DATE DUE			

A STORY THAT STANDS LIKE A DAM

A STORY THAT STANDS LIKE A DAM

GLEN CANYON
AND THE
STRUGGLE FOR
THE SOUL OF THE WEST

RUSSELL MARTIN

HENRY HOLT AND COMPANY

NEW YORK

Published by Henry Holt and Company, Inc.,
115 West 18th Street, New York, New York 10011.
Published in Canada by Fitzhenry & Whiteside Limited,
195 Allstate Parkway, Markham, Ontario L3R 4T8.

Library of Congress Cataloging-in-Publication Data
Martin, Russell.
A story that stands like a dam : Glen Canyon and the struggle for
the soul of the West / Russell Martin. — 1st ed.
p.   cm.
Bibliography: p.
Includes index.
ISBN 0-8050-0822-5
1. Glen Canyon Dam (Ariz.)—History.   I. Title.
TC557.A62G65   1990   2. Dams—Arizona   89-31619
CIP

Henry Holt books are available at special discounts
for bulk purchases for sales promotions, premiums,
fund-raising, or educational use. Special editions
or book excerpts can also be created to specification.

For details contact:

Special Sales Director
Henry Holt and Company, Inc.
115 West 18th Street
New York, New York 10011

First Edition

BOOK DESIGN BY CLAIRE M. NAYLON
CARTOGRAPHY BY DEBORAH READE
Printed in the United States of America

1   3   5   7   9   10   8   6   4   2

FOR GEORGIA

Contents

The canyonlands did have a heart,
a living heart, and that heart was Glen Canyon
and the wild Colorado.

Edward Abbey

The unregulated Colorado was
a son of a bitch. It was either in flood
or in trickle. It wasn't any good.

Floyd Dominy

A STORY THAT STANDS
LIKE A DAM

This was the third dam in Glen Canyon. The first, slowly shaped by sand and fed by a persistent stream, blocked a side canyon sometime before history began, forming a thin sweetwater lake that ultimately survived a civilization. The second dam belonged to the Anasazi, the people who forged that fragile civilization out of the rock and niggard soil. Their dam, built of sandstone blocks and sealed with clay mortar, stood in the canyon bottom at Creeping Dune. U-shaped and double-walled— its water gate a thin stone slab that could be adjusted to control the release of water into a system of ditches—it was a dam that trapped the flow from a spring bubbling to the surface of the sloping ground above the river, the water sustaining maize and spotted beans for perhaps two hundred years.

But the Anasazi abandoned the canyon in the thirteenth century. They left for still-mysterious reasons and never returned, their dam trapping water long after they were gone, until sand and eroding soil finally filled it up.

The first dam stood near the head of

1

little Lake Canyon until November 1915. A few prospectors, stock-men, and leather-hided hermits called desert rats occupied Glen Canyon that year—themselves members of a nascent civilization that was trying to come to grips with the most awesome and seemingly unreceptive country ever encountered—and they were there to witness when, after three days of heavy rains, the lake overflowed its sand dam, then breached it, and water thundered through the side canyon, tearing its way to the river.

Four decades after the Lake Canyon dam was severed, seven centuries after the Anasazi abandoned the dam at Creeping Dune, plans for the third dam were complete. The president of the United States had signed his name to legislation that would divert nearly $300 million from the federal treasury to build the structure just downstream from the mouth of Wahweap Creek, at a place where the canyon walls were sheer and narrow and almost 700 feet high. But unlike the Glen Canyon dams that preceded it—miniature and inconsequential impoundments by comparison, trapping only the slow, meager water from springs—this third dam would do some-thing that the Anasazi never imagined, something those grizzled, turn-of-the-century desert rats would have told you could never be done. This dam, ten million tons of it, would halt the relentless Colorado River itself. Wedged between the canyon walls like a keystone and rising 710 feet above the bedrock, this dam would utterly obstruct a river that collects the drainage of much of the western flank of the Rocky Mountains, a river whose watershed encompasses 244,000 square miles. Trapping the river between the canyon's serpentine walls, the dam would create a slackwater reservoir 186 miles in length, the longest in the world, covering Creeping Dune to a depth of 350 feet, reaching up Lake Canyon as far as that dune which itself was once a dam.

But this gargantuan dam in Glen Canyon, authorized in April 1956, and begun just five months later with appropriate presidential fanfare and a pig-tailed string of blasting sticks, wasn't the first audacious impoundment of the Colorado. Twenty-five years before, in 1931, work had gotten under way in Black Canyon, 400 miles downstream, on an enormous dam that would ultimately be named for Herbert Hoover—the largest dam in the world at that time, and the first time engineers had been able to test their convictions that a high concave wedge of concrete could successfully stop a river. Often called Boulder Dam during the desperate Depression years of its construction, Hoover Dam claimed the lives of 110 men before a swarm of workers topped it out, but it also captured the wonder and pride of the nation at a time when there were few palpable symbols of America's continuing might. It was a public work on the grandest scale imaginable, and its sweeping walls of concrete crowned by fluted, Deco-inspired intake towers attested to the fact that we as a nation knew we would be great again, signaled the certainty that our natural resources remained our secure and fundamental wealth.

Although the two projects took shape in eras separated by two decades and a world war, and although each was physically separated from the other by the great, wondrous gash of the Grand Canyon, Hoover and Glen Canyon dams had much in common: each dam and its companion power plant ultimately would generate 1.3 million kilowatts of hydropower, enough electricity to supply a city of a million people; measured at its crest, each dam rose precisely 587 feet above the Colorado's mucky, boulder-strewn bed. But there were subtle differences between them: Hoover was taller by 16 feet when the measuring began at bedrock; Glen Canyon contained 1.5 million cubic feet of additional concrete. The reservoir behind Hoover Dam held more water, but Glen Canyon's reservoir encompassed double the miles of shoreline and its length extended 76 miles farther upstream. Hoover Dam was wedged between hard, black walls of igneous andesite; Glen Canyon Dam abutted stained, striped, orange cliffs of spalling Navajo sandstone.

And there was an elemental, all-important distinction between the two gigantic structures. Hoover Dam was the first monumental project planned and executed by the Bureau of Reclamation, the western water-development arm of the Department of the Interior. It was with Hoover Dam that "BuRec"—as it became known in small-town weeklies throughout the dry and empty western quarter of the country—began a breakneck, thirty-year binge of dam building, an attempt to develop and populate the West and make it forever prosperous by harnessing its free-flowing rivers and putting their limited water to work. Although few people recognized it when Dwight Eisenhower telegraphed the signal to blow several ceremonial tons of rock out of a wall in Glen Canyon, the great dam that would rise at that spot during the subsequent decade would be a kind of swan song for the Bureau of Reclamation. The federal engineers and the wildcat heavy-construction companies would build more water projects, to be sure, but never again would they build such an enormous one. They could covet additional sites, but never again would they impound water in the canyons of the Colorado.

America was still on the make in the late 1920s when California water brokers convinced Congress of the need for Hoover Dam, two smaller downstream dams, and hundreds of miles of delivery canals that would regulate and supply the waters of the Colorado to the sprawling cities and fertile fields of the Golden State. In that era, possibilities for growth seemed unlimited throughout the Southwest—if only enough water could be pulled out of the canyons that held it captive—and Hoover Dam became concrete proof that America's engineering skill and industrial might together could work a kind of magic. Land and water existed only as rough raw materials to be manipulated, to be subdued, to be conscripted to the cause of the common good. The desert would bloom and great cities would sparkle with light if only we would set our machines in motion.

Thirty years later, by the time a high dam was rising in Glen Canyon, it had begun to dawn on many Americans that our technologies could diminish lives as well as enrich them, that the transformation of wilderness into something that represented the handiwork of humankind was an undertaking that ought to be considered very carefully. Although the West's water wizards, its politicians and farmers and cash-register entrepreneurs—clearly the lion's share of the region's slim and scattered population—saw in the construction of Glen Canyon Dam the same symbols of pride and prosperity that their parents had envisioned in Hoover Dam, there were others—many of them in many quarters of the country—for whom the project was nothing less than a crime. The dam and its power were not needed, they contended, and the canyon that would be lost—flooded up to its graceful, bald-domed rims—was a place unique on all the planet, so beautiful, so full of wonder that it should never have been disturbed.

As Glen Canyon Dam slowly grew up from the bedrock, and as more people began to mourn for the canyon that would soon disappear, America was undergoing a kind of sea change, an elemental shift in attitudes toward its land, its water, and its wild places. By the time the spare, buff-colored dam was completed, the notion that natural environments had great value in and of themselves, that *they* might matter as much as the march of progress, was becoming current throughout the country. And even those legions of people who remained convinced that the dam was entirely defensible argued that it *enhanced* its environment rather than somehow subdued it.

After three hundred years or so, we had finally conquered the continent. No doubt it was inevitable that our ongoing efforts to develop the wild regions that remained at some point would encounter the passionate opposition of those who were convinced we had done enough, that some of our best country simply should be left alone. That encounter, more specifically and dramatically than anywhere else, took place at Glen Canyon.

This book is the story in microcosm of that confrontation, the battle between those decent people who were committed to transforming the West for the collective good, so that it might be something more than a region with too little of everything but empty space, and those decent people who were equally certain that space—wild, open, undisturbed space—was the one resource our technology could never re-create, the legacy we must never entirely destroy. It is the story too of a singular spot on the American map—a slender, riverine cut in the rock and hardpan of the Colorado Plateau. Glen Canyon was tentatively named in August 1869, when John Wesley Powell, a scientist, Civil War hero, and kind of one-armed Columbus, first floated its quiet waters with a crew of eight men in four fragile and unforgiving wooden boats. For Powell and his men, the canyon was both a wonderland and a blessed relief, lying downstream from the harrowing rapids in Cataract Canyon, upstream from the disasters that awaited them in Marble Canyon and in Grand Canyon.

During the succeeding century, Glen Canyon continued to captivate and intrigue its infrequent visitors: first the industrialists who envisioned a canyon-bottom rail line connecting Chicago with Los Angeles, then the miners who somehow believed they could capture the powdered gold suspended in its sands. The federal dam builders first eyed the canyon in 1916. In 1940, Interior Secretary Harold Ickes proposed that it become part of a vast national monument, but by the early 1950s, conservationists themselves were eloquently arguing that Glen Canyon should be dammed—in order to prevent a similar fate for a canyon inside Dinosaur National Monument. By 1963, when a Sierra Club book called it "the place no one knew"—

perhaps expiating that organization's sense of guilt about its loss by fibbing about its anonymity—Glen Canyon had become somehow hallowed, the most potent symbol of environmental destruction in the nation, a symbol it long has remained. And in curious counterpoint, at the end of the 1980s, the reservoir that fills Glen Canyon—named to honor Powell—is now the most heavily visited of all the areas in the interior West managed by the National Park Service, receiving more visitors each year than the Grand Canyon, more even than Yellowstone, that extraordinary reach of land that the young nation first chose to save and protect.

What unfolds on the following pages is an account of a peerless and spectacular place—how it came to be dammed at the end of the nation's resolute and glorious dam-building days, flooded in the interests of water and power and play for the public good; how a fledgling environmental movement encouraged its inundation in the interests of compromise, then later made it a kind of battle cry in the struggle to prevent the building of similar dams and in ongoing efforts to protect the whole spectrum of wild places still extant. This book is a narrowly focused telling of what ultimately is the only real story in the American West—our collective and enduring quandary about whether we should develop this difficult and often majestic land to serve our material needs or protect it in the hope that it can sustain our spirits; it is a continuing query about how we should go about the business of surviving here, as the Anasazi somehow were able to do for a millennium or more. It is a story that winds like a red river in Glen Canyon, a story that stands like a dam.

1

SHUTTING THE RIVER OFF

The river ran free in Glen Canyon
for the first six years the dam was
under construction, bypassing the
cacophonous, swarming job site by
diving into a diversion tunnel dug
into the canyon wall. But by January
1963, with the vertical upstream wall
of the dam already soaring 600 feet
high and with the outlet pipes and
the eight steeply sloping penstocks
in place, it was time to shut the river
off.

In consultation with his superior,
Secretary of the Interior Stewart Udall,
and his chief construction engineer on
the site, an Arkansas-born bulldog
named L. F. "Lem" Wylie, Bureau of
Reclamation Commissioner Floyd
Dominy had sanctioned January 21 as
the day the diversion tunnel in the
west wall of the canyon would be
sealed. A team of ironworkers would
carefully begin to close the three steel
slide-gates that blocked the entrance to
the 41-foot-diameter tunnel; then
concrete masons, working in the wet
and now almost waterless hole, would
install a temporary concrete plug.

Finally, 400 feet of the tunnel would be plugged with solid concrete.

The entrance to a second diversion tunnel in the east canyon wall had been dug 33 feet higher than the one on the opposite side, and since the gates in that tunnel weren't scheduled to be screwed down for a couple of months, the river didn't have far to rise before it could escape again, its downstream journey only briefly interrupted. Yet there was something symbolic about those first 33 feet of reservoir, something that assured Wylie that he and his men actually would be able to harness the son of a bitch, something that proved to Dominy and Udall that the people's money and their own political energies had been well spent.

Those first 33 feet of the lake, high enough to back water upstream about 16 miles—sending it into the side canyons of Wahweap, Antelope, Navajo, and Warm creeks—were proof of a different kind to David Brower, evidence of the heart-wrenching certainty that the canyon soon would be submerged. Brower, executive director of a small, San Francisco–based conservation organization called the Sierra Club, had traveled to Washington, D.C., on the day Glen Canyon's west diversion tunnel was scheduled to be closed, hoping that he could convince Udall to forestall what he knew someday would be inevitable. In agreeing not to put up a pitched battle against the Glen Canyon construction back when it was authorized in 1956 as part of the Colorado River Storage Project, America's conservationists had won a single concession: no dam, reservoir, or related structure would be allowed to intrude on any national park or monument. In the case of Glen Canyon, that meant something very specific. When the Glen Canyon reservoir—now commonly being called Lake Powell—finally was full, it would send water far up Forbidding Canyon, then up a small tributary called Bridge Canyon and into the 160-acre Rainbow Bridge National Monument, a federal preserve since 1910 and the site of the world's largest and surely most spectacular natural stone arch. At the lake's maximum high-water elevation of 3,711 feet above sea level, water would sit 57 feet deep in the narrow inner channel of the canyon directly beneath the bridge. As far as Brower and his confederates were concerned, the situation was very simple: since Congress had mandated that the Glen Canyon project could not be built without adequate additional measures to protect Rainbow Bridge from encroachment, the lake legally could not begin to rise until those measures had been taken. If all of Glen Canyon could be spared in the interim, well, that would be a substantial, if temporary, additional victory.

What the conservationists wanted the Bureau of Reclamation to do was to build a dam downstream from the monument that would prevent the lake from encroaching upon it. A pumping facility would be required as well, one that would suck out the intermittent water that ran down Bridge Canyon from the slopes of Navajo Mountain—the water that had created the canyon and the bridge in the first place—and avoid the inevitable creation of a small but equally invasive reservoir on the upstream side of the barrier dam.

Dominy and the Bureau of Reclamation had been more than willing to cooperate. Bureau engineers were in the business of designing and building dams, after all, and Dominy—a charismatic kind of autocrat who liked to explain, with a grin, that he was a simple public servant—personally relished the irony that this time it was the conservationists who demanded a dam, a structure that otherwise they were invariably quick to despise. If Congress appropriated the money, Dominy would be glad to build the conservationists an impoundment they could call their own. For two years running, however, Congress had refused to approve legislation that would have made the construction funds available, and President Kennedy carefully had steered clear of the controversy, saying repeatedly that Congress should decide the issue.

As for Kennedy's cabinet member in charge of these enormously expensive federal water projects (as well as of all the nation's public lands), Udall was in favor of doing nothing, letting the lake slip under the enormous arch of Rainbow Bridge and lap beneath its massive sandstone abutments. Since Glen Canyon Dam had been authorized, Udall had made several trips to Rainbow Bridge, first as a congressman and then as secretary of the Interior. Despite his support for every western water project that had ever floated down the congressional stream, Udall was greatly enamored of the canyon country that formed the borderlands of his native state and neighboring Utah. In 1871, his great-grandfather, John D. Lee, a legendary Mormon pioneer, had settled briefly with his seventeenth wife, Emma, on the shore of the Colorado 15 miles downstream from the place where Glen Canyon Dam now stood, and his people had been proud and influential Arizonans ever since. As far as the young Interior secretary with the piercing eyes and the burr haircut was concerned, the rocky, rough, and very remote country surrounding Lee's Ferry and including Rainbow Bridge was *his* country, and it was the most captivating of all the desert canyonlands. And Udall was equally convinced that if a protective dam or dams were built near Rainbow Bridge, the cure would be far worse than the disease. Roads somehow would have to be blasted into an area that currently was traversed by only a single tough and primitive trail; mesas would have to be scraped to provide material for the structures; and electric lines to provide power for the pumps would have to invade the monument as well. Udall was convinced that Rainbow Bridge would have to be destroyed if it were to be saved.

Although in those first years of the 1960s, America's conservation organizations were not at all comfortable with Udall's record on water projects, in other respects his early stewardship of Interior had seemed enlightened compared with that of his Eisenhower-administration predecessors, Fred Seaton and Douglas "Give-Away" McKay. One thing that was very much in his favor, as the conservationists saw it, was that Udall wanted to add vast new acreage to the national park system. He wanted, in fact, to turn the postage stamp–sized Rainbow Bridge National Monument into a new 500,000-acre national park. But Udall's new park would have to include a foul little fjord of Lake Powell, and on that point, the conservationists so far had refused to compromise. It had been compromise, they reminded Udall, that had doomed Glen Canyon in the first place.

Early in the 1950s, the Bureau of Reclamation had proposed building a series of storage reservoirs on the Green and Colorado rivers in the upper Colorado basin, including a huge dam in Dinosaur National Monument up north on the Utah-Colorado border as well as the dam in Glen Canyon. Few conservationists were in favor of *any* of the proposed dams, but all were adamant in their opposition to the Echo Park project inside Dinosaur. As they saw it then, if any part of the national park system could be invaded, could be developed for whatever purpose, then the system itself had no integrity, and no part of the system was safe. The Echo Park proposal had to be defeated at all costs, and the logical, if lamentable, cost seemed to be withholding opposition to those proposed dams and reservoirs that did not affect or invade parklands.

After years of struggle, the West's water solons and the Bureau of Reclamation—where brash, often intimidating Floyd Dominy had not yet gained total power—agreed to scrap Echo Park in return for the conservationists' agreement that they would let the government get on with some dam building. The deal was set: Glen Canyon and five other dams would be built in areas throughout the upper basin of the Colorado River where they would not encroach on parklands; Dinosaur National Monument would be spared, and—there it was, written down in an act of Congress—"as part of the Glen Canyon Unit, the Secretary of the Interior shall take adequate protective measures to preclude impairment of the Rainbow Bridge National Monument."

On the morning of January 21, 1963, Dave Brower, who by default as well as his own force of will had become the unofficial chieftain of the nation's several conservation organizations, sat outside the office of the secretary of the Interior, hoping he would be granted a few minutes' audience. Brower planned to remind Secretary Udall of that very specific language within the Colorado River Storage Project enabling legislation, and to implore him not to let the diversion tunnel at Glen Canyon be sealed—not today and not until someday down the line when Rainbow Bridge had been blocked adequately from the rising water. But Secretary Udall did not meet with Brower that morning, not because he was entirely unsympathetic to his cause, but because that day—that day in particular—he had other fish to fry.

It was on the morning of January 21, by coincidence, that Stewart Udall had scheduled a press conference to announce his department's plans for an entirely new series of western dams, diversions, and delivery canals, a project that would dwarf what currently was being accomplished at Glen Canyon and elsewhere in the upper Colorado basin. With Reclamation Commissioner Dominy at his side, and with Brower now listening at the back of the room, Udall introduced with obvious enthusiasm and a certain public-relations flourish what Reclamation's imagineers had dubbed the Pacific Southwest Water Plan—a monumental dam, pump, and pipe system that would water vast new acreage in Arizona and California as well as deliver new supplies to the region's burgeoning cities. One of the beauties of the plan, according to the secretary, was that it would pay for itself in the end, not by the sale of water, but rather by the sale of electrical power, made possible by the deep canyons and steep gradient of the workhorse Colorado River. Hoover Dam had been generating hydropower for nearly thirty years now; Glen Canyon would begin its service in a year and a half or so; and sandwiched between them, the Bureau now planned to build two more hydropower dams, one at the head and another near the foot of the Grand Canyon. Together, the two dams would be able to produce enough electricity to pump the requisite water to Phoenix, plus surplus enough to sell to private utilities to generate the cash that slowly but surely would reimburse the government for its multibillion dollar outlay. Dominy, in fact, liked to refer to the Grand Canyon dams as "cash registers," and with them in place, the Colorado finally would come close to being entirely harnessed, a dream he had held fast to since he had joined Reclamation back in 1946.

Brower, the Sierra Club administrator, fifty years old and already silver-haired, was outwardly passive at the back of the room. He had known that something like this was in the works, and he knew certainly that Reclamation had had its eye on the two Grand Canyon dam sites for decades, but the timing of the announcement was particularly deflating. Exquisite Glen Canyon would begin to go under that day; Rainbow Bridge, a small but significant part of the national park system, seemed very unlikely—despite the Congress's stated wishes—to receive some sort of stay against encroachment; and now it was officially on the table: Dominy wanted to build two dams in the Grand Canyon, and Udall, the conservation-minded secretary of the Interior, wanted to let him do it.

It took two days to shut off the river, to get the slide-gates closed in the west diversion tunnel and to pour the concrete that would form the first of a series of plugs. Lem Wylie had never really worried that Udall would acquiesce to the dingbat conservationists and demand that he keep the Colorado running, but nonetheless, it felt good to him at long last to have the makings of a lake rising against the cofferdam that had kept the job site dry since February 1959. It wasn't that Wylie disliked the conservationists as individuals—at least not the few he had met. The Sierra Clubbers with whom he and the dam's chief designer, Louis Puls, had shared a boat ride up from

Lee's Ferry back in 1956 had seemed like decent sorts. But they couldn't see the big picture somehow. They didn't understand that a project like this one ultimately benefited millions of people, and surely people were more important than a hundred miles or so of canyon that didn't amount to more than a kind of crack in the ground. And another thing about those people, Wylie would aver over a highball or two during evenings down at the modest little Glen Canyon Country Club, they simply had no idea what it felt like to be a part of a project like this, to have a real hand in building something that would stand in that hole over there forever. "I feel sorry for people who have to make their livings other ways," he liked to say, "instead of going out there each day and putting another five feet of mud on the block."

By the time they closed the west-tunnel gates that January, Wylie had overseen the pouring of 617 vertical feet of mud. More than four million cubic yards of concrete—four-fifths of what would be the dam's total volume—were already in place, and Wylie had good reason to believe he'd get the thing finished on time. He had been on board for five and a half years now, since the very beginning. He had gotten roads pushed through the desert to access the site by playing politics with the governors of Arizona and Utah; he had secured a location for thousands of workers to live by playing politics with the Navajos. He had overseen the building of the world's highest steel-arch bridge across the river, reducing a 196-mile highway trip from canyon rim to canyon rim to a quarter-mile of suspended macadam, and he had watched that government camp on the treeless, sand-bedeviled mesa east of the canyon turn into a kind of town, metal prefabs and tin-sided shacks now giving way to supermarkets, motels, taverns, and pastel-painted bungalows—this city of Page, Arizona, with its thirteen churches and nine holes of golf and 6,106 people, presently the biggest community alongside the Colorado River for 330 miles upstream and for 390 down.

Wylie, who had been bestowed the Bureau's thirty-year pin the year before and who planned to retire within another, wasn't particularly worried about what would happen to the town that had been his personal fiefdom. Reclamation was estimating that Page would shrink to as few as a thousand people once the dam and power plant were finished, but he wasn't so sure. Likely as not, the Bureau would end up building the conservationists their barrier dam near Rainbow Bridge, and that work force would keep the population inflated a while longer. Then, once Marble Canyon Dam was under way a few dozen miles downriver, the new project and its companion payroll would surely keep Page open for business. And over the long term, Wylie was convinced, huge numbers of people would come to this remote stretch of Utah-Arizona border country not just to work but to play. Lake Powell would be the biggest recreational draw in the southwestern United States. Situated halfway between Phoenix and Salt Lake City, just a day's drive from Albuquerque and Denver, a long day's drive from Los Angeles, this reservoir couldn't miss. Although technically it wouldn't be as big as Lake Mead down on the Nevada border, it would seem bigger and it would be far more beautiful. Hell, they were already making movies here. Just a month ago, moviemaker George Stevens had brought in John Wayne and Charlton Heston and hundreds more Hollywood people to shoot a picture he was calling *The Greatest Story Ever Told.* Page had bulged at its seams, and the wife of every lowboy driver and jackhammer man on the job had put a sheet over her head to play the part of an extra. This country looked more like the Holy Land than the Holy Land did, Stevens had told a meeting of the Chamber of Commerce, and if he was right, well, this place had a very bright future indeed.

That visitors could become enraptured by the Glen Canyon country was something Art Greene had known since the early 1940s, back when he began operating commercial boat trips to Rainbow Bridge from his Marble Canyon Lodge near Lee's Ferry at the foot of Glen Canyon. Originally, Greene had hauled his boats, gear, and paying guests in trucks 400 miles overland to the settlement of Hite at the head of the canyon, then had floated them down a hundred miles of placid river to the mouth of Forbidding Canyon, at which point they had no option but to walk six miles up the winding creekbed if they were to see the spectacular bridge. By the late 1940s, the indefatigable desert rat was using jerry-rigged airboats to carry as many as 150 people a year *upriver* from Lee's Ferry to Rainbow Bridge, a route that cut the length of the average trip from ten days to three. It was on one of Art Greene's upriver excursions to Rainbow Bridge, in fact, that Wylie and the Denver-based dam designer, Puls, had hitched a ride as far as Wahweap Creek and made acquaintance with their first conservationists.

It didn't take Art Greene long to realize that if these Bureau boys really were going to build a dam somewhere between Lee's Ferry and Rainbow Bridge, they would play hell with his livelihood. In 1953, with word that Reclamation's engineers had their eyes on a dam site 15 miles upstream from Lee's Ferry, but with no assurance that the dam actually would ever be built, Greene and his family— his wife and four children and their spouses—decided to take a little gamble. The senior Greene journeyed down to Phoenix to visit the state land commission, and while there he nonchalantly leased six sections of state-owned land, 3,840 acres, near the junction of Wahweap Creek with the Colorado River, just a yard or two south of the Utah border—godforsaken ground from which the state of Arizona was glad to derive a bit of income. By 1956, when the dam had become a congressional certainty and Bureau engineers and prospective bidders from heavy-construction outfits were combing the mesa tops near the site, the Greenes were serving steaks in a small cafe nearby, renting out eight stone cabins by the night, and selling a little fuel every time a pilot set his plane down on the airstrip they had bladed out of the blowsand.

In January 1963, ten years after they had arrived on the scene, Art Greene and his clan were hardly worth Lem Wylie's worry anymore. Early on, although pleased that the Greenes had brought a vestige of civilization to that roadless and wind-scoured country, Wylie had dutifully informed them that the National Park Service would be managing the region as a national recreation area—in much the same way that it currently administered Lake Mead—and that they would have to pack up and move on because the park service had plans for a big marina there at Wahweap. Art Greene had responded that he appreciated the suggestion, but that he thought he'd stay, explaining that the state of Arizona had tendered him valid leases, and he had assured Wylie that it would all work out fine in the end because a marina was precisely what *he* had in mind for the place.

In the intervening years, no amount of subtle coercion or even veiled or outright threats had shaken the Greenes' resolve. The latest Wylie had heard was that Barry Goldwater, the Arizona senator who was making noises about running for president, had taken the time to go to bat for his old canyon country pal, Art Greene. Goldwater was pushing for the park service to resolve the whole hassle by making the Greenes its official Wahweap concessionaire, upgrading the de facto status the Greenes had enjoyed since 1959. And maybe that would be the simplest solution. On the day that they shut the river off, Wylie was well aware, a 200-foot-wide boat ramp already lay out on Wahweap Point waiting for water, and nearby was the Greenes' cafe and a general store, a campground, and a fancy new trailer park complete with hookups. By summertime, with all but a thousand cubic feet per second of the Colorado River languishing behind the dam, Art Greene planned to ferry tourists to Rainbow Bridge and back in less than a single day.

The use of the Colorado for regular commercial recreation had begun almost a decade before Art Greene's Canyon Tours boats first floated the ruddy waters of Glen Canyon. It was back in 1938 that Norman Nevills, an exuberant thirty-year-old from a remote hamlet called Mexican Hat, Utah, had collected $250 from each of four far-flung passengers to conduct them in wooden boats of his own unusual design down the Green River to the confluence with the Colorado, then on through Cataract, Glen, Marble, and Grand canyons to the headwaters of Lake Mead, a trip punctuated by rough water, meager rations, and more than a few rifts between the several strong-willed personalities in tow, but one that nonetheless convinced Nevills that a career awaited him on the Southwest's rivers. In 1940, Nevills organized a Green and Colorado expedition that included Barry Goldwater—at that time the thirty-one-year-old head of his family's Phoenix-based dry-goods business—and although World War II briefly interrupted Nevill's burgeoning enterprise, by 1944 he was back on the water, beginning to make a legendary name for himself and his wife, Doris, and training numerous boatmen, infecting them with similar, insatiable passions for riding the chocolate desert rivers.

Nevills was followed in the early 1950s by several friends and former employees—that second generation of river outfitters initially carrying as many Reclamation engineers and uranium prospectors through the desert canyons as tourists. River travel was still considered a kind of crazed adventure in those days, and the few people who spent their vacations riding rapids and sleeping on sandbars were assumed to possess a decided daredevil streak. But by the second half of the decade, by the time Glen Canyon seemed likely, then certain to go under, more and more people wanted to know what was hidden in that sinuous place, wanted to see it before it disappeared, and a launch or two a week from Hite at the head of the canyon was common during the summer months.

Beginning in the summer of 1957 and continuing through 1963, the canyon also had become a kind of laboratory, a field station for archaeologists, biologists, geologists, and historians. Garbed in khaki shirts and trousers, their heads protected by pith helmets, they worked under the auspices of the park service to perform what was called an emergency survey of the region's scientific resources—the most extensive such salvage project ever attempted. Professors and students from the University of Utah and the Museum of Northern Arizona had spent successive summers in small teams scattered throughout the 160-mile reach of the canyon's main stem, as well as in the canyon of the San Juan River, the Colorado's principal tributary in Glen Canyon, and in dozens of both rivers' side canyons—combing and cataloging, making site-specific evaluations, and attempting canyon-wide interpretations, endeavoring as best they could with adequate resources but with very little time to document for posterity what had been there before the flood.

And there were others in the canyon during those summers who were determined to make a similar but more emotional kind of record. Back in 1948, writing in *The Atlantic Monthly*, Wallace Stegner, a Stanford English professor, novelist, and the biographer of pioneer canyon explorer John Wesley Powell, had described his own 1947 trip with Norm Nevills through San Juan and Glen canyons. A decade later, contemporary accounts of Glen Canyon had yet to appear, but a burgeoning number of writers, photographers, and artists were visiting the canyon to try to glean some lasting memory, to describe its delights and lament its numbered days. California photographer Philip Hyde, a student of Ansel Adams, had discovered the canyon country on a Sierra Club–sponsored trip through Glen Canyon in the summer of 1955, and had returned to photograph it extensively in 1962. Eliot Porter, a physician and an innovator in the bold new medium of color photography, had photographed repeatedly in Glen Canyon since 1960, twice accompanied by artist Georgia O'Keeffe, who went on to make a half dozen trips of her own into the Glen. And a young novelist named Edward Abbey, a Pennsylvanian living in New Mexico at the time—a man who in later years would become the voice most readily identified with efforts to preserve the slickrock country of the Colorado Plateau— had floated Glen Canyon on a drugstore rubber raft in the sweltering summer of 1959, a languorous and reverential sojourn he had known he would want to write about someday.

David Brower, a writer, photographer, and editor in addition to his administrative duties for the Sierra Club, had never seen Glen Canyon until long after his role in the decision to dam it had ended, but his series of trips through the canyon during the years the dam was under construction—and his mounting horror at what was about to be lost—had led to plans for a lavish book of Porter's photographs, to be edited by Brower and titled *The Place No One Knew*. The book was being readied for publication on the day Lem Wylie shut the river off.

Fifty-two days passed before the lake ebbed high enough for its water to slip into the east diversion tunnel, where it met three temporary outlet gates that would control downstream flows until the power plant's penstocks—their intakes still high and dry on the face of the dam—were reached by the rising reservoir. On March 13, two of the gates were screwed shut; the third gate was lowered to precisely 50 inches, just enough to allow 1,000 cubic feet of water per second to escape downstream to keep distant Lake Mead from going dry.

The snowfall in the central Rockies was the slimmest in many years during the winter of 1963, and the spring runoff was commensurately light—almost as if the capricious river wanted to remind everyone just who held the ultimate hand. By June 15, the dam stood 650 feet high; a huge 345-kilovolt transmission line was being strung from the canyon rim south toward Phoenix, and the reservoir, despite the dry winter, had risen 226 feet and was continuing to gain about half a foot per day. The scientists and salvage crews, frantic to finish their work and traveling down the canyon by motorboat, encountered slackwater more than 110 miles above the dam. Throughout two-thirds of the canyon, the riverside sandbars were submerged; the saltbush and greasewood on the terrace slopes were going under as well, and the trunks of the enormous old cottonwood trees seemed to grow right out of the rising water. At the mouth of Lake Canyon, 90 miles upstream from the dam, the water was 30 feet deep, and—a mile and a half away from its old channel—it lapped at the cliff wall below Wasp House, an Anasazi habitation built eight hundred years before.

By the middle of July, the conveyor and cableway operators, signalmen, and cement finishers had added a few more feet of mud to the block, and the end was almost in sight. Fifty miles upstream from the dam, the reservoir had backed up Forbidding Canyon far enough to spill into Bridge Canyon, and the tourists whom Art Greene guided to Rainbow Bridge had to walk little more than a mile to enter the national monument. The reservoir had risen 14 additional feet during the month, but, with the runoff receding, and with the water climbing the high, widening canyon walls and reaching ever farther up the side streams, this implausible new lake in the desert now was gaining only an inch a day.

2

WATERS THAT RAN
TO WASTE

The logical place for the first high dam
was in Glen Canyon, E. C. LaRue was
convinced of that. As chief hydrologist
for the United States Geological Survey
in 1916, it was clear to LaRue—a fellow
with a gruff demeanor, a thin and oily
mustache, and a singular mind when it
came to matters riverine—that the initial way to calm and regulate the torrents of the Colorado was to build a
gravity dam near Lee's Ferry, Arizona,
just south of the Utah border, a dam
that would tame the wild brown river
throughout the 600 remaining miles of
its descent to the Sea of Cortez. But by
1922, instead of the 244-foot-high dam
immediately downstream of John D.
Lee's old river crossing he had proposed
six years earlier, LaRue wanted to construct a similar dam four miles upstream. Then, in a paper published a
year later, the hydrologist announced
what had evolved into a truly grand
plan: a 780-foot-high dam could be
built at that second site simply by
"blasting in the canyon walls." It would
be relatively inexpensive: it unquestionably would dampen the flood, and

it would create a 50-million-acre-foot reservoir that would back water almost 250 miles upstream. As far as E. C. LaRue was concerned, if you were going to bother to try to contain the Colorado, you might as well envision something spectacular.

Since 1901, when the Colorado had torn through an ungated diversion channel near the Mexican border, plowed through the thick soil of California's Imperial Valley, and turned the sub-sea level Salton Sink into the brimming Salton Sea, there had been a growing consensus among Californians and Arizonans that the mountain-born river that formed their common border—its flows drastically fluctuating each year, ranging from a trickle to an absolute torrent—had to be impounded somewhere in its upriver canyons if it ever was going to be put to dependable, beneficial use. Most Californians, including Senator Hiram Johnson, Imperial Valley Congressman Phil Swing, newspaper magnate William Randolph Hearst, and William Mulholland, influential head of the Los Angeles Water and Power District, were fierce advocates of a dam in either Boulder or Black canyons on the Arizona-Nevada border, a dam which, they argued, should be built at public expense by the newly formed federal Reclamation Service. Arizona's senators Henry Ashurst and Ralph Cameron, George Hebard Maxwell, a lawyer and founder of the National Reclamation Organization (a private group lobbying the government to jump into the water business with both feet), and E. C. LaRue, among others, were equally adamant that the big dam belonged in Glen Canyon.

As far as the Californians were concerned, the dam should be built downstream, near the locations where the river already was being diverted and put to use, nearer to the coastal cities that would consume the electricity the dam would produce. The granite in Boulder and Black canyons was unassailable, and the small Mormon town of Las Vegas, Nevada—a few miles west of either of the two dam sites they preferred—could serve as a ready supply center. In contrast, they contended, the Glen Canyon site was far too distant from the civilized regions of the Southwest, there were no towns or roads or rail lines within hundreds of miles, and the sandstone walls of the canyon would be of dubious reliability in supporting the weight of a massive dam and reservoir.

What the Arizonans liked best about the Glen Canyon site was that it was entirely within the boundaries of their state. Given a requisite system of tunnels, siphons, and canals, water stored at Glen

Canyon could flow by gravity to the dry valleys and population centers to the south. The Glen Canyon geology offered a perfect reservoir pool, and a number of experts had gone on record affirming that a dam of 400 feet or so in height would be entirely stable. But perhaps most persuasive of all, people like Maxwell and LaRue vehemently argued, was that building the Glen Canyon dam *first* would make it much easier to construct a series of dams farther downstream—a virtual staircase of dams and reservoirs descending through Marble and Grand canyons, through Boulder and Black canyons, and on down to the river delta. Glen Canyon shouldn't be the location of the Colorado's only dam, they averred, but it should certainly be the site of the first.

It wasn't until 1921, however, that the building of a dam *anywhere* on the Colorado began to seem distinctly possible. It was in that year that the seven bickering states of the Colorado River basin—Arizona, California, Nevada, Utah, New Mexico, Colorado, and Wyoming—agreed to form a commission, to be chaired by Secretary of Commerce Herbert Hoover and to be charged with determining how the river's waters should be divided among them. It was in that year as well that the U.S. Geological Survey began a several-summers' effort to survey accurately and comprehensively the canyons of the Colorado and its tributaries, to accumulate a trove of geological and topographical data, and to assist the Reclamation Service in determining where the dam or dams eventually should be constructed.

By the end of 1922, the San Juan and Green rivers had been examined by teams of engineers, geologists, and surveyors navigating the canyons in small wooden boats; the survey of the main stem of the Colorado above Lee's Ferry also had been completed, and only one more summer would be required to complete the mappings and investigations downstream as far as Needles, California, where the deep canyons finally gave way to riverside dunes, benches, and desert lowlands. The members of the Colorado River Commission, meanwhile, working independently of the USGS survey and already having met for months with no success in their attempt to devise an agreeable way to divide the river, decided to convene in Santa Fe, New Mexico, at the end of the year to try once again to hammer out an accord; there really wasn't any need for the commission to wait for the USGS report on the lower canyons before it completed its own business.

For his part, E. C. LaRue was delighted that the downstream canyons had yet to be studied. He had invited several members of the Colorado River Commission to journey with him to Glen Canyon en route to their Santa Fe meeting, to familiarize themselves with the river and its most desirable canyon, as well as to publicize the river's wonderful resources and the importance of their deliberations. If plans for a dam in Boulder Canyon were scrapped in the process, well, LaRue could live with that.

In late autumn, with the river running low and peacefully, LaRue and a crew of boatmen took their own wooden survey crafts out of the shacks where they had been in storage at Lee's Ferry, equipped them with outboard motors, and traveled upriver to Hall's Crossing, a similar and equally remote ford 110 miles upriver. At Hall's, the official party was waiting—the director of the Reclamation Service, the chief engineer of the topographic branch of the USGS, Secretary Hoover's representative in his regretted absence, the heads of a variety of railroads and power companies, a press contingent, and, of course, several commission members.

Although the low water caused the outboards to clog periodically with silt, the downstream trip went smoothly. The group visited Anasazi ruins, which in those days more commonly were called Moqui sites, and an abandoned gold dredge rotting away in mid-river, and some of the party even hiked the six miles to the impossibly large and lovely stone arch known as Rainbow Bridge. Later, LaRue gave a grand tour of his dam site, although there was little to show but the high canyon walls. Nonetheless, the hydrologist heralded how much sense it would make to build the first dam right there, speaking nonstop about the beauty of the plan before the flotilla traveled four more miles to its landing at Lee's Ferry, the commission members making their way by rough road to Flagstaff, then by train to Santa Fe.

It was the Reclamation Service's 1921 announcement that it was recommending construction of the so-called All-American Canal to carry water from the lower Colorado to California's Imperial Valley—as well as proposing the expeditious building of a large storage dam somewhere upstream on the main stem of the river—that really had set the seven states to quarreling. As far as the headwater states of Colorado, Wyoming, Utah, and New Mexico were concerned, California plainly was planning to steal *their* water, and the federal government wanted to help them do it. As far as Arizona was concerned, the feds were welcome to lend a hand to the thievery, as long as they made sure Arizona got as much water as the big state to the west. Nevada, too sparsely populated and too little affected to raise a fuss, simply sided with California.

At the time, California was the only state of the seven that had a need for any of the water, and its need was real. Southern California was growing faster than any place on the planet; the city of Los Angeles in particular would soon outgrow its water supply from California's own Owens Valley; and the big, flat, and potentially lush desert valleys to the east, filled with eons of the Colorado's rich silt, could spawn a virtual revolution in the nation's farm production—if only there was water. The state's agricultural elite, in concert with its servile politicians, originally had assumed that getting hold of the Colorado was a rather straightforward matter of taking it. Since all western water law was based on the doctrine of prior appropriation—meaning it belonged to whoever swallowed it first— California would simply lay legal claim to most of the Colorado by putting it to use.

But although their combined populations didn't come close to the size of California's, the other Colorado River states did have enough collective congressional clout to make the case successfully that California was entitled to *none* of the river's water except by the clear consent of the several states where the river rose. After all, nearly half the river's total volume originated in the state of Colorado; Wyoming and Utah contributed a third; even Arizona and New Mexico were the points of origin of a few paltry streams. California and Nevada, in comparison, contributed nary a drop.

So California reluctantly had agreed to join the river commission and search for some sort of compromise. And if Secretary of Commerce Hoover was getting nowhere after several months of meetings in his efforts to see these seven states through to some kind of compromise, the reason seemed to be that no single state wanted to risk locking itself into a specific share of the river's resources if it could possibly avoid it. But prior to the Santa Fe meeting, Denver water lawyer Delph Carpenter, representing the state of Colorado, had had an idea: why not simply divide the river between *basins*— upper and lower—then let the states in each basin ultimately divvy up that basin's share? All the upriver states were immediately game.

They wouldn't be using—or needing specifically to allocate—their shares for decades perhaps. The two-basin division at least would guarantee them roughly half the river's water, and that, for now, was all the guarantee they needed.

Once the group gathered at Bishop's Lodge, a resort outside Santa Fe, California's Commissioner W. F. McClure made it clear that his state might be able to live with such a solution if the lower basin could use upper basin water until it was needed upstream—the bet being that a few of these hayseed states would never amount to much. And also, if the United States were ever to enter into a treaty with Mexico allocating a specified share to the neighboring country, any necessary reductions in the two basins' allotments would have to be borne equally. Arizona, still deeply suspicious of California's motives, was skeptical, but Carpenter's plan started to snowball.

Based on some rather sketchy data, the Reclamation Service had estimated that the Colorado's average annual flow was 17.5 million acre-feet. (An acre-foot is enough water to cover an acre of ground a foot deep: 325,851 gallons; enough water, Reclamation engineers long have answered in response to queries from nonhydrological minds, to supply a family of five for a year.) Hoover, a practical man and an engineer himself, proposed to the seven commissioners that each basin be allocated 7.5 million acre-feet annually. A million and a half acre-feet would be reserved for Mexico. The remaining million would serve as an unallocated reserve.

When California announced that it could abide by Hoover's formula only if the lower basin were granted that final million acre-feet—since it, any fool could see, was the only state that *needed* any of the water—the other states reluctantly agreed to the demand rather than risk having the meeting come to naught. On a raw November day in 1922, the official representatives of the river's seven states joined Hoover in signing the Colorado River Compact, a document that would be of enormous import during the coming decades, one that effectively divided the river in two. The agreed-upon point of division was Lee's Ferry, Arizona, at the mouth of Glen Canyon, 28 miles downstream from the Utah border, a logical enough spot to rend the river if you were determined to, but one that played hell with E. C. LaRue's plans for a grand Glen Canyon dam.

Arthur Powell Davis, head of the Reclamation Service and nephew of John Wesley Powell, had had his eye on the Colorado River since first assuming his post. Part of his interest, to be sure, stemmed from his storied uncle's explorations of the river and his instructive championing of the cause of reclamation—a rather obscure term that, in the West at least, meant only a single thing, the "reclaiming" of unproductive land by saturating it with irrigation water. Yet much of Davis's interest in the Colorado was entirely personal. He was enchanted with the river's sheer, rocky canyons; they virtually begged to be dammed. And what a career he would build for himself if his fledgling Reclamation Service indeed could create an agricultural empire with water impounded behind such dams. The signing of the Colorado River Compact had been the political accord Davis needed in order to set his engineers to work on the Colorado; it paved the way for the construction of the high dam and storage reservoir that would begin the process of bringing the river under control. Now the seven states needed to ratify the compact to make it law, and his own outfit needed to determine precisely where the dam ought to be built.

One by one, six states ratified the compact. The Arizona legislature, however, put up its back and firmly refused to budge. As far as the Arizona solons were concerned, the compact asked their state to sacrifice much in order to gain very little. Arizona would have to battle it out with gargantuan, water-gulping California to secure itself a share of the river that flowed through its own territory for hundreds of miles before it began to form the boundary with California; and the power-producing potential of Arizona's own canyons was apparently being put up for grabs, the state's special geography garnering it nothing in the end. If the compact had expressly defined the several states' intentions to build a high dam in Glen Canyon, or even one at the site commonly called Bridge Canyon near the mouth of the Grand Canyon, that might have been enough to sway the Arizonans. Either of those two dams would be entirely within the state, and, with the addition of tunnels, irrigation water could be channeled out of either and sent south toward Phoenix. But the compact, of course, had mentioned nothing of the sort.

The real nightmare for the legislators in the Grand Canyon State was the possibility that the dam and storage reservoir would be built in either Black or Boulder canyon on the Arizona-Nevada border.

Not a drop of water from that potential impoundment realistically could be delivered to Arizona's agricultural valleys, and the power that dam would produce could be transmitted far too readily to California. No, the great new state of Arizona was signing nothing.

On a separate front, Arthur Powell Davis's engineers were finding much to like on the Arizona-Nevada border. The USGS survey in the summer of 1923 had turned up nothing in either Black or Boulder canyon that would make a high dam there seem problematical, and the "Bureau of Reclamation"—fresh with a name change—had made its own downstream reconnaissance in 1920, 1921, and again in 1924, Bureau field men each time liking very much what they saw in the dark, igneous canyons along the Nevada border.

The 1924 report on Glen Canyon's potential as a dam site was a rather more mixed bag. The federal engineers liked a lot of things about the canyon: its upstream regulation of the river would facilitate future downstream developments; it would not interfere physically with any other proposed developments; it would trap the river's awesome annual load of silt; and the uninhabited canyon was "of no particular value so far as is known." On the down side, the engineers noted that the canyon's sandstone, although probably adequate for the physical demands of the dam, was not ideal as a weighty dam's foundation and abutment material; the dam site was too far from existing agricultural and power-consuming markets, meaning California; the dam and its reservoir would not be able to control floods from the Paria, Little Colorado, and Virgin rivers that lay downstream from it; and, perhaps most telling, it was terribly inaccessible in comparison with other proposed sites. You just couldn't get there from anywhere.

Zealots such as George Hebard Maxwell and E. C. LaRue continued to fight the good fight, repeatedly making what they perceived as the sound and simple case that in developing a river system the only smart way to proceed was to start upstream and work your way down. But more than anything else now, the Glen Canyon site was the victim of the Colorado River Compact. LaRue's public-relations junket had backfired. Seemingly the only thing the commissioners and their cronies had remembered from the trip was its termination at Lee's Ferry. That remote river access point at the base of the Vermillion and Echo cliffs, John D. Lee's redrock hideout, was the last place for travelers to leave the placid river before it entered the forbidding Grand Canyon, and with Lee's Ferry fresh in their minds, the commissioners had deemed it a suitable place to divide one river basin in two. And this was the consequent rub: even if the Glen Canyon site hands down had seemed to be the best of the several dam sites—better topographically, better geologically, better hydrologically—the dam and reservoir now would be part of the *upper basin*. There simply wasn't any way that the powerful Californians would allow this key structure, this great plug that would make many things possible, to be built in what would amount to hostile territory. With virtual faucets in Glen Canyon, the upper-basin states could send downstream precisely the water the compact required them to, but no more. They could effectively *use* all of their water by keeping it in storage, and California had agreed to the compact in the first place only because it seemed to allow the big state to borrow as much of the river as it wanted to, at least for the foreseeable future. A dam in Glen Canyon plainly would be disastrous.

Arthur Powell Davis found the lower basin's arguments persuasive, but more than anything, he simply was itching to build a big dam *somewhere* on the Colorado—a dam that surely would be the largest ever constructed, and he had encouraged his crony, California Congressman Phil Swing, to introduce his Boulder Canyon Project bill into the House of Representatives, where, for two years running, it died quietly in committee. Even after the Bureau of Reclamation officially proposed building a monumental concrete gravity-arch dam in Black Canyon, 30 miles southeast of Las Vegas (the Boulder Canyon site now deemed less desirable by the Bureau), Swing's bill—which, to complicate matters, he continued to name after Boulder Canyon—languished undebated for two years more. California Senator Hiram Johnson had had no better luck with his own version of the bill until finally, in February 1927, it made it to the floor of the Senate, only to be beaten back by a three-day filibuster led by Arizona's two senators, who were wild-eyed in opposition. "I'll be damned if California will ever have any water from the Colorado River as long as I am governor of Arizona," George W. P. Hunt proclaimed in solidarity with the senators.

Assuming that these Arizonans weren't bluffing, the members of the Colorado River Commission quietly went back to work to make the compact a *six*-state agreement. Arizona could continue to object as loudly as it cared to, but if Congress could be assuaged by a fully ratified compact, then perhaps the dam finally could be authorized.

In May 1928, Swing's bill passed the House with an impressive majority, but in the Senate, Carl Hayden, Arizona's young new senator, mounted a marathon filibuster of his own, warding off any action on the bill until the session adjourned. But when the Senate reconvened in December, a quick motion to limit debate on the bill passed overwhelmingly, taking away the only real weapon the Arizonans had left. On December 14, the Senate passed the Boulder Canyon Project Act, authorizing the construction of a structure still strangely named Boulder Dam but located in Black Canyon, as well as Imperial Dam on the lower reaches of the river, and the All-American Canal, which would deliver water from Imperial Dam to the Imperial Valley. The bill limited California's share of the river's annual flow to 4.4 million acre-feet, plus any surplus, and Nevada was guaranteed 300,000, implying but not specifically stating that Arizona's share was 2.8 million acre-feet. It did assure Arizona and Nevada a paltry annual royalty from the sale of hydropower at the high dam, but other than that, Arizona took a true whipping. There would be no dam in Glen Canyon, none in Bridge Canyon, and no water would run toward Phoenix.

On the December day President Calvin Coolidge signed the Boulder Canyon Project Act into law, George Hebard Maxwell, ever the irrigation zealot, simply refocused his efforts. Returning to Phoenix from his sojourn in Washington, he was fighting now to win a series of smaller reclamation projects that would belong solely to Arizona. E. C. LaRue, however, was out of work, fired from the USGS because of his strident and seemingly endless criticism of the Bureau of Reclamation and its Black Canyon site.

After twenty-six years in operation, the Bureau of Reclamation finally was going to build the big one. Its engineers, the sharpest technological minds in the nation by reputation, had designed more than fifty dams and irrigation systems by the time Boulder Dam was authorized, yet none of them had had this kind of scope, this kind of challenge and appeal, and more than a few of their projects plainly had been flops—not engineering failures but rather farming fiascoes. These clever young fellows knew a great plenty about load factors and coefficients of thermal expansion, but some of them didn't know a tractor from a truck, a shovel from a shoehorn. And that was a problem. Reclamation had been founded not as an engineering outfit, whose raison d'être was to build, but rather as a kind of federal irrigation company, whose job it was to see that fields were fully watered.

Although it was Francis G. Newlands, congressman from Nevada, whose name had been attached to the Newlands Reclamation Act of 1902, it had been Theodore Roosevelt—president for less than a year following William McKinley's assassination—who convinced a skeptical Congress that western reclamation made sense. "The western half of the United States would sustain a population greater than that of our whole country today if the waters that now run to waste were saved and used for irrigation," Roosevelt had proclaimed in his State of the Union speech in December 1901. "The reclamation and settlement of the arid lands will enrich every portion of the country."

Newlands, assisted by George Hebard Maxwell and buoyed by his National Reclamation Association, first had proposed an autonomous reclamation agency that would develop the West's water resources as it saw fit, without having to submit to congressional oversight. Not surprisingly, Congress had rather firmly turned him down. Roosevelt, however, had been able to convince Newlands to delete that and a few other controversial aspects of his proposed legislation—as well as to tone down significantly its blatantly socialist rhetoric—and largely because of Roosevelt's ability to coerce as well as inspire, Newlands's bill had become law in June 1902.

On paper, the Reclamation Act seemed simple enough. Initial projects would be financed by a reclamation fund, made up of monies received from the sale of public lands in the sixteen western states and territories that would be eligible for reclamation projects. The fund subsequently would be replenished by the repayment, *without interest*, of the costs of the various projects by the farmers who benefited from them. The fund would be self-sustaining, and the West, in effect, would develop itself. But in practice, Reclamation's first few years were years of disillusionment. The new agency was flooded with requests for projects, and within five years it was at work on two dozen of them, yet apart from constructing dams, dikes, and ditches, it did little to see that its water was delivered to suitable soil in agreeable climates or to train farmers, many of whom had no experience with this strange new practice of irrigation. The result often was fields turned to salty swamps, crops successfully harvested but bereft of markets, and farmers gone bust. By 1910, the government had to loan the reclamation fund $20 million to keep it from going bankrupt as well, and in 1914, the act was adjusted to allow farmers twenty years instead of ten to repay their shares of their projects. A decade later, with 60 percent of the Reclamation irrigators defaulting on their obligations, Congress again stretched the repayment period, this time to forty years, still without interest. Twenty-two years into the reclamation era, only 10 percent of the money loaned from the reclamation fund had been returned to it.

Ironically, it was the engineers—the boys in the khaki shirts and the shiny, broad-brimmed hard hats, most of whom knew nothing about farming—who devised a way to bail out the farmers as well as to keep their agency open for business. If the irrigated farms couldn't pay for the expensive waterworks, the dams themselves would have to generate some income—and to do so they would have to generate electricity. Hydropower—the ability of falling water to crank dynamos to produce clean and reliable kilowatts—became the catchword of the twenties. Wherever possible, Bureau dams would be fitted with power plants. The electricity they produced, wholesaled to cities and private utilities, would provide the cash to keep the reclamation fund afloat. Farmers still would be required to pay their shares, but only when and if they were able to, and the fund would regain some basic stability.

Farmers and the small local irrigation districts that represented them were thrilled by the possibilities of hydropower, if for no other reason than that the money it would provide would take a little of the economic pressure off their backs. The private power companies, however, decried the potential "socialization" of their industry and were mollified only when they were assured by Reclamation officials and members of Congress alike that they could become effective partners in the plans, providing local service and transmission lines and, in some cases, actually owning the power plants at the toes of the power dams.

It was more than simply coincidental that the Boulder Canyon Project was authorized at a time when the Bureau of Reclamation was planning to get into the electricity business in a very big way. The canyons of the Colorado offered hydropower potential on a stunning scale. The Bureau's engineers had the technological wherewithal to build enormous concrete plugs that could impound oceans of water, and they knew precisely how to turn that water into electricity. And the Colorado, the river that was otherwise wreaking havoc, whose waters were running to waste, was perhaps the best place in the world to erect those sheer, sleek, majestic dams with their wonderful spinning turbines.

This river that had carved Grand Canyon, that drained fully one-twelfth of the landmass of the United States, was actually a piddling thing in terms of its water-carrying capacity. It was the third longest river in the continental United States—either 1,450 or 1,750 miles long, depending on where you marked its source—yet its annual flow matched only that of the meager Delaware River. The Colorado was nonetheless a superstar in three particular categories. It fell 14,000 feet in its journey from the Continental Divide to the Gulf of California; the Nile, in contrast, dropped only 6,000 feet in its entire 4,000-mile trek to the Mediterranean. And the Colorado might well have won the prize as the world's siltiest river, carrying 160 million tons of silt each year past Yuma, Arizona, a few miles upstream from the river's mouth—*11 tons* of silt for every acre-foot of water—the river flowing in a semisolid state, seventeen times more silt-laden than that so-called muddy Mississippi. It was the Colorado's steep gradient and the awesome amount of earth it hauled away to its delta that pointed to its third riverine claim to fame: no river on earth had cut such canyons.

Once upon a time, a languid river without a name wound through a well-watered plain, so reluctant to leave that lush country that in many places it bent back on itself in a series of oxbows and indecisive meanders. Then slowly, pressures within the earth began to push the plain upward, so slowly that the river was able to cut into its bed instead of changing course, the river carving deeper and deeper until the plain, a plateau now, stood suspended and dry above it. It was an event that geologists have labeled the Laramide Orogeny, and it created the Rocky Mountains and the high Colorado Plateau—actually a kind of lofty and irregular basin—beginning about seventy million years ago. But it wasn't until Pleistocene glaciers swept down the continent, then began to recede some six million years ago, that the river ever carried enough water to get truly serious about its erosive capabilities. Far upstream in its Green River tributary, Lodore, Desolation, Gray, Labyrinth, and Stillwater canyons were cut; along the main stem, Gore, Glenwood, Ruby, Horsethief, and Westwater canyons were carved by the raging waters. Below the broad confluence of the two upriver branches,

Cataract Canyon was next in line to be trenched, followed by Glen Canyon, Marble Canyon, the mile-deep gash of Grand Canyon, then dark Boulder and Black canyons, and lastly, narrow little Topock Gorge—all of them, more than a thousand miles of canyons, scoured by the relentless work of the river, the earth that once filled them finally deposited in an enormous alluvial basin, the uppermost 200 miles of the Sea of Cortez ultimately filled in by the dust and the soil from the canyons.

By the time Europeans first saw the canyoned river in 1540, its work was essentially complete. In his wanderings in search of the Seven Cities of Cíbola, Francisco Vásquez de Coronado had heard tales of a huge river that lay somewhere off to the west, and he had dispatched trusty García López de Cárdenas to lead a reconnaissance party that would try to determine whether it truly existed. Hopi guides, who had no doubts about the river's existence, led Cárdenas and his men across the Painted Desert, through scrubby piñon and juniper forest, and finally to the south rim of the Grand Canyon, where the Spaniards looked down into an astonishing subterranean world. At the bottom of this ghastly and impossible pit, almost three miles down by line of sight, Cárdenas could see a sliver of water. But, sadly, the stories hadn't been true. He rejoined Coronado with a report that he had encountered a terrifying chasm, but the stream at its nadir, he said, appeared to be little more than five feet wide.

Sixty years passed before Spaniards returned to the river, this time to give it a name: el Río Colorado, "the Red River," the river laden with silt. Juan de Oñate and his party of explorers first encountered the river in 1604 near its confluence with a tributary now known as the Little Colorado. Confusing the two rivers, Oñate actually named the tributary the Colorado; the main stem of the river he called el Río de Buena Esperanza, "River of Good Hope." But it was the bigger, redder river that he meant to get a measure of, and after bestowing a name upon it, his party followed its southwesterly course by marching across the plateau lands that flanked it, traveling as far as the flat delta at its mouth, where Oñate ceremonially took possession of the Colorado in the name of the king of Spain.

The river's name and its proper course were the subjects of minor confusion and occasional controversy for three subsequent centuries. In the territorial era, cartographers often confused this Colorado with an insignificant but much better known stream of the same name in Texas, and the legislature in Arizona Territory often toyed with the idea of renaming the river something, well, like Arizona. Additionally, it seemed to people in several other states that it made no sense whatsoever for the Colorado seemingly to get its start in the canyon country of southeastern Utah.

Above its confluence with the Green River, the main branch of the river had always been called the Grand—if indeed you could call it the main branch, that too being a disputed subject. The Green River, born in Wyoming's Wind River Range south of Yellowstone, journeyed 730 miles before it joined forces with the Grand; the Grand, whose headwaters tumbled out of Colorado's Never Summer Mountains in Rocky Mountain National Park, traversed only 430 miles before it reached the confluence. So why wasn't the longer Green the *true* river? boosterish newspapermen in Utah and Wyoming often asked. Because the Grand carried more water, one and a half times more, as a matter of fact, countered the partisans in the state of Colorado.

With a fight for the river's water looming in the early years of the twentieth century, it seemed obvious to members of the Utah legislature that a political as well as rhetorical advantage could be pocketed if the Green were renamed the Colorado, the true river then officially bisecting their beloved state. Wyoming politicians sanctioned the plan, but before either legislature bothered to enact the change—and while both were out of session—the jealous Colorado legislature, miffed in principle that the Colorado didn't originate in *Colorado*, and also mindful of how rhetorically valuable it would be to make it plain to the downriver states exactly where all this water came from, jumped into action. In the winter of 1921, a bill was introduced, debated, and passed under the cover of legislative darkness. The Grand River forevermore would be called the Colorado River, at last making it a unified waterway, 1,440 miles long, coursing from Colorado's Continental Divide to the Sea of Cortez. Congress cooperated and settled the matter by confirming the rather crafty change of appellation.

Almost four hundred years had passed between the time Europeans first saw the river and the time its current name was definitively attached to it; Indian peoples had discovered the river and given it a variety of names during the preceding thousand years, since their own migrations into the desert Southwest had commenced. Yet in all that span of time, the river and its canyons had changed hardly at all. In the 1920s, the canyons were no more or less imposing than they had been in the 1820s, when fur trapper James Ohio Pattie had decried "these horrid mountains, which so cage [the river] up, as to deprive all human beings of the ability to descend to its banks." And the river itself still roared down the canyons in awesome torrents or trickled politely across baking sandbars, much the way it had done in 1857, when Lieutenant Joseph Christmas Ives powered his shallow-draft steamboat *Explorer* from Fort Yuma, near the river's mouth, 300 miles upriver to Black Canyon, where the boat grounded on a submerged rock. Although the damage was minor, it was enough to give the navigator pause. Rather than risk further obstacles upriver, Ives and his men freed the *Explorer* and steamed back out of the canyon. In his official report on his exploration, Ives assumed that he and his subordinates had been "the first and will doubtless be the last party of whites to visit this profitless locality. It seems intended by nature that the Colorado River, along the greater portions of its lonely and majestic way, shall be forever unvisited and undisturbed." Yet only seventy-five years later, the rock that had halted Ives's upstream voyage was excavated by a growling, diesel-driven dragline to make way for the highest dam in the world, the largest construction project ever attempted.

When Henry J. Kaiser, a California highway builder who had been stirred to frenzied excitement by the very idea of building so big a dam, proposed tackling the project to W. A. "Dad" Bechtel, another small-time California contractor, Bechtel responded with the great understatement of the era: "I don't know, Henry," Dad said after hearing out the ebullient Kaiser. "It sounds a little ambitious."

The dam in Black Canyon—confusingly called Boulder Dam until 1930, when, while it was still in the planning stages, President Herbert Hoover's secretary of the Interior announced that thereafter it would be known as *Hoover* Dam—would be the largest structure of any kind ever built and, accordingly, the construction contract that would be let to the lowest bidder would be the largest single contract ever entered into by the United States government. That was the exciting part. But the problem with Hoover Dam, as Dad Bechtel had cautioned, was that it was going to amount to a good-sized piece of work. There were several construction companies in the East that potentially were big enough to take it on, but to most of them, a desert canyon near someplace called Las Vegas,

Nevada, sounded like a hell of a hardship post. Dozens of western contractors, on the other hand, wanted to get in on the action in the worst way, but virtually all of them were such small fry that it would be impossible to get surety companies—to say nothing of the Bureau of Reclamation—to take them seriously. The only solution, Kaiser figured out, was to form a breathtakingly big and powerful partnership.

After Kaiser convinced Bechtel what a terrific project this was going to be, the two were able to come up with only $1.5 million of the $5 million or so surety companies were expected to want as a cash guarantee before they would agree to underwrite the deal. But, Kaiser knew, there were a few other fellows out there who were trying to land this fish. Principal among them were two Mormon brothers, E. O. and W. H. Wattis, whose Utah Construction Company had specialized in building railroads until it had won the contract to build the Hetch Hetchy Dam in Yosemite National Park in 1917. Then there were Harry Morrison and Morris Knudsen, two former Bureau of Reclamation engineers, who lately had formed a construction company in Boise, Idaho; they had employed Frank Crowe, a former Bureau general superintendent who, if it were possible, was even more enthusiastic than Kaiser. "I'm wild to build this dam," Crowe confessed to the Wattis brothers and Morrison and Knudsen. Between them, the two companies could muster $1.5 million, but Morrison knew where else to turn. The J. F. Shea Company of Los Angeles, tunnel and sewer specialists, joined on for half a million; the Pacific Bridge Company of Portland, Oregon, contributed the same amount; and MacDonald and Kahn, Inc., of San Francisco, builders of hotels and office buildings, added $1 million to the potential pot. When Kaiser and Bechtel joined them, the western consortium was set.

In February 1931, at the Engineers Club in San Francisco, Six Companies, Inc., was incorporated. Seven firms were actually part of the deal, but Kaiser insisted on the name, borrowed from the tribunal that settled the grievances of the tongs, the Chinese crime families. It was Kaiser's private joke, and even though it was bad arithmetic, it was a good, simple appellation and Six Companies the seven became. The next step was somehow to secure the contract.

The group had less than a month to ready its bid, and William H. Wattis was in a hospital bed in San Francisco, where he was dying of cancer. There being little alternative, the several executives simply set up shop in Wattis's hospital room; Frank Crowe wheeled in the working model of the dam he had constructed; and in a series of marathon sessions, the men haggled over, then hammered out down to the dime precisely what this thing would cost. They tacked on 25 percent for contingencies and profit, and, two days before the Bureau of Reclamation's deadline, Crowe put a final polish on the Six Companies proposal. At $48,890,000, the partnership's bid was $5 million under its closest competitor and just $24,000 over Reclamation's own estimate when bids were opened in Denver on March 3. Six weeks later, Six Companies had a job.

The Bureau of Reclamation had engineered fifty concrete dams by the time Hoover Dam was on the drawing board, so it wasn't new to this audacious sort of enterprise, but the *scale* of Hoover Dam was something else again. It would contain more concrete than all those previous dams combined, and to amplify the challenge, it would be erected in one of the most inhospitable places in North America. Summertime temperatures inside Black Canyon commonly reached 130 degrees, and the setting process of the concrete was expected to generate enough heat to add appreciably to the air temperatures. That much heat, in theory at least, had seemed to make a concrete dam impossible.

Assuming that a gigantic form, as big as the dam itself, could have been built, and assuming that the form somehow could have been filled to its rim with concrete, the dam would have needed roughly 150 years to cure. During that lengthy wait, it would have been unusable, its vast interior the consistency of custard. And, as it ever so slowly cooled, the dam would have shrunk and cracked and buckled to the point that it ultimately would have been little more than the modern, technological version of the boulder that had grounded Lieutenant Ives—a mid-channel obstruction that could aggravate the river but never stop it. The solution, as devised by the boys at the Bureau, was to assemble the dam from a series of little blocks—230 of them, and little only in relation to the overall size of the dam. The smallest of the multisized blocks would be as big as a house. A network of one-inch copper tubing, enough pipe to reach from the dam site to San Francisco, would be embedded in the concrete blocks at five-foot intervals, and refrigerated water continually flowing through the pipes would turn the rising dam into a kind of gigantic ice chest, cooling the concrete uniformly and ultimately dropping the temperature on the water-contacting upstream surface of the dam to 43 degrees, on the exposed downstream face to 72 degrees. Bureau engineers were confident that instead of requiring a century and a half to cure, their cooling system could complete the task in nineteen months, give or take a day or two.

This dam would be a mongrel breed, an innovative cross between the basic gravity dam, which had been in use for four hundred years, and the continuous arch dam, which had been used successfully only since late in the nineteenth century. The way a gravity dam worked was to resist water pressure by turning it downward, toward the dam's wide and heavy base and its bedrock foundation below. A gravity dam was pyramidal, thick and substantial at the bottom, thin and light at the top, whether made of rock and earth-fill, as was most common, or concrete. Arch dams, on the other hand, could be built only of concrete or masonry, and they were suitable solely to deep canyons. Sometimes called eggshell designs, they were incredibly thin in comparison, and they resisted water pressure simply by their action as arches, transferring the pressure to thrust that was carried by the canyon-wall abutments. Arch dams required far less mass, but they required rather more specific topographics.

What the Bureau of Reclamation's dam designers had done for Hoover Dam was to devise a structure that would carry its load in part by weight and in part by arch action. It employed the best of both basic types, and, they were willing to guarantee, it would last. Vertical on its upstream face, the dam's downstream face would slope dramatically. Along its narrow crest, the dam would arch 1,244 feet, almost a quarter mile, from one canyon wall to the other. Yet, although that crest would be a mere 45 feet of horizontal concrete, the base, 726 feet below, would be a whopping 660 feet thick. Just upstream from the dam, and mounted midway up the opposing canyon walls, two enormous intake towers, each as tall as a forty-story building, would admit water into 30-foot-diameter steel penstocks and send it plunging to a total of sixteen turbines in the power plant at the toe of the dam. Before the project was finished, 3.25 million cubic feet of concrete, 3 million board-feet of lumber, 662 miles of copper pipe, and untold tangles of electrical cables, conduits, hoses, adits, and drains would be pressed into government service, a conscription whose length, Bureau engineers knew, would be measured in centuries.

"The problem" with getting this thing built, Frank Crowe explained, "which was a problem in materials flow, was to set up the right sequence of jobs so they wouldn't kill each other off." And by "they" Crowe meant his partners as well as the men who were building the blocks. "Kaiser thought the job should be run like an army, with a general in supreme charge," Crowe explained. "The idea got nowhere because no one, least of all Henry himself, wanted to be a private." What the Six Companies fellows finally worked out was a four-person executive committee, with Charlie Shea heading up canyon construction; Felix Kahn in charge of money, legal matters, and room and board for the crews; Dad Bechtel's son Steve overseeing materials and transportation; and Henry Kaiser serving as chairman of the group—a CEO if not quite a supreme commander.

Within a month of the awarding of the contract, Six Companies crews had begun blasting the four diversion tunnels that would allow a potential flood of 400,000 cubic feet per second to pass around the dam site—*double* the river's maximum known flow, but then, with the Colorado, no one wanted to take any chances. And on the canyon rim, Six Companies was at work on a construction camp dubbed Boulder City—750 wooden bungalows equipped with a blessed innovation called air conditioning, eight men's dormitories, several mess halls, and a huge company store. Boulder City seemed to take shape overnight, but it took two full years to dig the tunnels, build cofferdams that would divert the river and protect the site, strip the canyon walls to expose a solid new surface for the abutments, and excavate the riverbed 40 feet down to rock.

In June 1933, with a new president in the White House and an executive order stating that this would indeed be *Boulder* Dam, not something named after the Republican who had presided over the onset of the Great Depression, concrete finally poured into the canyon—220 cubic yards of it an hour, twenty-four hours a day, enough concrete each day to form a stream 20 feet wide, a foot deep, and a mile in length, all of it batched in a single concrete plant perched on the Nevada edge of the canyon. With 3,000 men working now, bucket by relentless bucket the forms were filled; then new forms were erected on the concrete blocks below, the work proceeding at a breakneck pace, frantic enough that it resulted in thousands of minor and major injuries and ultimately in 110 deaths—awful enough work that early on laborers had struck to raise their base pay from $4 to $5 a day. But with hundreds of Depression-battered men camped in a "Rag Town" near the site, dozens were desperate to take the place of every striker—and with the Six Companies' charge that it was Communists and Wobblies who were fomenting trouble—the strike had been broken in only a few days. "They will have to work under our conditions or they will not work at all," W. H. Wattis had proclaimed from his hospital bed, and for four more years no one struck again.

One of the thousands of men who had gone first to booming Las Vegas, then on to Boulder City to find work in the depths of the Depression was a stocky, confident kid named Lem Wylie, who had received an engineering degree from the University of New Mexico in 1931, and who had been lucky enough to find work hauling a wheelbarrow for the company that held the subcontract to build Boulder City's sewers, curbs, and gutters. Living in a tent on the edge of town in April 1932, Wylie had approached Ralph Lowry, the Bureau of Reclamation's field engineer, explaining that he, too, was an engineer—a good one—and that he'd appreciate a job better suited to his talents, if one was available. Lowry, somehow impressed with the young man but having nothing much to offer him, told Wylie that he could come aboard as a rear chainman on a survey crew down in the diversion tunnels. It was hot, dusty, suffocating work, but it was a *government* job, and nothing was more secure than that, Wylie knew. He jumped at the offer, and by the time the dam began to rise, he had been promoted to a junior inspector, overseeing minute details of the work Six Companies did. Wylie, now rising quickly up the ranks, was a Bureau of Reclamation assistant engineer on March 23, 1935, the day the final bucket of concrete was poured on the crest of the dam.

In just twenty-two months, the greatest structure on earth had taken shape. It was a spellbinding achievement, and Boulder Dam was unquestionably the biggest story in the country throughout that spring and summer. By September, the powerhouse was completed; Frank Crowe was ready to "saw the job off," as he liked to say; and President Franklin Roosevelt, battling the Depression as much with rhetoric and ceremony and a sense of national pride as with jobs programs, made the difficult journey west personally to dedicate this magnificent structure, this proof of American ingenuity, strength, and resilience. Twelve thousand people crowded onto the crest of the dam on September 30 to hear Roosevelt laud its visionaries, its designers and builders, and to memorialize the men who had given their lives to see it through to completion.

Neither W. H. Wattis, E. O. Wattis, nor Dad Bechtel lived to celebrate its collective achievement, yet Six Companies, even without them in the end, had accomplished something extraordinary. It had finished the job ahead of schedule; it had produced a profit of more than $10 million, 21 percent of the contract price, and in a fitting kind of irony, these several industrialists were heralded as populist heroes, as men who put other men to work, and who, with the help of their dedicated labor, finally had subdued the Colorado.

The reservoir behind the dam, named in honor of Elwood Mead, Reclamation commissioner during the long years of the dam's construction, had begun to fill in February. But it was not until October 1936 that water rose high enough to spill down the penstocks to begin generating power. Then, in 1942, in the midst of a war that had become the next enormous project for these industrialists and working people alike, the reservoir actually filled, doing so in only seven years, backing as far upriver as the Grand Canyon, reaching to within inches of the top of the great gray dam as poet May Sarton rhapsodized, "Not built on terror like the empty pyramid, / Not built to conquer but illuminate a world: / It is the human answer to a human need . . . It proves that we have built for life and built for love / And when we are all dead, this dam will stand and give."

By 1946, with the war over and gas rationing ended, hordes of Americans were on the road, seeing this nation's sights in a kind of patriotic frenzy, and the dam in Black Canyon hosted hundreds of thousands of visitors, most of them peering over the rim of the canyon at the giant monolith first in a kind of speechless awe, even reverence, then in giddy delight. Among the pilgrims that summer were Wallace Stegner, his wife, Mary, and their young son. Stegner, at work on a biography of John Wesley Powell—a biography that would applaud Powell's claim that while, yes, you could reclaim some small portion of the Southwest's arid lands, you could never do so on an enormous scale—nonetheless was as captivated by the dam as were his sunburned contemporaries who stood beside him and stared. Writing about it in *The Atlantic Monthly*, Stegner affirmed that "nobody can visit Boulder Dam without getting that World's Fair feeling. It is certainly one of the world's wonders, that sweeping cliff of concrete, those impetuous elevators, the labyrinth of tunnels, the huge power stations. Everything about the dam is marked by the immense smooth efficient beauty that seems peculiarly American." And so it seemed in the heady, hopeful euphoria that followed the terrible war.

Early in 1947, the first Republican Congress in fifteen years passed a bill certifying that the structure blocking Black Canyon of the Colorado forever after would be known as Hoover Dam. These Republicans, in congressional power again at long last, were not about to allow Herbert Hoover—engineer, builder, architect of the Colorado River Compact—to be kicked around anymore. President Harry S Truman made no objection, made no comment of any kind about the bill, and simply signed it into law.

Early in 1947, thirty years after E. C. LaRue first had pointed out its possibilities, Glen Canyon was under scrutiny again, the focus of a Bureau of Reclamation grown bold and confident of its dam-building prowess, the focus of Colorado River men in need of another canyon.

3

THE BATTLE FOR
ECHO PARK

It was an article in *The Atlantic Monthly*
that first alerted Theodore Roosevelt to
the majesty of the canyon country. The
1898 essay was the product of a California mountaineer and transcenden-
talist named John Muir, a man who
saw unspoiled nature as a "window
opening into heaven, a mirror reflecting
the Creator"—a kind of Eden miraculously still extant. Muir had founded
an organization of like-minded San
Francisco–area folk called the Sierra
Club six years before. Although California's Sierra Nevada, the "range of light"
as he called it, held a special fascination
for him, he was convinced that pristine
lands throughout the country deserved
some sort of federal protection, much
the way the Yellowstone plateau had
been declared a "national park" in
1872, as had the region surrounding
Yosemite Valley in 1890. But it was the
canyons of the Southwest, the canyons
cut by the Colorado River, that were
the subject of the article in the *Atlantic*,
and Muir argued that they were no less
deserving of protection than the alpine
and sylvan kinds of country that most

Americans envisioned as places of natural beauty. Muir wrote glowingly of the special grandeur and serenity of the desert, and affirmed that this spectacular canyon landscape was one of "God's wild blessings" and therefore should become the nation's newest park.

Although eager to have a look at these desert canyons, Roosevelt, governor of New York when the article appeared, had had to wait until 1903 to journey west to see them for himself. On a summer trip to visit Yellowstone and to tour Yosemite with Muir, Roosevelt, by then president of the United States, stopped first at the Grand Canyon, where he took time to join a mountain-lion hunt and to remark that the canyon "was the one great sight every American should see." Counseling the nation to leave it as it is, he cautioned, "You cannot improve upon it. The ages have been at work on it, and man can only mar it."

When he met Muir in Yosemite and gave him a rapturous account of the "manliness" of the Grand Canyon and his days hunting inside it, Roosevelt was admonished by the conservationist both to "get beyond the boyishness of killing things" and to do something concrete to protect the canyon environs. Roosevelt, though taken aback, told Muir that perhaps he was right about hunting, and he assured him that Grand Canyon would be preserved. Five years were to pass, however, before Roosevelt signed an executive order creating Grand Canyon National Monument, a 25-by-50-mile preserve in the rough center of the Grand Canyon, including about 75 of the 280 miles of the Colorado's course from one end of the canyon to the other. As the nation's chief executive, Roosevelt was empowered to create national monuments by simply signing his name to documents so ordering them. National parks, on the other hand, had to be authorized by Congress. And although Roosevelt said he would have preferred to see the region become a park, he knew that Arizona mining, timbering, and agricultural interests, as well as the congressmen they kept in office, would have battled him furiously on the issue.

Yet within another decade, with the completion of the Santa Fe railroad's spur line to the canyon's south rim and its attendant increase in tourism, as well as growing national sentiment to acknowledge formally that the canyon was something very special, Arizona Senator Henry Ashurst and Carl Hayden, still a congressman, agreed to shepherd a park-creating bill through their respective houses of Congress. Neither man *wanted* to see a park created, but each understood that whatever clever obstructions and delaying tactics they used, a Grand Canyon bill would have the votes to succeed before long. If there had to be a park, it might as well be a park that Arizonans could live with. The two men made sure that their Grand Canyon bill did its best to remove mining claims from inclusion in the park, to keep the boundaries largely inside the canyon itself and away from the valuable timber on its rims, and Hayden was careful to hammer language into the bill specifically allowing future use of the preserve for federal reclamation projects. With the long-standing Arizona opposition to the park now removed, the bill easily wound its way through Congress, and Grand Canyon became the nation's seventeenth national park in February 1919.

Herbert Hoover was still president and the dam named in his honor was almost ready to rise in 1932 when the next stretch of the river corridor was set aside—319 square miles of land, including 39 miles of river frontage, adjacent to the western boundary of Grand Canyon National Park, and again called Grand Canyon National Monument. Later that year, when voters replaced the genial engineer with a patrician from New York State named Franklin Roosevelt, and when Roosevelt appointed Harold Ickes—stolid and surly and known as the "Old Curmudgeon"—to be secretary of the Interior, the nation began a virtual parks-acquisition spree. The park service took control of the land surrounding what would become Lake Mead—the new reservoir behind Boulder née Hoover Dam—which was administered under a novel classification called "national recreation area," a designation applicable to public lands that were neither wild nor undisturbed but that nonetheless offered particular opportunities for outdoor enjoyment. Then Ickes recommended to this second President Roosevelt that he create dozens of new national monuments, including Cedar Breaks, Death Valley, Saguaro, and White Sands in 1933, Joshua Tree in 1936, Capitol Reef in 1937, and a major expansion of Dinosaur in 1938—all of those in the scarcely populated and still largely unvisited desert Southwest. Roosevelt took pen in hand and obliged.

In 1940, Ickes offered the president a grandiose proposal, based on a recommendation made three years earlier by National Park Service Director Arno Cammerer: the Roosevelt administration would establish an enormous new preserve straddling the Colorado River and reaching from Lee's Ferry, near the Utah-Arizona border, north and east all the way to the town of Moab, Utah, on the main stem of the river, and up the arm of the Green almost as far as the town of Green River, Utah. It would be called Escalante National Monument, and it would encompass 280 miles of the Colorado's winding canyons, including all of Glen and Cataract canyons, 150 miles of the canyons of the Green, and, along its north bank, more than 70 miles of the San Juan River, 4.5 *million* acres in all. Only a single rough and primitive road cut across the proposed monument throughout its length, and where this two-track road met the Colorado, at a settlement called Hite, there wasn't a bridge or even a ferry to connect the opposing shores. No place in the entire United States was more remote than the Escalante region—named for one of the two Spanish priests who had traversed it in 1776—and perhaps there was none more unique. It was a land of bald rock striped in rainbow hues, of standing rock—bare buttes and spires and haughty monoliths—of sheer and rocky canyons that cut deeply into the earth, endlessly twisting back on themselves, their rivers and intermittent streams cutting sweeping overhangs, plunge pools, cool and mossy grottoes. If the Grand Canyon was overwhelming—and it was—the scale of the Escalante country was more intimate some-how, more accessible, an inviting maze of stone and sky and, sometimes, water.

This new national monument would be a bold stroke, there was no question about it, yet Ickes knew his chief executive had vowed that he would be known as a great conservationist. Unlike his cousin Teddy, however, who had been captivated by his introduction to the desert canyons, Franklin Roosevelt hadn't been much impressed by his own first glimpse of the Grand Canyon. "It looks dead," he had commented, peering into the canyon from Yavapai Point. "I like my green trees at Hyde Park better." Roosevelt hadn't seen Glen Canyon, nor any of the Escalante country, and it bore no more relation to his home in the Hudson River Valley in Duchess County, New York, than did the moon, yet Ickes was confident that the president would not be able to resist creating what would be the nation's largest preserve, larger even than Yellowstone.

But much of the world was at war in 1940, and Franklin Roosevelt had many things on his mind. Rather than acting quickly on Ickes's proposal, he put it on the back burner, requesting comments from Utah's congressional delegation as well as from Marriner Eccles, a Mormon Church leader, banker, and businessman, now chairman of the Wattis brothers' Utah Construction Company and FDR's chairman of the Federal Reserve. To a person, the Utahns who got wind of the proposal were outraged. Didn't Ickes realize that there were important mineral reserves in southeastern Utah? There had been substantial gold-mining activity in Glen Canyon in the years surrounding the turn of the century, and no doubt there would be again, not to mention the grazing leases that blanketed the country on both sides of the Colorado. Too little of Utah belonged to the people of Utah to begin with, and the last thing those good, hardworking people needed was to have more of the lands administered by the General Land Office removed from *use*—from mining and grazing and like activities that fed people and produced a few tax dollars. No, Utah most definitely could not abide some sort of butterfly-chasing sanctuary astraddle the Colorado River.

Ickes was undeterred. He had dealt with these hinterland hayseeds before. He knew that with patience and perseverance, and, if need be, with the threat that the federal money-faucet might run dry, the Utahns could be convinced that Escalante National Monument would be of benefit to them. His plan to put time on his side was still in place in early December 1941, when Japanese planes screamed out of the sky at Pearl Harbor, Hawaii, and America went back to war. Harold Ickes remained secretary of the Interior until shortly after Franklin Roosevelt's death in 1945, serving in that position longer than anyone before or since. But he never again proposed to the president the creation of a national monument in that strange and wonderful Escalante country encompassing and surrounding Glen Canyon.

If Interior's park-acquisitions program came to a standstill during the war years, its big-budget and now high-profile Bureau of Reclamation remained decidedly busy. The Roosevelt administration early on had seized the Bureau and its new power-producing, dam-building fervor as the perfect means to put people to work building vital, if enormously expensive, public-works projects. Hoover Dam had been a tremendous success, and the Bureau had taken little more than a deep breath before it had launched into California's gargantuan Central Valley Project, into Shasta Dam—bigger by half than Hoover was—on California's Sacramento River, then into Grand Coulee Dam on the Columbia—bigger in concrete volume than Hoover and Shasta combined. By 1941, the United States possessed what many critics considered a ludicrous and altogether wasteful amount of surplus, publicly owned hydropower, electricity that wouldn't find a market for fifty years or so, by some estimates. Yet only a year later, with the nation desperate to arm itself, every bit of the Bureau of Reclamation's hydropower was being swallowed up by defense plants, and the Bureau was scrambling to come up with more. By the time the war was over, 60,000 aircraft had been built with Bureau electricity—a contribution that arguably was the single most important one in the defeat of the Axis powers—and Reclamation's stature seemed secure for decades to come.

Although no major projects were begun during the war, Reclamation planners and designers were determined not to be caught flatfooted at its conclusion. There was much work to be done in the Columbia and Missouri river basins, in California's bounteous rivers and rich valleys, and the work of harnessing the Colorado had only just begun. In 1946, under the leadership of Mike Straus, a wealthy New Englander, a disheveled, unmade bed of a man who had a remarkable flair for project salesmanship—also Reclamation's first nonengineer commissioner—the Bureau released its seminal study, *The Colorado River: A Comprehensive Report on the Development of Water Resources*. Nicknamed the "Blue Book," and subtitled *A Natural Menace Becomes a Natural Resource*, the report outlined 134 potential projects, one in virtually every river canyon and farming valley throughout the upper and lower basins of the Colorado. But it also issued a warning: there wasn't nearly enough water in the river to make all the projects feasible, and worse, there wasn't even as much water in the river as the seven states had been told there was way back in 1922 when they first had signed the Colorado River Compact.

The years of the teens and twenties had been extraordinarily wet ones throughout the West, Bureau hydrologists now knew. In the case of the Colorado watershed, the long-term annual flow, which had been estimated at 17.5 million acre-feet, more accurately was something like 13, *maybe* 14 million acre-feet. Yet the compact had apportioned 7.5 million acre-feet to each basin; California was already using nearly 5 million; and in 1945, the United States had signed the Mexico Water Treaty, guaranteeing that country 1.5 million acre-feet annually. The Colorado, with only one major storage dam in place, and three small diversion dams in operation below it, was already a "deficit river," a stream that couldn't quite meet its demands.

Acknowledging the hard hydrological facts, the Blue Book made it clear that some nasty decisions would have to be made regarding *which* projects actually would be built, and it strongly encouraged the states within each basin to decide their intra-basin shares. That would be the critical first step, and it would have to precede any construction. But in the lower basin, an agreement seemed problematic at best. Arizona was still whining about how it had gotten nothing from Hoover Dam, still complaining about how the Glen Canyon and Bridge Canyon dam sites had been scrapped because of California's clout, still decrying the water thievery that the Bureau of Reclamation was sanctioning. Arizona was threatening to settle the whole business in federal court, if it had to, and that was fine with California. The state's lawyers could keep a water suit tied up in the courts for *years* if they had to, during which time the Golden State could use all the water it wanted.

In the upper basin, things were decidedly more amicable, and with Reclamation's encouragement, the four states executed their own compact in October 1948, agreeing to allocate 51.75 percent of the river's resources to Colorado, 11.25 percent to New Mexico, 23 percent to Utah, and 14 percent to Wyoming. The four states seemed to understand very clearly that they *had* to develop some significant water storage in the upper basin. California very soon would begin to "borrow" upper-basin water, if it flowed downstream unchecked, and with millions of immigrant Californians dependent upon it a few years down the line, how would these Rocky Mountain states ever get it back? Their compact signed, the four states began to court the Bureau of Reclamation with open and unbridled passion. Mike Straus and his engineers and economists soon responded in kind.

In 1949, a Reclamation planning report suggested several potential storage-reservoir sites: one in the lower basin at Bridge Canyon near the western end of the Grand Canyon, largely as a means of mollifying the Arizonans; another in Glen Canyon in the upper basin, a recognized dam and reservoir site since 1916; and two sites far up-basin on the Green River, one in Flaming Gorge on the

Wyoming-Utah border, the other at Echo Park near the Colorado-Utah border, inside Dinosaur National Monument. These possibilities weren't firm proposals by the fellows at Reclamation; they were mere sink-or-swim suggestions. The Bureau needed to know which projects would float politically before it started designing in detail how to impound the water.

The Sierra Club, 6,000 members strong that year, long had considered itself to be a guardian of the West's national parks—since John Muir had led the club in a losing battle in 1913 to save Yosemite's Hetch Hetchy Valley from inundation. Not long after the turn of the century, the city of San Francisco had been in need of a larger and more dependable water supply, and it had proposed the construction of a dam on the Tuolumne River to the federal government, which owned the land inside the park and therefore controlled all its potential uses. Muir had argued publicly with Congress and San Francisco's municipal officials, and privately with President Theodore Roosevelt—with whom he had remained in contact in the years since their Yosemite wilderness trip—that while there were many possible reservoir sites in the Sierras, there was only one Yosemite, one Hetch Hetchy Valley, and the nation had already decided that they should be preserved in their natural state. Roosevelt was sympathetic with Muir's position—he fancied himself a man who had been shaped by his experiences in wild country—yet he was equally swayed by the widespread notion that the conservation of natural resources meant their *wise use*. When Gifford Pinchot, influential chief forester of the U.S. Forest Service, endorsed the Hetch Hetchy project as conservation at its best, Roosevelt formally sided with him and Congress gave the dam its approval. Seventy-six-year-old John Muir died in 1914 as the Wattis brothers of the Utah Construction Company were raising a dam across the Hetch Hetchy, leaving Muir's successors in the Sierra Club to vow that no other national preserve would ever suffer a similar fate.

In 1949, the Sierra Club remained a regional organization—a confederation of conservative businesspeople and academics, most of them men, who liked to load mules with a week's provisions and wander into the high Sierra. The summer "high trips," in fact, were the club's principal endeavor. It did pay attention to political issues that affected the Sierras in particular, and wilderness in general, but many club members were decidedly uncomfortable with the interest of a few young firebrands in openly fighting for conservation causes. So when the Bureau of Reclamation's proposals for the Colorado River basin were made public, there was a mixed reaction among club members. Some, who had been paying attention to the Colorado and its potential development, were quickly convinced that Reclamation's announcement boded ill indeed for the Grand Canyon, yet at least there was the certainty that the Arizona-California confrontation would fester indefinitely, long postponing the building of any dams. And as for the dam site in Dinosaur National Monument, former club president Walter Huber actually had visited the remote site, and he reported that it was nothing to worry about—just rock and sagebrush, certainly nothing special, although, yes, it was a national preserve. Except for Huber, novelist and historian Wallace Stegner, photographer Ansel Adams, and a few others, club members were rather exclusively oriented to the alpine out-of-doors, and none had visited the desert Southwest. The Grand Canyon, everyone knew, was unique in all the world, but otherwise, the region as a whole sounded like a *lot* of rock and sagebrush.

The club's official response to the Bureau's proposals was to ignore the dam in Dinosaur altogether. And by vote of its board of directors, the club went so far as to *endorse* the proposed Bridge Canyon Dam, despite the fact it would impound water throughout the length of Grand Canyon National Monument, despite the fact it would push slackwater 13 miles along the border of Grand Canyon National Park. Club director Bestor Robinson—a former club president, attorney, and chairman of a conservation board advising President Truman's secretary of the Interior, Oscar Chapman—had argued that the responsible position for the club to take was to support the dam *if* it was built to a low height that would minimize the size of its reservoir pool, if it did little to detract from the beauty of the inner gorge of the canyon, and if the reservoir would be made available for recreational use. When board member David Brower suggested that, additionally, the club should favor Bridge Canyon Dam *only if* a dam in Glen Canyon were in place first as a means of trapping the Colorado's enormous silt load, the organization passed a resolution offering its qualified support.

In 1950, a bill sponsored by Carl Hayden, now an Arizona senator, authorizing solely Bridge Canyon Dam successfully wended its way through the Senate, but companion legislation was trounced soundly in the House, where representatives were of no mind to approve anything until Arizona and California settled their neighborhood squabble.

With the fate of Bridge Canyon Dam decided, at least temporarily, conservationist concerns about Reclamation's threat to the national park system became focused specifically on Dinosaur. And while the Sierra Club had no interest in the fate of the remote little monument, other conservation organizations, most of them national in scope, were interested indeed. The National Park Service, also alarmed, reissued its 1946 report on recreational resources within the Colorado River basin, a report that had highlighted the region's unique character and the recreational potential of its wild, unspoiled canyons. A separate park service pamphlet identified the Bureau of Reclamation as the parks' particular nemesis: "The greatest peril to the parks from dam proposals comes from the plans and programs of the governmental dam-building agencies themselves and the pressures which their activities generate in the various sections of the country." In plainer language, if the Bureau wouldn't tout its pie-in-the-sky projects so loudly and so early on, the people out west wouldn't scream that they had to have them.

Faced with mounting pressure from his subordinate Newton B. Drury, the director of the park service, as well as from numerous conservationists, Interior Secretary Chapman decided to call an informal hearing in his office in April 1950, to listen to both sides on the issue of whether a dam in Dinosaur would be compatible with the principles and goals of the park system.

The conservationists' only motive, they wanted the secretary to know, was to preserve Dinosaur and the rest of the nation's parklands. "We recognize thoroughly the importance of water," the Izaak Walton League's William Voight told Chapman. "No one in his right mind can be opposed to sound and logical development of that prime resource." On the other side of the issue, Utah Senator Arthur V. Watkins tried to be equally conciliatory, claiming that no one was more appreciative of "rugged scenery and the preservation of nature's great wonders" than he was, yet you could only take scenery so far. "To my mind," stated the senator, "beautiful farms, homes, industries and a high standard of civilization are equally desirable and inspiring."

Chapman duly deliberated on what he had heard. Then in June

1950, confident that the issue now would be settled, he sent a memorandum to the heads of both Reclamation and the park service outlining Interior's approval henceforth of the idea of a dam at Echo Park in Dinosaur, yet assuring them that the decision would "not be a precedent for tampering with the inviolability of our national parks and monuments," and instructing the two agencies to stop their silly bickering.

The conservationists, shocked by this Interior secretary who seemed to believe the nation could have it both ways, and equally adamant that the Echo Park issue was *not* settled, were buoyed a little by the appearance of an article in the *Saturday Evening Post* soon thereafter by Bernard DeVoto, novelist, historian, and a native of Utah. The angry article, entitled "Shall We Let Them Ruin Our National Parks?"—a rather inflammatory one for a staid and decidedly mainstream publication like the *Post*—generated widespread attention and commensurate public outrage. *Reader's Digest* promptly condensed and reprinted the article under the same title, and soon millions of Americans were aware of a little monument that theretofore they had never heard of, and they were hopping mad about what this outfit called the Bureau of Reclamation was planning to build there.

Mike Straus's reign at Reclamation was about to come to an end. During his seven years as commissioner, the Bureau had been transformed from a kind of administrative dynamo making electricity for the war effort into an agency that built dams for dams' sake—not just to provide irrigation water or flood control or hydropower, but to do all these things as part of a hammer-and-tongs technological assault on the whole of the arid and undeveloped West. But in November 1952, the Democrats had lost their twenty-year hold on the White House, and despite his best efforts to make himself appear indispensable, Straus knew this Republican, Eisenhower, almost surely would ask him to resign.

The Democrats had campaigned in the West in the summer and fall of 1952, warning that if the Republicans came to power, farmers and ranchers could kiss reclamation good-bye. The Republicans were tight as ticks to begin with, and Eisenhower in particular had made it plain that he would direct the nation away from the socialism of the Roosevelt-Truman decades. As far as Straus and his boss Oscar Chapman were concerned, the situation was clear: Congress obviously would remain receptive to Reclamation's endeavors, and, God knew, there was much good work left to be done, particularly in the upper Colorado basin. If the two men could get the massive upper-basin project that had been in the planning mill for several years officially on the table before they left office, then surely cooperative senators and representatives would introduce authorizing legislation as soon as Congress reconvened in January 1953, and the project might well win approval before Eisenhower and company could put on the brakes. Their legacies at Interior and Reclamation would long outlast their terms of service because this one, dubbed the Colorado River Storage Project, was going to be a whopper.

By the early 1950s, the Bureau had given up thinking in terms of single dams, single irrigation projects—it was just too difficult to juggle the figures to make them economically feasible. The new Reclamation math called for basin-wide calculations of costs and benefits, thereby making it possible to come up with a way to pay for a series of irrigation projects, which otherwise would lose big money, with the power proceeds from dams hundreds of miles upstream or down from the lands that would receive Bureau water. And now Mike Straus's swan song was going to do precisely that. The cornerstones of the CRSP would be two massive storage reservoirs and power-producing dams, one at Echo Park on the Green River, the other in Glen Canyon on the main stem of the Colorado. The project would include three more storage facilities: on the Gunnison in western Colorado, on the Green in northeastern Utah, both also power producers, and on the San Juan in northwestern New Mexico. Twelve different irrigation projects would be built in four states from the sale of CRSP power—1,622,000 megawatts' worth of power, the combined total more than ten times the output of Hoover Dam's power plant—supplying, repaying rather, 85 cents of every dollar spent for irrigation, saddling the farmers and municipalities with only 15 percent of the cost of their bountiful new federal water supply. For a projected total cost of $1.4 billion—real money back in those days—the project would store 48.5 million acre-feet of water, almost *four years* of the river's total flow. Three decades after Hoover Dam, the upper river, too, would begin to resemble a faucet. And a natural menace would become a natural resource as Mike Straus bade his beloved agency farewell.

Politicians and water planners in the upper-basin states were ecstatic, and the Bureau's career engineers were hopeful that the plan had been unfurled in time to guarantee them good work throughout the 1950s, yet there were others in government service who weren't so sure about this Colorado River Storage Project. The U.S. Geological Survey, for instance, worried that gigantic underground basins beneath the Kaiparowits Plateau adjacent to Glen Canyon might draw water from the proposed reservoir, inadvertently creating two additional subterranean reservoirs that would store water, to be sure, but which would make it impossible to retrieve. The Bureau of Mines wondered whether it made sense to build enormous hydroelectric facilities in the midst of a region containing the largest known coal reserves in the world. The Fish and Wildlife Service expressed concern that the proposed reservoirs would destroy 700 miles of riverine fisheries, game, fur-animal, and waterfowl habitat. The Public Health Service pointed out that the average population density in the upper Colorado basin was 2.6 people per square mile, compared with a nationwide average of 44.2 people. The largest "city" in the entire region was Grand Junction, Colorado, population 12,000. Clearly, the basin's current health-services infrastructure could not support the population influx the project would bring. (As a kind of afterthought, the Health Service added that with the creation of all that motionless water, people could expect a few more mosquitoes.) And just prior to his resignation, Park Service Director Drury authored his agency's official reaction to the CRSP, declaring that the proposed dam in Dinosaur National Monument, in particular, "would result in nationally significant scenic, scientific and other recreational values being irreparably damaged or lost all together."

But if one branch of the Department of the Interior, the park service, was willing to set itself at odds with another branch, the Bureau of Reclamation, a third, the Bureau of Indian Affairs, supported Reclamation enthusiastically. The BIA pointed out that 38,000 American Indians lived in the upper Colorado basin. Those indigenous people, it said, were utterly dependent on the natural resources of their lands, resources which were "inadequate for their minimum needs." CRSP water and power, the Indian agency was certain, would do much to help sustain the several tribes.

The governors and legislatures of the four upper-basin states were quick to endorse the Bureau's grand plan. But in the lower basin, the

Arizonans equivocated; they were pleased that at long last they might be getting Glen Canyon Dam, but they noted in their response to the proposal that *no water* and *no power* from the project was specifically allocated to them. California and Nevada in turn were plainly nonplussed. Development in the upper basin on the scale of the CRSP would drastically shrink the amount of water that would flow downstream and be available to be "borrowed."

Oscar Chapman at Interior and his man at Reclamation, Mike Straus, indeed had been relieved of their positions by President Eisenhower by the time the first CRSP bills were introduced in the House and Senate in April 1953. The new Interior secretary was Douglas McKay, owner of an Oregon Chevrolet dealership and an ardent supporter in Eisenhower's election campaign. Sitting in the commissioner's office at Reclamation was Wilbur Dexheimer, an engineer who had spent his life building Bureau dams. Neither man had any particular problem with the CRSP proposal inherited from the Democrats; in fact, both were anxious to prove that sound reclamation projects would not go begging in a fiscally responsible Republican administration. And both were secure in the knowledge that Eisenhower himself, a product of the dry, short-grass prairies of Kansas and a kind of honorary resident of his wife's home state of Colorado, thought the CRSP sounded dandy. Had the Democrats still been in office, it might well have proved to be evidence of creeping socialism, but with Republicans at the helm, the CRSP was simply sound development, the government giving capitalism a helping hand out on the hardpan.

Despite the president's support, despite staunch advocacy by the Bureau of Reclamation and an influential block of western senators and congressmen, CRSP legislation wasn't in for an easy time, and everyone knew it. It wasn't the project's cost—at least $10,000 in federal expenditures for every man, woman, and child in the upper basin—that was the sticking point, however. It wasn't the decision to emphasize hydropower in the midst of vast coal fields that bothered people, nor was it the possible loss of an entire reservoir if the subterranean pools that flanked Glen Canyon somehow sucked the water down. What really had people riled about the CRSP was that one of its dams would be built three miles downstream from the confluence of the Green and Yampa rivers in northwestern Colorado, two miles upstream from the Utah border. The reservoir behind the dam would reach 64 miles up the Green, 44 miles up the canyon of its Yampa River tributary, and an extremely remote, almost unvisited national monument would be transformed into a kind of water tank. The Bureau of Reclamation, the region's water boards and booster organizations were heralding the water storage, the new irrigation and electricity, the terrific recreational opportunities the reservoir would provide. There would be "BENEFITS FOR ALL!" the brochures said, but somehow, so far at least, the American people weren't buying it. A coalition of conservation organizations was succeeding in its attempts to convince Americans everywhere that this place few people knew actually mattered enormously and ought to be left alone, and that was why passage of the CRSP legislation was anything but certain.

It was President Woodrow Wilson who in 1915 had designated 80 acres of land on the Utah-Colorado border as a national monument, his sole purpose being the protection of an important paleontological resource—whole dinosaur skeletons embedded in a shale and sandstone bluff. With Harold Ickes's encouragement, Franklin Roosevelt had enlarged the monument two decades later, increasing its size by 2,500 percent, preserving in the process more than 100 miles of the canyons of the Green and Yampa.

Although Howard Zahniser, executive secretary of the Wilderness Society, had never heard of Dinosaur National Monument before the Bureau of Reclamation announced its intention to inundate much of it, he had done much of the quick, frantic organizing in opposition to the plan that preceded Interior Secretary Chapman's hearing in 1950, and in the intervening months since then he had assembled seventeen member groups into a large and rather unwieldy organization calling itself the Citizens Committee on Natural Resources. Additionally, Zahniser had enlisted the support of Edward Mallinckrodt, Jr., a wealthy St. Louis chemicals manufacturer and, like himself, a passionate defender of wilderness. The fight to save Dinosaur would be very expensive, Zahniser was well aware, and with Mallinckrodt, he had secured an important patron.

Mallinckrodt, a member of the Sierra Club, had been disappointed that the club hadn't been involved in the early opposition to the Echo Park Dam proposal. Regardless of whether people like Walter Huber *liked* the rough desert country of Dinosaur, it was a national monument and it deserved protection. When Mallinckrodt took his case to David Brower, the club's new executive director—its first full-time staff member—he discovered he already had a committed ally.

Soon after he had left the Sierra Club's board of directors to become its paid staffer at the end of 1952, Brower had published an article in the *Sierra Club Bulletin*, which he edited, called "Folboats Through Dinosaur," an account by club member Stephen Bradley of a family float trip through the monument on the waters of the Green, and he had been invited to view the home movie of the journey filmed by Bradley's father, Harold. The article and the short film were an inspiration to Brower; he was a mountaineer, an alpine man, and had little experience in boats or in desert country, yet the Bradleys had made the river and the canyons come alive for him. The trip sounded like a terrific adventure, and surely Dinosaur must be a marvelous place. Brower vowed that the following summer he too would float through the Canyon of the Lodore, through wide and pastoral Echo Park, where the Yampa spilled into the Green, through Whirlpool and Split Mountain canyons; he would see and experience Dinosaur for himself, and in the meantime, the Sierra Club—at least in the body of its executive director—would begin to work to save it.

Brower had no background in political organizing, but he did have a lifetime of experience in coordinating group expeditions into the back country, so he began his battle to save Echo Park by doing what he knew how to do. By the early summer of 1953, he had scheduled three Green River trips through Dinosaur, each one including sixty-five people, most of them club members. Five dozen or more people was an enormous number to send down the river in a single boating party, but with the assistance of Bus Hatch, operator of a commercial river-running business in nearby Vernal, Utah, the hearty adventurers from Berkeley and Bolinas, from Mill Valley and Menlo Park were ecstatic when the river washed them out— sunburned and mud-caked—at Split Mountain campground at the end of the 44-mile run. On one trip, seventy-three-year-old Harold Bradley, a Green River veteran and the recently retired head of the University of Wisconsin medical school, was along as proof that this strange sport of river running wasn't limited to daredevil adolescents. On another, Brower was joined by two of his sons, still children. And on that trip, Brower himself got into the movie business, working with photographer Charles Eggert on the raw footage of what would become an invaluable visual aid, a twenty-eight-minute film entitled *Wilderness River Trail,* demonstrating what a dam and reservoir in Dinosaur would destroy.

In the summer of 1953, the Green and Yampa rivers were floated by far more people than had ever done so before, and tourists arriving in cars across the rutted dirt roads that led to Dinosaur—people who had been told by the *Saturday Evening Post* and by the dependable *Reader's Digest* that Dinosaur was a terrific place—crowded the monument's meager campgrounds. But Dinosaur's visitors did more than make curious kinds of pilgrimages. When they returned home, they wrote letters, thousands of letters. They showed slides and snapshots to their friends and neighbors and encouraged them to write letters. The publications of the Wilderness Society, the Audubon Society, the Izaak Walton League, the Wildlife Federation, and the Sierra Club encouraged their members—those few who actually had visited Dinosaur and the many who had not—to write letters! By the time its hearings on the CRSP were scheduled to begin in January 1954, the House Subcommittee on Irrigation and Reclamation was receiving hundreds of missives a day, arguing not so much against the CRSP itself as against the idea, the *very idea* of building a dam in Echo Park. Although the chambers of commerce in Vernal, Utah, Grand Junction, Colorado, Farmington, New Mexico, Flagstaff, Arizona, and dozens of other towns did their best to counter with letters of their own—letters that fairly pleaded for the gargantuan federal project—the congressional mail was estimated to be running 80 to 1 in opposition.

Western congressmen were dumbfounded. How had all these people from watery Massachusetts and humid Ohio gotten bent out of shape by some canyon out in Colorado? Bureau of Reclamation engineers and administrators were mystified, and some of them were a little hurt. Heretofore they had been heroes; they had constructed great dams that had helped to feed and shelter the burgeoning nation; their dams had helped this country weather a depression and win a war. But now, as they tried to do no more than continue their wonder work, they were being made out to be destroyers, killers of landscape. It was difficult to understand, but it didn't make sense to panic. The hearings would put Echo Park into perspective and would clarify the critical importance of the project to the states of the upper Colorado basin. The hearings, much like dams themselves, surely would calm the conservationists' rough rhetorical water.

David Brower had never testified before a congressional committee or battled the Bureau of Reclamation. The House CRSP hearings were going to be a baptism by fire, no doubt, and it made sense to him to observe them from gavel to gavel rather than just to arrive in time to make his own statement. At Howard Zahniser's invitation, Brower checked into Washington's Cosmos Club and prepared to stay a while as the hearings were convened on January 18, 1954.

A parade of administration witnesses opened the nine-day proceeding, beginning with Under Secretary of the Interior Ralph Tudor, an engineer who had designed the San Francisco Bay Bridge before accepting the call to government service. And rather than commence with an overview of the proposed project, Tudor jumped directly to the defense of Echo Park Dam, claiming it was fundamental to the success of the project as a whole, explaining that alternative sites would result in unreasonable water losses from evaporation. When sympathetic California Congressman Clair Engle asked him if taking Echo Park out of the CRSP would be "like taking the engine out of the automobile," Tudor said it would be "like taking the pistons out." But the under secretary also wanted to make it clear that Interior's endorsement of Echo Park did not mean that it would allow a developmental assault on the national parks. After all, Conrad Wirth, a respected park service administrator since the days of the Civilian Conservation Corps, was currently its director, and Wirth wouldn't sanction any attack on the system.

Tudor was followed by Ole Larson, a Reclamation regional director based in Salt Lake City. Larson agreed that Echo Park Dam was "the wheel horse of the upper basin," and stated that despite its best efforts, the Bureau had been unable to find an acceptable alternative to the "remarkable storage vessel" created by the canyon topography of Dinosaur. Then came in long and sometimes tedious succession Reclamation numbers-men arguing in support of the project's economics, and congressmen, ranchers, and civic leaders from Colorado, Wyoming, and Utah claiming that it was dry as dust out west and that the only antidote to drought was a CRSP that included Echo Park. G. E. Untermann of Vernal explained that as far as "saving" Dinosaur was concerned, the people who lived nearby didn't want to be saved. "We want to be damned," he declared.

After six days of supporting testimony, it was time to hear from "the abominable nature lovers," as Utah Senator Arthur Watkins had termed project opponents during his own testimony before the House subcommittee. Yet if the subcommittee members were expecting to hear next from a bunch of sob sisters, they were indeed surprised when gentlemanly General U. S. Grant III began his testimony. Grant, an officer of the Army Corps of Engineers for forty-three years, pointed out that the Federal Power Act of 1920 prohibited private power companies from developing projects inside national parks and monuments. It made no sense, he continued, for a government agency to be allowed to do what private enterprise could not, "namely to do irreparable damage to a scenic area set aside after careful study." The CRSP, *in toto*, was "not a well-worked-out, balanced project," the general declared, his physical presence, his courtly manner, as well as his name and his professional renown, lending a special credence to his statement. "The hard fact is that the Bureau of Reclamation has already made site surveys and other preparatory work—including propaganda—to such an extent that it does not want to make similar studies of alternative sites," he charged.

The succession of opposition witnesses was long, if not necessarily always as impressive or eloquent as General Grant. Joe Penfold, speaking for the Izaak Walton League, said it seemed duplicitous for Reclamation engineers to reject alternative sites as too wasteful of water when an estimated 25 percent of all the water the Bureau delivered was currently lost through seepage in unlined irrigation canals. Howard Zahniser curiously chose to give a dramatic reading of the Robert Southey poem from which the Canyon of the Lodore had derived its name. And when it was David Brower's turn, the rookie witness, representative of the lowly and largely unknown Sierra Club, began by comparing the danger to Echo Park with the one great developmental disaster his organization could never forget.

Cranking up a still-montage film called *Two Yosemites* to show to his captive audience, displaying before-and-after photos of the Hetch Hetchy Valley with a show-and-tell kind of flourish, Brower got right to the heart of his concern. "If we heed the lesson from the tragedy of the misplaced dam in Hetch Hetchy, we can prevent a far more disastrous stumble in Dinosaur National Monument," he told the members of the subcommittee. Then he issued a kind of challenge: "I know, and I will bet Reclamation knows, that if the river disappeared in its course through Dinosaur, or was somehow unavailable, a sound upper Colorado storage project could be developed elsewhere." That was the strategy Brower, Zahniser, and the confederate conservationists had agreed to pursue—not to oppose the CRSP as a whole, but rather to acknowledge the need for water and power in the isolated upper basin in tandem with their claim that other reservoirs in other canyons would be as beneficial as the one proposed for Echo Park. But Brower astonished the assembled congressmen—who tended to view Bureau reports, with their complex and seemingly impenetrable jumble of computations, as semisacred texts—when he went to a blackboard and pointed out that, in the case of these critical evaporation rates they were talking about, the engineers couldn't correctly add and subtract.

In arguing against alternatives to Echo Park—in particular, a conservationist proposal to increase the height of Glen Canyon Dam and thereby store the water now slated for Echo Park—Reclamation engineers had claimed that such an alternative would result in an annual loss by evaporation of 165,000 acre-feet of water more than would Echo Park. Yet Brower, who coyly defined his arithmetic as the simple, ninth-grade variety, pointed out that in making the comparison, the engineers had forgotten to subtract the evaporative amount caused by Echo Park (assuming it wasn't built) from the evaporative total of the proposed alternative. Instead of causing far more evaporation than would the dam and reservoir at Echo Park, Brower demonstrated with a piece of chalk, a higher Glen Canyon Dam would result in 2,610 acre-feet less evaporation and, he paused for emphasis, the resulting Glen Canyon reservoir would store 700,000 acre-feet *more*.

It was startling testimony, but it was true. Under Secretary Tudor subsequently admitted to the subcommittee that *somehow* Reclamation's figures had been in error. In nine days of testimony pro and con Echo Park, Brower's had been the most powerful. The neophyte had successfully chipped a hole in the rhetorical Echo Park dike. Perhaps the dam in Dinosaur *wasn't* the pistons, the engine, the whole damn drivetrain of the CRSP. Perhaps the way to bring this legislation on board, wondered some subcommittee members, was to forget about Echo Park.

President Eisenhower had held a press conference in March to recommend officially the construction of the Colorado River Storage Project, including both Echo Park and Glen Canyon dams, so by the time the Senate Subcommittee on Irrigation and Reclamation held *its* hearings at the end of June—most of the same people saying much the same thing as they had in January—the CRSP steamroller appeared to be moving inexorably forward. Letters in opposition were still arriving by the bundle, but the House Interior Committee nonetheless had reported its CRSP bill favorably to the full House. The Senate, where western senators had far more power by sheer force of numbers, was widely seen as a kind of home ground for wayward water projects. And as far as Utah Senator Arthur Watkins was concerned, the principal goal of the Senate hearings was to crush this higher Glen Canyon alternative that the conservationists had hoisted before the House. Watkins, often acting as temporary subcommittee chairman, often the only senator present in the Senate Office Building hearing room during the six long days of testimony, had planned to ferret out what seemed to be obvious about this fellow Brower: he wasn't an "honest conservationist" as the others were. He was a *Californian*, and as such was surely in cahoots with the Colorado River Board of California, the Los Angeles–based organization that was fighting the CRSP simply to keep as much of the Colorado's water as possible flowing downstream for as long as possible. Those water-stealing bastards weren't fighting a fair fight, Watkins was convinced. The San Francisco conservationist was nothing more than a pawn of the L.A. water boys, and their opposition to Echo Park was simply a smoke screen.

Although Watkins found that Brower was very willing to admit that he had had one hour-long meeting with Northcutt Ely, attorney for the Colorado River Board, he otherwise withstood the senator's challenge. No, neither he nor the Sierra Club was in regular communication with the board; no, they had never discussed common strategy; no, the board had made no contribution to the club or to its efforts in opposition to Echo Park; and no, Senator, the Sierra Club and all the allied conservationists were interested only in protecting Dinosaur National Monument, not in sabotaging the whole of the CRSP. Watkins finally surrendered and left Brower alone to state his case, a second time, that the Bureau had a better alternative, that as long as little Rainbow Bridge National Monument, tucked high in a side canyon 50 miles upstream from the dam site, was protected, Reclamation ought to build Glen Canyon Dam to the very rim of the canyon walls.

Realizing that Brower and the conservationists, in their push for the higher Glen Canyon Dam, had found a real opening, Interior Secretary McKay sought in the fall of 1954 to diffuse their arguments. The Interior committees of both houses had approved Echo Park and the rest of the CRSP in the spring and summer; in late July, President Eisenhower, in Colorado on vacation, had flown over Dinosaur and pronounced that on visual inspection there seemed to be no reason not to dam it. Neither body of Congress had voted on the CRSP prior to adjournment, but if the conservationists could be held at bay until early in the next session, and if this Glen Canyon business could be put to rest, it still appeared that Echo Park could be built.

What the secretary did was to write privately to the Sierra Club representative, explaining, almost as if he were sharing a secret, that while, yes, raising the height of Glen Canyon Dam was an interesting possibility, the Bureau of Reclamation had discovered sadly that it wouldn't be feasible. In his November letter, the secretary stated that the sandstone that would support the dam's weight was "poorly cemented and relatively weak in comparison to most high dams." It simply couldn't be trusted to support a 700-foot-high structure, and therefore Echo Park still would be needed.

In his response to the secretary, Brower reminded him of an earlier Reclamation report that had dispelled the U.S. Geological Survey's concern about the underground basins beneath Glen Canyon and that had endorsed the dam's foundation material as "remarkably free of structural defects." Brower implied, but didn't go so far as to charge, that Reclamation and the Interior Department were again playing fast and loose with the truth. Then, having learned very quickly how you fought these political fights, he made the letters public.

Elsewhere on the public relations front, the conservationists were planning something extraordinary. With the assistance of Alfred Knopf, a New York publisher and chairman of the Interior Department's Advisory Board on National Parks, they were preparing to publish a book devoted entirely to Dinosaur National Monument. It would be a piece of propaganda, of course, but it would be an elegant one. Entitled *This Is Dinosaur*, and containing dozens of black-and-white and color photographs, the book would be an eloquent explanation of why Echo Park and all of Dinosaur were so important. Wallace Stegner had agreed to edit the text, which would include essays by himself, Knopf, geologist Eliot Blackwelder, wild-life biologist Olaus Murie, archaeologist Robert Lister, pioneer river-runner Otis "Dock" Marston, and journalist David Bradley. Together, the several essays would be a kind of celebration of the book's subtitle—*The Echo Park Country and Its Magic Rivers*. A book like this, an unabashedly conservation-minded picture book, had never been published before, but in the face of what threatened Dinosaur, and with Knopf's generous underwriting, it was certainly worth the risk. If it made enough money to help continue the political battle, so much the better; but even if the book lost money, if it alerted more Americans to what was in danger, well, that would be a success in itself.

Although David Brower's name was in no way associated with the book when it appeared in January 1955, its pages nonetheless bore the watermark of his increasing commitment to the Echo Park struggle. The book had been his idea, and it was he who had approached Stegner about playing a major role in its creation. Stegner, a Sierra Club member, had been strongly in favor of the organization's forays into the world of power politics under Brower's leadership, and he supported the book as a tool in the campaign. He had once been enamored with Hoover Dam, but twenty years later, he had begun to regard desert dams with great suspicion, and now he cautioned Brower that there really wasn't any way to win this battle. In recent years, he had twice boated through Glen Canyon. The trips had been overwhelming experiences, and Glen Canyon, he could attest, was even more remarkable than Dinosaur. A high dam in Glen Canyon, even if it somehow spared Echo Park, would be a tragedy.

Brower's first trip floating the Green through Dinosaur had been the most profound wilderness experience of his life, and he told Stegner that he couldn't imagine that *any* place could be more beautiful, more wondrous than Dinosaur. Yes, Glen surely was another glorious canyon, but unfortunately, it had never received federal protection and that was what this battle really was all about. It was too much to hope that they could scuttle the entire upper-basin project, but if together they could save Dinosaur, that in itself would be something very significant. They just didn't have the resources, the time, or the political muscle to fight on every front.

Understanding the sad practicality of that argument, Stegner willingly devoted his talents to the Dinosaur project, and in the foreword to *This Is Dinosaur* he stated its purpose: "If the American people, through Congress, choose to authorize the proposed dams and take the losses, they can do so. Our book attempts no more than to show—so far as words and pictures can show a region so varied and colorful—what the people would be giving up, what beautiful and instructive and satisfying things their children and their grand-children and all other Americans from then on would never see."

Wayne Aspinall knew when he opened his copy of *This Is Dinosaur*, sent to him in January 1955, courtesy of Alfred A. Knopf, that things were going from bad to worse. What four or five years ago he had presumed would be a brief whimper of opposition to the Echo Park dam site had turned into a roar, a carefully orchestrated howl of protest that showed no sign of abating. Following the House subcommittee hearings a year before, his staff had counted the letters commenting on the CRSP, and only 53 of 4,731 of them were in support of the dam. With no little amount of maneuvering, the western Colorado congressman had succeeded in favorably reporting the bill out of the full Interior committee, but only by a tenuous 13 to 12 vote. And now, the conservationists had published a book all about how terrible a dam in Dinosaur would be, and a copy had been sent to every member of the House, every Senator, most of the high-level people over at Interior, and to every newspaper editor between Boise and Albuquerque. Colorado Senator Eugene Millikin had assured Aspinall that the bill was secure in the Senate, but in the House, where the West's influence was spread mighty thin, it still looked like a hell of a scrap, and this book wasn't going to help matters a bit.

Aspinall, a small, bespectacled ex-schoolteacher—quick to grin but testy, nearly venomous at times—had grown up beside the Colorado River, the Grand back in those days, in the small fruit-growing community of Palisade, Colorado, and nothing had been of more import to him during his many years of public service than reclamation. Now, as chairman of the House Subcommittee on Irrigation and Reclamation, he wielded the kind of power that could get dams built where dams were needed, yet it was the kind of power that had been nurtured by compromise, by a kind of legislative realism that was legendary among his colleagues.

Early in the CRSP struggle, Aspinall had issued a dire warning about the conservationists: "If we let them knock out Echo Park, we'll hand them a tool they'll use for the next hundred years." But now, with his House head count a little worrisome, and with the conservationists clearly appearing to be gaining strength, he was beginning to wonder whether Echo Park Dam might have to be sacrificed to save the CRSP. Echo Park was in Aspinall's district—it would be *his* dam—and he certainly did not want to lose it. Neither did he care for this David Brower, this cocksure young Californian who was constantly disputing the best men in the Bureau of Reclamation, and who must have played a part in this fancy coffee-table book he had been sent. Yet maybe Brower would have to win this one. Increasing the height of Glen Canyon Dam— assuming the boys at the Bureau could make it work—would accomplish much; the upper basin would get the gigantic storage facility it had to have, as well as a colossal hydropower dam that would pay for a large supporting cast of irrigation projects. If it came down to it, Aspinall already knew heading into a second round of hearings in March 1955, designed to counter the swelling opposition to the project, that he could live without Echo Park.

The Senate's second CRSP hearings were showcased by the Bureau of Reclamation, its administrators and engineers making it as plain as they could that Glen Canyon was a good and safe place to build a dam. Yes, the Interior secretary had made statements about possible limitations on the height of the dam, but those concerns couldn't be addressed in detail until the project was authorized and extensive field tests were completed. The conservationists monitoring the testimony were incredulous. Reclamation was claiming that first Congress should approve the CRSP and *then* it would figure out exactly what kind of dam it could build in Glen Canyon, was that correct? That was indeed correct, responded Reclamation Commissioner Dexheimer, and it was long-standing Bureau procedure. And the senators seated on the dais plainly were undisturbed by what seemed to the conservationists to be the prospect of buying a pig in a poke.

But when the House hearings opened four days following the

close of the CRSP proceedings in the Senate, southern California Congressman Craig Hosmer—no fan of a CRSP in any form—pounced on the subject of Glen Canyon. He had it on good geological authority, he told the committee, that the sandstone forming the walls of Glen Canyon was so porous it was a virtual sieve. As water backed up behind Glen Canyon Dam, it would leak so swiftly into the rock that the reservoir might never fill and "the country might be stuck with the most enormous white elephant in history." Chairman Aspinall winced as Hosmer continued his assault, announcing that other experts had proved that electricity generated by atomic power soon would be the cheapest electricity to produce, so cheap it might not even be metered, and certainly far cheaper than the hydroelectricity that would come from Glen Canyon. And as if that weren't enough, Hosmer next hauled out the Bureau's own project cost-benefit statistics demonstrating that although the four upper-basin states would receive $137 million in benefits from the CRSP for only $75 million in revenue contributions, every other state in the nation would lose money, a deficit totaling more than $3 billion when you factored in the loss of investment interest. "For instance," Hosmer shouted, "the state of New York will receive, according to the Bureau, $77,398,000, but the taxpayers of New York will have to fork out $493,600,000 for the project. Who does the Bureau of Reclamation think it's kidding?"

Congressman Aspinall was beside himself. Why was he worried about the wacky conservationists when a member of his own subcommittee—a Californian whose plain purpose was to reserve the whole of this great river for his own and his neighbors' swimming pools—was attacking the bill with a tomahawk?

The Senate Interior Committee favorably reported the Colorado River Storage Project bill on March 30, and after three days of debate on the floor of the Senate—including fervent, often eloquent oration in opposition by Senators Paul Douglas of Illinois, Hubert Humphrey of Minnesota, and John Kennedy of Massachusetts—Senate bill S. 500, authorizing the construction of six dams and storage reservoirs, including those at Glen Canyon *and* Echo Park, and twelve participating irrigation projects, at a cost of $1 billion and change, passed by a vote of 53 to 28. It was a resounding victory for Reclamation and for the senators from the Rocky Mountain states.

Wayne Aspinall didn't celebrate, however. He knew, by doing no more than counting noses, that the House version of the bill the

Senate had just passed couldn't even survive the Interior Committee. The *New York Times* editorialized against Echo Park, but not against the whole of the CRSP, in early June, and a few days later, Aspinall reluctantly did the same in a subcommittee executive session. Although the Californians on the subcommittee kept up their efforts to kill the entire project by voting to *keep* Echo Park, Aspinall prevailed, and the dam in Dinosaur National Monument was deleted from the bill. Then, on July 8, the House Interior Committee favorably reported it, still *sans* Echo Park, by a vote of 20 to 6.

But even with that kind of showing in the Interior Committee, Aspinall was reluctant to bring the bill to the floor of the House for debate. His current estimate was that House Democrats would split about evenly on the legislation, but Republicans would vote it down three to one. "I would rather have such an important bill given full debate and rational consideration when the emotions of the members are not as high-strung as they are," he explained. But there was more to his reluctance. The conservationists now were claiming that deleting Echo Park from the House bill was nothing more than an end run to get it passed. Plans were already laid, they contended, to resurrect Echo Park Dam in a House-Senate conference committee. A little time was needed to calm down the conservationists and get them behind this bill of their own making, Aspinall understood. And time also would help the old general in the White House. He still staunchly supported the project, and he, no doubt, could eventually persuade the recalcitrant House Republicans to shape up, but even Ike could do nothing overnight. So, on August 2, Congress went into recess without taking further action on the CRSP, and the battle for Echo Park stretched out a little longer.

During the five years that Echo Park had been under ever-increasing national scrutiny, Americans had poured into the place that theretofore had been frequented only by a handful of dinosaur buffs and scientists. In the summer of 1955, 45,000 people visited the monument, an estimated 1,000 of them traveling through it on the bucking backs of the rivers. Members of conservation organizations, journalists and newspaper people, congressmen and their staffs, and even the odd rancher or uranium miner determined to demonstrate his open-mindedness floated the Yampa and the Green during the runoff-swollen months of early summer. Dinosaur, for the first time since those great, fated creatures themselves roamed the region, had become a busy place.

But Glen Canyon, on the other hand, remained a relative refuge for silence. Reclamation geologists and survey crews had made occasional forays into the canyon, taking core samples in the riverbed at the two sites proposed for the dam, drilling drift tunnels into the looming walls. A few boatmen who had worked for Norman Nevills had escorted paying customers of their own down through the canyon each summer—some of whom actually tried to generate a bit of congressional enthusiasm for making the place a national park—and Art Greene had ferried a few more upstream to Rainbow Bridge. Yet although the dam proposed for Glen Canyon would be far bigger than the one at Echo Park, its reservoir vastly larger, no one fighting the political fight from either side had paid any personal attention to the place. Neither the Interior secretary, nor the high-ranking appointees at Reclamation, nor the solons who ultimately would decide the canyon's fate, nor more than a smattering of the legions of conservationists who were encouraging them to dam it had ever traveled to Glen Canyon.

During the congressional recess in December 1955, California Congressman Craig Hosmer finally went to have a look. Accompanied by John Terrell, public-relations man for the Colorado River Board, a couple of independent geologists, and a cameraman, Hosmer set out to prove that nobody had any business building a high dam in such a geologically suspect setting. Despite the Senate vote, the conservationists were doing a good job in their efforts to scuttle Echo Park. If Hosmer could lead a last-ditch effort to impugn the rocky integrity of Glen Canyon, perhaps neither dam would be built and the Colorado would continue to flow unchecked to California.

Using Art Greene's Marble Canyon Lodge as a base, for three days the four men helicoptered in to the dam sites, poked about with pick hammers, photographed everything that captured their interest, and hauled back out of the canyon gunnysack loads of sandstone and shale. In the evenings, warmed by Art Greene's piñon fires and his inexpensive whiskey, they surveyed their booty and pronounced it precisely what they had come looking for. By the time President Eisenhower had delivered his State of the Union address on January 5, 1956, Congressman Hosmer had mailed a small sample of Chinle shale to the editor of every daily newspaper in the nation, and each of his colleagues in the House was a recipient as well, the congressman attaching a note with every chunk suggesting that fun could be had if the bearer dropped it into a bit of water.

David Brower, Howard Zahniser, Joe Penfold, and other leaders of the confederation now known as the Council of Conservationists paid little attention to the congressman's largess. They were in the midst of negotiations with Wayne Aspinall that would result in the prohibition, plain and simple, of a dam in Echo Park. In late October, the conservationists had gotten wind of a secret meeting scheduled in Denver at which governors, congressmen, and water-board members from the upper-basin states would try to figure out how to pass a House bill without Echo Park, then make sure that the Senate's CRSP bill (Echo Park included) prevailed in the conference committee. In order to head them off, Zahniser had come up with some quick cash, flown to Denver, and taken out a full-page ad in the October 31 edition of the *Denver Post*, the "open letter" making it very clear that unless Echo Park *irrevocably* were deleted from the CRSP, the conservationists and their allies would launch an all-out attack, this time not just on the Echo Park proposal, but on the entire upper-basin project.

It was no bluff, the politicians recognized, and they knew by now that they were dealing with formidable foes. Despite the objections of Senator Arthur Watkins and the rest of the Utah delegation, who still clung to Echo Park as if it had been ordained by Brigham Young (which, from a particular Mormon perspective, it *had*), Colorado Representative Aspinall declared to his assembled colleagues that he could strike a deal with Brower and his bunch if they would capitulate on Echo Park. On November 2, the Upper Colorado River Commission, with the blessing of the several states' elected officials, formally rejected the proposal to build a dam inside Dinosaur National Monument. Then, following a series of meetings and correspondences, and the revision of the house CRSP bill—now containing language stipulating that "no dam or reservoir constructed under the authorization of this act shall be within any national park or monument," and mandating "adequate protective measures to preclude impairment of the Rainbow Bridge National Monument" from the reservoir behind Glen Canyon Dam—Brower, Zahniser, and company drafted a letter to Aspinall on January 23, 1956, withdrawing their opposition to the Colorado River Storage Project.

On Wednesday, March 1, President Dwight Eisenhower held a press conference to make two brief announcements. First, he would be a candidate for a second term in office. And second, Congress should "get busy and get on with" passage of the CRSP. The following day, however, California Representative Hosmer held the floor in the House of Representatives and held the president at bay, announcing that the geologists with whom he had journeyed to Glen Canyon had discovered that, yes indeed, water backed up by the dam would flow into gargantuan underground basins capable of hording 350 million acre-feet of water; the Navajo sandstone that formed the canyon walls was so porous that 15 million gallons of water a day would flow around the dam's abutments; and the Chinle shale that underlaid the Navajo formation would literally disintegrate when it came in contact with water, causing the canyon walls to avalanche into the reservoir pool, filling Glen Canyon with wet rock instead of water.

Congressman Hosmer did not paint a pretty picture, and he had graphic show-and-tell assistance from Florida Representative James Haley, who poured water into a glass containing a piece of Chinle shale and quickly produced mud, warning the House of the dire consequences the Chinle formation would bring about if the dam were built. But Representative Stewart Udall, a first-term congressman from Arizona and therefore a staunch supporter of the Glen Canyon site, had his own piece of stone in tow. As he rose to speak in support of the grand benefits of Glen Canyon Dam, he dropped a little cylinder of sandstone—part of a core sample from the dam sites, he explained—into a water glass. At the conclusion of Udall's glowing remarks, he drank the clear water from the glass to great guffaws and hearty applause. Moments later, he and his colleagues voted 256 to 136 in favor of the CRSP and its centerpiece dam in Glen Canyon.

Despite the deletion of Echo Park Dam, David Brower had been tempted to keep up the opposition. The Bureau of Reclamation's indecision about the specifications for a dam in Glen Canyon, the geologists' accusations, and the *cost* of the whole thing seemed reasons aplenty to kill the entire project. But the Sierra Club's board of directors had informed him in January that, with the Aspinall-engineered compromise, *its* fight, and therefore his fight, was finished. And so Brower, certainly not as jubilant as some of his fellow conservationists, but nonetheless satisfied with the certainty that they had accomplished *something* of enduring value, left the Cosmos Club for the last time and returned to San Francisco. Howard Zahniser, who would quickly shift gears to begin congressional lobbying for a law designating and protecting wilderness lands throughout the nation, pronounced the final version of the CRSP an unqualified victory. Joe Penfold of the Izaak Walton League said the bill now was a sound one, and he praised Aspinall and other congressional realists "for their successful efforts in reaching an accord with us." Ansel Adams, head of a Sierra Club–affiliated lobbying group called the Trustees for Conservation, said his organization gladly could now support the upper-basin bill because of its "reaffirmation of the national park principles." Newton Drury, who had been director of the National Park Service back when the struggle began, and who had prophesied that "Dinosaur is a dead duck," expressed his astonishment at the outcome, and his great admiration for the people who had proved him wrong.

It had taken more than half a decade, but the band of mountaineers and bird-watchers, tree-huggers and river rats had blocked the concerted efforts of the Eisenhower administration to turn the canyons of Dinosaur into a slackwater lake. Forty years following John Muir's failure to save Hetch Hetchy, they *had* preserved Echo Park—forever, it seemed—and they successfully had made the case that preservation itself could be a kind of wise use. But they had done something perhaps even more astonishing. They had called into question the competency and the integrity of the Bureau of Reclamation, an agency that theretofore had been a symbol of government service at its finest. They had sown seeds of suspicion about precisely how the federal bureaucracy regarded the public's lands, and they had demonstrated that conservation battles could be waged and won within the political arena. In the process, too, they secured the overwhelming transformation of Glen Canyon.

On March 28, the House-Senate conference report on the CRSP was approved, the ultimate version of the bill authorizing the Curecanti Dam on the Gunnison River, Flaming Gorge Dam on the Green, Navajo Dam on the San Juan, and Glen Canyon Dam on the main stem of the Colorado at the far downstream edge of the upper basin. Eleven irrigation projects were included in the final package, and its estimated total cost was scaled down to a frugal

$900 million. "This represents something I believe in," President Eisenhower declared as he signed the bill into law on April 11, 1956.

Later that year, Colorado Representative Wayne Aspinall introduced legislation that would have made Dinosaur a national park, deciding in his practical-minded way that the people of western Colorado deserved to get some sort of economic spoils from the place, but the congressmen in the Utah delegation—who would have had to cosponsor the bill to give it much chance of passage—vowed that, despite their recent setback, they would see Echo Park submerged long before they would acquiesce to a further lockup of the canyonlands on the Colorado border. And with the passage of the CRSP, the small group of people—a disenfranchised and thoroughly disgusted band of conservationists and river runners from Salt Lake City, as well as a few others scattered about the West—who had been pressing, albeit improbably, for a revival of Harold Ickes's plan for a national monument down in the Glen Canyon country—abandoned their efforts, the canyon's future now otherwise defined.

4
INTO THE HOLE

The call came while Lem Wylie was at
work in Amarillo, Texas. The brass in
Denver had a proposition for him. They
wanted to offer him the biggest job in
the country, if he'd take it, so he was
soon on a Bureau plane headed for
Colorado to hear them out. As assistant
regional director of Reclamation's Region Five in Amarillo, Wylie had been
office-bound for less than a year, yet he
was eager to get out and get his hands
dirty again. But he wasn't at all sure
why Commissioner Dexheimer and his
subordinates had singled him out for
this new assignment. Maybe it was
because Wylie had been on board way
back when Hoover Dam was built and
had firsthand knowledge of how you
went about erecting a dam in a desert
canyon. But maybe it was just because
nobody else wanted the job. Maybe
everybody else was too damn intelli-
gent to leave behind willingly the com-
forts of Denver or Salt Lake City, or
even Amarillo, for that matter, to spend
ten years in some hellhole in the back
of beyond. Whatever their reasons, they
wanted Lem Wylie to be the construc-

tion engineer for the Glen Canyon Project, to be the head man on the site, to endure a hundred headaches a day for as long as it took to get the blocks stacked and the turbines spinning. Wylie listened to Dexheimer's pitch, heard chief concrete design engineer Louis Puls explain precisely what kind of structure they were going to raise down there in Arizona, thought it over for ten or fifteen seconds, then took the job.

If it was a surprise to Wylie to be shifting gears so suddenly, he was sure it was an absolute shock to the fellows in the Bureau's regional office in Salt Lake City, who, he knew, had presumed that the construction engineer would be picked from among their immediate ranks. Wylie had gotten his start in the canyons of the Colorado, all right, but after Hoover was finished he had been transferred to Irrigation, working on the All-American Canal down on the Mexican border until the United States entered the war and Wylie went with the Marines to the south Pacific. He had been based in Amarillo, working on the Platoro Dam in southeastern Colorado, from 1946 to 1952, then had gone to Alaska for three years to head up the Eklutna Project near Anchorage prior to being sent back to Amarillo and a desk job in 1955. Nobody in Salt Lake could quite figure out why Wylie had seemed so right for Glen Canyon, but Ole Larson, Reclamation's regional director in Salt Lake, assured him they were glad to have him join them out there in God's country and would assist him in every way they could.

Wylie and Louis Puls had flown from Denver to Salt Lake early in July not only to make acquaintance with the people in the Region Four office, but also to arrange transportation down to the dam site. Wylie wanted to see the place with his own eyes just as soon as he could, and he informed Larson—who was his boss in the hierarchical scheme of things—that the first way he could help would be to get him down to Glen Canyon. Aboard a Bureau plane the next day, Wylie and Puls and a couple of field men who had made an earlier reconnaissance flew south to the town of Kanab, a burg of about a thousand people, most of whom were descendants of the Mormon pioneers who had scratched the place out of the rock a hundred years earlier. In Kanab, just two miles north of the Arizona line, the group rented a Jeep and a driver and bounced 73 miles across dunes and slickrock and dry arroyos east to the canyon rim of the Colorado.

Wylie had read the geological reports on the two potential dam sites—analyses that had made it clear that one site was far superior to the other—and he knew he would see a lot of sandstone. Yet although he admitted to being mildly apprehensive about the rock (it was as different from the granite in Black Canyon as aspen is from ebony), he wasn't worried about the prospect of this dam washing away. Sure, he'd prefer a granite foundation, but since they plainly didn't have one, Louis simply would have to design for what they had, and design the bastard to last for a thousand years or so.

What Wylie didn't expect to see, however, was a round sandstone butte, shaped like a beehive, standing on the west rim right where the dam would abut the canyon wall. There were several beehive buttes scattered throughout the canyon country, and Utah was the "Beehive State," the hive a long-standing Mormon symbol of industriousness. It would have been stretching it to say this beehive or any of the others was sacred, but they were *there*, after all, and they had always received a kind of de facto protection. Yet this beehive, a few miles south of the Utah border, stood right where a cableway tower would have to stand; there wasn't any option unless you moved the dam to a wider spot in the canyon and added millions of dollars to its price tag. Wylie knew, but he didn't say that day—didn't bother Puls with it because neither cableways nor beehives were Louis's responsibility—that before much else was under way in Glen Canyon, he'd have to shave that beehive back a bit, and he'd do it under cover of darkness if he had to to get it done.

The engineers spent the night back in Kanab. Both Wylie and Puls were dissatisfied with what they had seen. They needed to know what they were up against down in the hole itself; they wanted to have a look at the gravel-mining site the geologists had found on the low bench above nearby Wahweap Creek; and they were also in the market for a town site. They hadn't seen one they liked on the west rim of the canyon, and Wylie had insisted that they see if the east rim held any promise. There was a fellow at Marble Canyon, 70 miles away by paved road, people in Kanab told them, who hired himself out for boat trips from Lee's Ferry, near his little lodge, to Rainbow Bridge, 65 miles up the Colorado. Art Greene would take them up the canyon, you bet he would.

The following morning, Wylie and Puls drove to Marble Canyon by themselves, found Art Greene, his ever-present cigarette sticking out of his mouth like an appendage, preparing to take a small group upriver in an airboat he had designed specifically for these journeys, and the two Bureau men promptly hitched a ride. Greene had ferried geologists and survey crews several times and had known for years that Glen Canyon was a glint in Reclamation's eye, but now they were getting serious. Congress had authorized a gigantic Glen Canyon project in the spring; then the Bureau of Reclamation had settled on a specific dam site, 15 miles upriver from Lee's Ferry, and now here were two Bureau boys who weren't just out "scoping," as the earlier engineers had described their nosing around. These two fellows in pith helmets, wearing khaki pants tucked jauntily into their Wellingtons, claimed they were actually going to build the son of a bitch.

At Wylie's request, Greene pulled his boat ashore at the place up-canyon where, the engineer explained, huge diversion tunnels would emerge from the canyon walls, spilling the river back into its bed after it had skirted the dam site. Greene's scheduled passengers— a group of Sierra Club members who were eager to see fabled Rainbow Bridge—roamed the sandbar where the boat was beached while Wylie and Puls poked around in the rocks at the base of the cliffs, scrambled 300 yards upstream, then stood shoulder to shoulder, gesturing at the canyon walls, nodding, Puls taking notes.

The dam was a certainty by now, the sightseers knew that, but the canyon was so serene in that spot, so splendid; once they were all onboard the boat again, they confessed to the two government men that it seemed such a shame to destroy it. Wylie made a reluctant, unimpassioned remark about how the dam would turn Glen Canyon into something useful, something that would benefit everyone, but he was of no mind to say anything more. Puls was silent, and before he started the deafening engine, Art Greene sought some sort of compromise by saying that the one thing he was sure of was that no one had consulted him on the matter.

The boat pulled ashore again at the mouth of Wahweap Creek, Wylie and Puls shaking hands all around before they stepped onto the sand. Art Greene pulled his boat back out into the current, and the two men began the hike up the moist but waterless creekbed to the place in the sloping side canyon where Bureau geologists had found a deposit of gravel that was plenty large enough for a ten-million-ton dam, and where a Jeep was scheduled to meet them. As they walked, the quiet canyon rich with the aroma of cottonwood trees, Puls, puffing on the corncob pipe that was his constant companion when he was out in the field, asked Wylie what he thought about those folks with whom they had shared the ride.

"Seemed like nice enough people," the construction engineer averred, looking a little like Teddy Roosevelt under his helmet, except that instead of the big handlebars favored by the Rough Rider, Wylie's mustache was more the trim Ernie Kovacs kind. "But I didn't pay any attention to them. We just have different views of such things, I suppose." Then he stopped for a moment and turned to speak directly to Puls to impress the point. "Louis, as long as they get me the money to do it, I don't think there'd be anybody in the world who could stop me from climbing those walls and building this thing."

On the third day of their sojourn, Wylie and Puls drove south from Kanab again, crossing the river on the highway bridge spanning Marble Canyon, then on to the Navajo settlement of Bitter Springs, where an Arizona Highway Department crew was working to blast a road through the Echo Cliffs to connect Flagstaff with the dam site—the states of Arizona and Utah already in a furious contest to see who could get a road to the rim of Glen Canyon first, hoping thereby to secure the lion's share of the project's economic impact for their respective sides of the river. Sure, said the foreman on the highway job, he'd be more than happy to take the two men the 25 miles to the dam site in his four-wheel-drive truck—anything to assist the bountiful Bureau of Reclamation—and the three of them soon bounced and rocked and rolled their way north across the high, sandy, treeless sweep of land on the northwestern edge of the Navajo reservation, passing an occasional round, log-roofed, mud-stuccoed hogan that was home to a Navajo sheepherding family, passing drifting dunes with sensuous, curving ridge lines, passing rather more stable ground spotted with saltbush, the high, dark hump of Navajo Mountain visible off in the east.

At the canyon rim again, separated by just a quarter-mile of air from the beehive on the opposite side, Wylie and Puls made another casual reconnaissance, Wylie peering from the precipitous edge itself until Puls begged him to back away. The two men walked north along the rim, then south, then circled the sloping ground that flanked it on the east, but they couldn't find what they were hoping to find, and Wylie was getting disgruntled. Finally, before they gave up and went back to the truck, the two climbed the slope up to the mesa above them in the east that looked like it might hold some promise. They were winded when they got to the top, and neither spoke while they surveyed the flat expanse, but each man immediately knew that they had found what they were looking for.

"Lem," Puls asked at last, still puffing on his pipe, "what's the matter with this?"

"Not a thing. Not a damn thing," Wylie responded, letting a grin slip onto his face. But before they left the mesa, Wylie made his partner promise that he wouldn't mention it to anyone—not to his wife, not to Ole Larson, and certainly not to the commissioner of Reclamation. "If anybody asks you if we found anything we liked down here, just tell them it was awfully discouraging," Wylie instructed. The mesa belonged to the Navajos, after all, and in these modern times you couldn't just take it from them; you couldn't even do it if you were the federal government. And besides, it wouldn't pay for the fellows at Reclamation to show their hands on a town site too soon. If both Utah and Arizona wanted to work breakneck to get their highways built, presuming that the town-site selection would hinge on who did the faster job, then Wylie was more than willing to string them along and let them lay some road. Hell, now that he'd found this mesa, the whole business was beginning to look operational.

Wylie waxed confident as he and Puls started back down to the truck. It looked as though he had a nice narrow hole to work with—sheer walls that would clean up fairly easily and seamless sandstone that was serviceable if not spectacular. That beehive would be a bit of a bother, and transportation from rim to rim would be a bitch until they got a bridge built; he'd have to have plenty of powwows with the Navajos, no doubt, before they would let him build a town on their ground, and the governors of two states would be breathing down his neck for God knew how long, but all things considered, this looked like country he could raise a dam in.

This region where the dam would rise had been listed by the U.S. Geological Survey as the most isolated area—the fewest people, fewest roads and frowsy settlements—in the nation. (No doubt there were places in Alaska that were even more remote, but Alaska was just a territory, so its status was rather different.) In all the forty-eight states, no place came nearer to the end of the earth than did this far northwestern fringe of the Navajo reservation—the Navajo *Nation*, as the tribe described its particular kind of sovereignty—bounded on the north by the San Juan River, spilling out of the craggy mountains of the same name in southwestern Colorado, and on the west by the ruddy Colorado, already a thousand feet deep in Glen Canyon, en route ultimately a mile into the earth in the Grand Canyon.

The Navajo Nation as a whole was by far the biggest tribal reserve in the country—its total land area of 24,000 square miles roughly the size of the combined areas of Massachusetts, Vermont, and New Hampshire—the reservation consuming the northeastern quadrant of the state of Arizona and stretching north and east into Utah and New Mexico as well. And although the estimated 75,000 Navajos comprised the most populous tribe in the country, they couldn't come close to filling up the high-desert terrain that had been their home for more than two hundred years.

Descended from the Athapaskan peoples who had migrated southward from subarctic Canada, the first identifiable Navajos had settled in northwestern New Mexico around 1500, then slowly spread west and south as their numbers increased. Profoundly *American* in their ability to adapt to geographical and economic demands, as well as in their knack at incorporating the cultural traditions of other tribes, the Navajos had flourished until the early 1860s, when they were very nearly destroyed by Colonel Christopher "Kit" Carson and troops of the U.S. Army under his command. Carson and his men, convinced that the Navajos were responsible for ongoing Indian violence in the Southwest, had burned Navajo farms and settlements, slaughtered livestock, and finally forced tribal members to undertake the "Long Walk," a 300-mile march to Fort Sumner on the Pecos River in eastern New Mexico. Following a terrible four-year imprisonment and the eventual signing of the Navajo Treaty in 1868—an agreement that established the initial boundaries of a Navajo reservation in Arizona—the Navajos had been allowed to trek homeward again. A few extended families, or "outfits," had traveled beyond the previously settled tribal lands near the Arizona–New Mexico border and continued on to the sand-swept heights of the Kaibito Plateau, which dipped southward from Navajo Mountain, one of several peaks that served as spiritual landmarks in the Navajo world, and which lay just east of what they called "Old Age River," the crooked Colorado in Glen Canyon.

Although grass was sparse between the buttes and wind-sculpted spires that dotted the plateau, its highlands were lush in comparison with much of the rest of the Navajo country, and sheep and goat flocks grew large, the majority of the animals belonging to the four outfits that eventually controlled the region—the Tsi'najinii and Bigishii outfits initially, later joined in that outpost country by members of the Yazzie and Bidootlizhi clans. By the 1950s, the 2,500-square-mile overgrazed and overlooked plateau was inhabited by as many as 3,000 Navajos, all of them living without electricity or running water, without medical clinics or schools, without paved or even gravel roads, most of them traveling horseback or by buckboard—a wealthy few the envied owners of Ford and Chevy pickups—the only trading posts in the whole region located at the scattered settlements of Shonto, Kaibito, and Coppermine.

Manson Yazzie long had watered his sheep at a spring a few miles north of Coppermine, the small seep tucked under exposed rock at the north end of an empty mesa. The spring wasn't limited to his use; it was used by herders and headmen from all the outfits, yet because he tended it, cleaned it, and kept it excavated, it bore his name, and the mesa too—Manson Mesa—was named after him. The name didn't appear on any map or public record, but everyone who lived on the Kaibito Plateau knew it—knew that Manson Mesa was way over there, beyond Leche-E Rock, above the river—and somehow the tribal leaders in faraway Window Rock knew the name as well when they sent a telegram to Washington, D.C., on April 18, 1956, just a week after President Eisenhower had formalized the federal decision to build a dam in Glen Canyon. The wire, sent to Commissioner of Reclamation Wilbur Dexheimer—with copies going to Arizona's senators Carl Hayden and Barry Goldwater and congressmen John Rhodes and Stewart Udall—explained that the tribe had completed a preliminary survey of its lands surrounding the dam site and had "LOCATED IDEAL TOWN SITE ON COOL SOILED MESA SOUTH OF DAM." And the message assured the men in Washington that the "TRIBE WILL COOPERATE TO MAXIMUM WITH PLANS TO LOCATE TOWN SOUTHEAST OF RIVER." The telegram was signed by Paul Jones, chairman of the Navajo Tribal Council, and if the message tried to minimize the temperatures on the mesa top, if it rather imprecisely defined sand, it was nonetheless enthusiastic.

Yet for some reason, Commissioner Dexheimer didn't bother to mention the communication from the tribe to Lem Wylie six weeks later when he sent him out to take a firsthand look at Glen Canyon and the desert plateaus that rolled away from it. Wylie had had to find Manson Mesa on his own, after puffing his way up its steep western slope with Louis Puls. But he had pronounced it equally ideal, if not so cool and soiled, when he saw it for himself.

The Bureau of Reclamation's preliminary geological report, filed before the CRSP had passed Congress, hadn't been so charitable about the "upland plains" that flanked Glen Canyon, pointing out that there was good reason why the area was "uninhabited except for scattered Indian families on the reservation. Travel over land through this area is difficult and slow," said the report. "Water is scarce and only a very scant desert growth, developed on the scattered patches of blowsand and alluvium, covers the otherwise continuous rock exposure. The entire area," it concluded, "is a vast expanse of wasteland." But wastelands had never deterred the Bureau before. It was in the business, after all, of transforming wastelands, of converting deserts into wonderlands of water, and although the geological report hadn't been kind about the plateaus, it had greatly approved of the canyon.

Reclamation geologists originally had examined six potential dam sites in the 16 miles between Lee's Ferry and the mouth of Wahweap Creek, then had quickly concentrated on two sites where the canyon walls were narrowest: one at Mile 4 (measuring up-stream from the gauging station at Lee's Ferry) and a second at Mile 15. The Mile 4 site, the very spot where E. C. LaRue had wanted to erect a dam forty years before, at first had seemed the most promising. By locating a dam in the upstream half of a tight horseshoe bend in the river there, diversion and power-intake tunnels could be drilled laterally through the rock spur to a powerhouse and spillway in the downstream end of the bend, avoiding the construction of tunnels through the segments of the canyon walls that would form the dam's abutments. The canyon was at its most narrow at the Mile 4 site, and a reservoir that was dammed there would have stored significantly more water than would a reservoir impounded at Mile 15. Yet Mile 4 had a couple of serious flaws. Beneath the silty riverbed, the thick-bedded Navajo sandstone that formed the canyon walls gave way to 30 feet of the Kayenta formation, a fractious and less massive sandstone than the Navajo, beneath which were the thin Wingate stratum and 100 feet of crumbly and highly suspect Chinle shale—foundation materials that would result in huge water losses at best, and in a collapsing dam if the Chinle gave way with saturation. When they examined the canyon walls, the geologists also found four joint systems cutting diagonally across the canyon, the joints extending from rim to riverbed on either wall. They were simply cracks in the massive block of the Navajo formation; they weren't faults and none of the blocks had been displaced, but they weren't the sort of cracks into which you cavalierly pressed the abutments of a dam, and Mile 4 began to look like a loser.

The Mile 15 site was in a narrow, straight-walled section of the canyon where the rims stood 650 feet above placid river. The site lacked the tunneling advantages of Mile 4; it would be harder to reach with roads and powerlines and access tunnels, and it would store a lot less water. But on the positive side, it was only seven miles from an enormous aggregate deposit lying along Wahweap Creek, and at the dam site itself, the Navajo formation extended at least 434 feet—and maybe much farther, the drills hadn't gone any deeper— beneath the bed of the river. Navajo was a fine-grained and only moderately well cemented sandstone; you could easily chip it with a hammer, and it sucked up water like a sponge, yet 400 or more feet of it would make a remarkably solid foundation for even a ten-million-ton dam. When you compared the two sites, there really wasn't any choice, the report concluded. You could build a dam at Mile 4, then worry, or you could build one at Mile 15 and walk away for a thousand years, knowing it would be standing strong when you came back to check on it.

The Bureau geologists had no patience with the so-called experts who had told Congress that the walls of Glen Canyon would collapse when they came in contact with water, and they didn't believe for a minute that water would escape into giant underground basins. They did acknowledge the basins' existence, and they admitted that the Navajo sandstone would absorb water into "bank storage," but they could find no evidence of faults that would lead water deeply underground, and they were certain that water would readily flow back out of the sandstone when the reservoir pool was drawn down, and at least that water wouldn't evaporate while it was captive in the rock.

By the time President Eisenhower set the CRSP in motion, by the time the Navajo Tribal council telegraphed its offer of Manson Mesa and Lem Wylie had his first look around the region, the choice of the dam site was set. The Californians and the conservationists could continue to cry all they wanted to about how a dam in this place wouldn't hold, would collapse in the biggest federal fiasco of all time, but neither the Bureau administrators, the field geologists, nor Lem Wylie were much concerned. Wylie himself had far too much to get on with to try to second-guess the core samples, and besides, he boasted—putting his preference for granite out of his mind for the next few years—he wished he could be as sure of the sun coming up in the morning as he was of that 400 feet of foundation.

By the time Lem Wylie had set up shop in an abandoned school building in Kanab in early August, every con artist and snake-oil salesman in the western half of the country had descended on the little Mormon town—every one of them out to pocket a share of the $200 million or so the government was planning to spend in the immediate vicinity. Wylie's secretary quickly got to be good at protecting her boss from the steady stream of two-bit contractors and unemployed dozer operators who wanted to get some work. But one sweltering morning, a man marched past her despite her imprecations, opened the door to Wylie's office, and introduced himself as Arthur Watkins, saying he was mighty happy to see this thing get going, telling the man behind the desk that he looked as though he could handle the job. Wylie recognized the name, but it took a few seconds of conversation before he could place the white-haired man with the bold demeanor. Why, Christ Almighty, this must be Utah's Senator Watkins, Wylie realized at last—the selfsame fellow who was still fighting mad about losing Echo Park, but who nonetheless recognized that his state would gain no small benefit from the dam that was going to rise in Glen Canyon. As Wylie settled back in his chair, Watkins made it clear that he thought the construction camps and supply depots belonged on the Utah side of the river, and he told the construction engineer that they were going to need to put together one whale of a ground-breaking ceremony. As Watkins explained it, the Republicans finally were going to be able to make some political hay from Reclamation, and there wasn't any point in waiting until Glen Canyon Dam was completed to make it clear to the citizenry exactly who was at the helm. In fact, the senator was so excited about the prospects for a Glen Canyon ground breaking that he picked up Wylie's phone and got Dick Nixon on the line.

The senator told the vice-president that he was calling from Kanab, Utah, temporary headquarters for the Glen Canyon Project, Reclamation's biggest job since Grand Coulee, and he wondered if the vice-president remembered how Harry Truman seemingly had dedicated every bucket of concrete they'd laid up in the Columbia basin. Nixon remembered, and he agreed with Watkins that this time around they certainly would get President Eisenhower involved— maybe not go so far as to send him out to a place that sounded desperately desolate, but at least get him in some photographs, some newsreel footage, make it plain that Ike was the CRSP's catalyst. Nixon said he would talk to the president, and Watkins assured him that he would take care of things on the Utah end. After Watkins hung up the phone, he took care of things quickly, telling Wylie they'd need to make some noise and kick up some dust in the canyon in a month or two. He'd leave the details up to him, he said, making an equally abrupt exit.

Work was already in progress on the section of road from the Utah state line to the dam site, and bids were out on the contract to bore the west diversion tunnel, but it would be six months or more before the job had a general contractor. Wylie supposed that this "ground breaking" ought to involve something a little flashier than a Cat scraping a road-cut, but it took him a while to come up with a suitable plan. When the bids for the diversion tunnel were opened on September 11, and when the representative of the low bidder, Mountain States Construction Company of Denver, said sure, his crews were ready to go to work, Wylie struck on something he thought would suit Senator Watkins and probably the president as well.

The idea was to have two high-scalers—fearless young fellows willing to work the cliff face suspended only by a bosun's chair and a thick manila rope—swing off the rim above the spot where the tunnel outlet would be, drop down almost to river level, set several tons of explosive in a crack behind a house-sized slab of rock, wire it with primer cord, then string a line up to a plunger perched near the rim. From the White House, President Eisenhower would tap a telegraph key, the signal bound for Kanab, where Bureau employees would relay it by radio to a flagman at the dam site, whose wave would mean that it was time to lean into the plunger and blow out a bit of rock. The blast would powder the canyon with dust, and its reverberations would make the requisite noise.

All was ready on Monday morning, October 15. Over the weekend, the two high-scalers had set what would be a sizable blast indeed. The cables were checked, the two-way radios tested, the phone lines cleared from Washington, D.C., to Kanab. Reporters were watching the canyon wall; radio stations from Salt Lake City and Phoenix were on hand to broadcast the blast; Bureau of Reclamation film crews were poised on the opposite rim, and 2,000 miles away, the president, joined by his newly appointed secretary of the Interior, Fred Seaton, was set to depress the ceremonial key.

But when the president pushed, nothing happened. He pressed the key a second time, then waited. Secretary Seaton whispered into his phone, telling Kanab that the president was waiting. The radio message was frantically sent to the dam site, to Senator Watkins himself, who proudly shouted the chain-reaction command into a walkie-talkie to a flagman stationed at a high spot a half-mile back from the rim. But the flagman wasn't sure if he heard "go" or "no-go" amid the static. When he called back for clarification, the *GO!* he heard was emphatic if none too friendly, and he waved his flag. The plunger dropped and the canyon was filled with a torrent of rumbling, crescendoing sound. Great boulders arced into the air and fell away to the river; a billow of grit and dust lifted above the rim and wafted dramatically away, and the report of a successful blast began the reverse journey from Senator Watkins to Washington. The president let out a little shout, then, seeming slightly befuddled, said, "I guess it takes electricity a long time to travel out West," when Secretary Seaton informed him of the successful detonation. The secretary simply nodded before he got on the phone to tell Arthur Watkins he'd done a splendid job, and to tell Lem Wylie to proceed with building the project.

The job had to move forward on several fronts at once. Tunnels had to be built to bend the river around the job site; roads and airstrips had to be bladed out of the blowsand so men and materials could get to work; a bridge had to be built to connect one rim with the other, the two now separated by a quarter-mile of yawning space; the sheer walls of the canyon had to be stripped of loose rock, then great keyways, the gaping slots in the walls that would hold the dam in place, had to be chiseled into them; hundreds of miles of electrical and telephone lines had to be strung to serve the project; and temporary construction camps—and later a permanent town—had to rise up from the saltbush. In addition to the gigantic prime contract for the construction of the dam and power plant themselves, and including the years of preparatory work that would precede construction, Reclamation would let literally hundreds of contracts in the coming years—contracts for everything from ten-ton hoists to handrails, from eight hydraulic generators at $1 million apiece to 10,000 gallons of lubricating oil at less than a dollar a gallon. Wylie would not have to attend to every piddling contract personally, to meet with the president and purchasing officer of every bolt-manufacturing business in the Southwest, but he would have to keep himself savvy, well apprised of each new contract, and, at least in the beginning, at least until he had a prime contractor whose project manager he could keep his thumb on, he would have to watch every hard hat like a hawk.

Mountain States Construction had won the job on the 2,778-foot west diversion tunnel with a bid of $2.4 million and beer money, then had promptly subbed out the drilling and blasting to North-wood, Inc., a Vancouver outfit, and the mucking to Theo. Wood of Salt Lake City. When those jobs were done thirteen months down the line, the prime contractor for the dam would begin to line the tunnel with concrete, producing in the end a cylindrical tunnel 41 feet in diameter that could carry 100,000 cubic feet of water per second, large enough in tandem with the similar east-wall tunnel (which would be started when this one was finished) to handle a huge flood if it had to, up to 200,000 cubic feet per second that might, just might, come barreling round the bend in a very bad mood some June.

Beginning in October 1956, just as soon as the ceremonial shoot was out of the way, Mountain States crews began blasting the wall at the tunnel portal, setting ring supports to shore up the entrance, placing roof bolts to secure suspect rock. When the portal was adequately opened, the two subcontractors joined the fray, working in leap-frog fashion to drill, shoot, and muck an average of 14 feet each day. Working from a tire-mounted, four-level scaffold called a jumbo, as many as fifteen drillers would attack the butt end of the excavation each morning, drilling up to 250 15-foot holes in the dark rock face during the course of the day shift, each hole packed with explosive during the subsequent swing shift by a crew of powdermen. Six hundred cubic yards of pulverized, graveled sandstone would be produced when the nightly shot—its muffled growl barely leaking out into the canyon—went off sometime before midnight. Then the mucking crew, just fourteen men working the graveyard shift, would load the rocky debris into dump trucks that would haul it out of the dust-clogged, dimly lit tunnel to the riverbank.

There was nothing novel about boring a tunnel in Glen Canyon, except perhaps that the tunneling took place at the bottom of an inaccessible 700-foot hole, the dump trucks, shovel-loaders, and a small army of air compressors lowered from the rim by a highline cable, the men sent to work each shift in a caged, cable-suspended elevator they called a monkey-slide. And neither was the highway work anything new; engineers like Wylie, like Louis Puls, considered highway building to be among the most prosaic of all the engineering arts. To build a road you simply surveyed a route, calculated acceptable grades and curve radii, then cranked up the Cats. Even the necessary road-cuts were straightforward enough; much like the tunnelers did, you simply drilled the rock that was in your way, shot it, then scraped the rubble. It was true that on the 73-mile stretch of road connecting Kanab to the dam site, road-cuts through the ridge called the Cock's Comb reached as deep as 250 feet, and on the Arizona side of the river, nearly 400 feet had to be cut out of the Echo Cliffs monocline in order to punch the 25-mile road north from the Navajo community of Bitter Springs, where Highway 89 wound south to Flagstaff. But with a little patience and a dose of dynamite you could build a road through anything, and already the 98 miles of new roads—graded, subsurfaced, and graveled but still in need of an asphalt seal coat—were taking shape, making the dam site a place you could get to from here, from anywhere.

Yet if all you needed to do was to travel from one rim to the other—just 1,271 feet as the ravens and turkey vultures did it—you currently had to drive a 192-mile circuit, heading south through the Echo Cliffs, crossing the river on the Marble Canyon Bridge, visiting Cliff Dwellers, Jacob Lake, Fredonia, and Kanab en route, crossing a moonscape of rough, incredible country before you ended up just a shout from where you started. A bridge across the canyon there at the dam site was absolutely essential, and the sooner it was built and open for traffic, the better.

Unlike the mundanity of boring a tunnel or blasting a roadbed, however, the bridge the Bureau engineers had designed to span Glen Canyon was going to be something special. This bridge in the bleak outback of the Colorado Plateau was going to be the highest steel arch ever constructed—its centerpoint 680 feet above the brown river—and its massive arch connecting the canyon walls would be the second longest span in the world, exceeded only by the Bayonne bridge that joined Staten Island to the New Jersey mainland. It was precisely the kind of project that could get Fran Murphy excited, and its remoteness simply didn't matter. You could live anywhere for a year or two, he knew, as long as you were doing good work.

Murphy, son of the president of Judson Pacific–Murphy Company, had had an earlier hand in two terrific jobs—the 37-span bridge across the Columbia River at The Dalles, Oregon, and the three-mile-long Richmond–San Rafael Bridge across San Pablo Bay northeast of San Francisco. When Murphy, his father, Phil, and Tom Paul, vice-president of Peter Kiewit Sons' Company, had flown out to take a look at Glen Canyon ten days following the first blast in October 1956, they quickly knew they wanted the job. They informally agreed that day that the two companies would form a partnership to secure bonding and to bid on the bridge, and they correctly guessed that Reclamation wouldn't exactly be overrun by competing offers to build it. On December 19, the California-based partnership, awkwardly called Kiewit–Judson Pacific–Murphy, submitted the lower of the two bids the Bureau received, and for a fee of $4.14 million, Murphy and company went to work.

Fran Murphy, who would become the project manager, didn't need his engineering degree or his upbringing in the construction business to realize that when you were building a bridge 600-some feet in the air, you couldn't support it by falsework from the bottom. The only way you could suspend the arch trusses as they slowly curved out from the canyon walls was by means of tieback cables running through tall towers on the rims, then anchored in concrete deadmen tied into the plateau rock. But Murphy did need his own expertise, plus that of project superintendent Bill Choate, to devise a means of placing successive triangular-braced sections, each one riveted to the next, out over the gaping hole. Traveling cranes, riding on the arch itself, might have done the job, but they would have added 150 tons or more to the enormous weight already supported by the tiebacks. A separate cableway system, with which a cable moving between towers on either rim could swing each new section into place, seemed to be a logical solution. Yet the completed bridge would be composed of two parallel truss arches 40 feet apart, one beneath either edge of the roadway, the two tied together to form a single structure. To move materials into position along that 40-foot width, you could build a cableway with towers mounted on tracks that would allow the cable to be moved laterally from arch to arch, or—and this, of course, sounded like the exciting thing to do—you could construct a cableway with *tilting* towers, the towers anchored smack between the centerlines of the two arches and which, by means of hinges and side guys, could "luff," or lean, the 20 feet necessary to move the cable into position directly above either arch or anywhere in between. The luffing towers would be efficient; they would be comparatively inexpensive; and they'd be very slick, the young Murphy surmised.

With the Kiewit Sons end of the operation in charge of excavating the skewbacks in the canyon walls, as well as executing all the concrete work, and with the Murphy branch overseeing the fabrication and erection of over 4,000 tons of structural steel, the bridge got under way in February 1957, when high-scalers backed off the canyon rims and dropped down 170 feet to begin drilling holes for the explosive charges that would blast enormous, slotlike caves into the cliff face. The caves, called skewbacks, would serve as canyon-wall abutments for the two arches once they were fully excavated, reinforced with 30-foot-long anchor bolts, and finally filled with concrete—and it would take almost a year to construct them. In the meantime, the luffing cableway, capable of moving a 25-ton cargo, would have to be erected, along with a smaller cable system that would carry crews out onto the span.

A thousand miles away in Emeryville, California, in the spring of 1957, workers at the Judson Pacific–Murphy yard began fabricating steel for the giant arch, every member in each section having to be milled to tolerances within $\frac{1}{10,000}$ of an inch if the spans sweeping out from each canyon wall ultimately were to meet in the middle. But before the steel was shipped by rail to Flagstaff and north to the job site by truck, it was first assembled into four half-arches in the Emeryville yard, the sprawling pieces inspected, measured, and measured again before they were disassembled, sandblasted, painted, and loaded on flatcars headed for Arizona.

When Lem Wylie first had spotted the mesa, he had assumed it would be a hell of a long time before the Navajos were willing to give it up, before the state of Utah would quit squalling that it once again was getting the short end of the stick. But you never quite knew how things were going to go in this line of work, and on March 22, just nine months since Wylie's first visit, Commissioner Dexheimer announced that the permanent town adjacent to Glen Canyon Dam would be located on the east side of the river. The Navajo Tribal Council had agreed that it would trade a 24.3-square-mile piece of its reservation—the sandy tableland on Manson Mesa as well as the land sloping west to the river—for a similar-sized chunk of McCracken Mesa over near Aneth, Utah, land the Navajos long had coveted and claimed was rightfully theirs. Anglo stockmen in Utah who held grazing leases on McCracken Mesa, as well as the oil companies that had exploratory drilling leases there, were outraged by the plan, but although Senator Arthur Watkins was doing his best to defend their interests (and in the process, trying to make a last-ditch effort to secure the dam town for the Utah side), a bill sanctioning the swap seemed certain to glide swiftly through Congress.

Less controversial was Dexheimer's announcement that the town would be called Page Government Camp in its earliest incarnation, then Page, Arizona, once it was a going concern. Suggestions for the name for the town had come in from many quarters since the CRSP's approval a year before. "Glen Canyon City" was a popular and obvious choice. Some people favored "Navajo City" as a means of acknowledging the aboriginal inhabitants of the area; others who were more conquest-minded liked "Powell" or "Powell City," as a tribute to the early explorer; a Phoenix newspaper columnist wanted to call it "Hayden's Crossing" in honor of Arizona's senior senator; and one wag thought the town simply should be named "Dam Site," claiming the West had a history of humorous if unadorned place names.

But Dexheimer and Floyd Dominy, the associate commissioner of Reclamation who seemed rapidly on the rise these days, owed their respectively waning and waxing careers to the Bureau, and it seemed only fitting to the two of them to honor the Bureau by naming this new town after one of Reclamation's early leaders. John C. Page had been the commissioner of Reclamation from 1937 to

1943, and had died just two years before. Page had been a project engineer, struggling with the egos and eccentricities of the Six Companies, on the Boulder Canyon Project, then had been at the Bureau's helm during its early development of the Columbia River basin, as well as during the chaotic power-producing years of World War II. Plans for the Colorado River Storage Project had not been hatched before his retirement, it was true, but Page had long averred that Glen Canyon ought to be dammed, and the quiet, methodical man was a true champion of the development of the Colorado. When Dexheimer and Dominy ran their proposal by Secretary of the Interior Seaton, he made no objection—it seemed like a decent enough name to him—and neither did he see any reason to bother President Eisenhower with the trivial detail that the new Glen Canyon community was going to be named in honor of a vaunted New Deal bureaucrat, so the name soon was official.

Fran Murphy and his crew, at work on the construction of the cableway towers that one day would lower the bridge into place, were living in metal barracks on the west side of the canyon when the announcement about this new place called Page was made in Washington—their primitive camp separated from the deep hole in the ground only by the high hump of the beehive. The tunnel-boring crews were housed beside them, and the two groups of men shared a mess hall fashioned from a circus tent, the food—surprisingly good food for the middle of gaping nowhere—catered by Willard Wood, a man whose culinary talents would draw rave reviews for years to come, ever after Page had mess halls of its own, and later, cafes and restaurants.

The Bureau of Reclamation personnel who had been assigned to the project—Lem Wylie and already dozens of subordinates—still commuted to the job site from Kanab, daily driving the gravel road with their boots bearing heavily on the accelerators of the Bureau's sedans and station wagons, negotiating the 73-mile stretch in under an hour when the traffic was light, when the road was watered by thunderstorms and the dust didn't nearly blind them. But the office work, the small mountain of paper that already was piling up, was attended to in the old Kanab high school building, and it would remain project headquarters for another year and a half or so, until Page had a sewer and water system, a school and a hospital, warehouses and retail stores, until at least a few hundred trailers had been set down on blocks for the contractors' crews, and a few dozen permanent homes had been built for the Bureau's employees. Until then, and while all this government largess lasted, little Kanab looked like a gold camp.

It was in Kanab, in fact, on the promising morning of April 11, 1957, with the cottonwoods leafing and the hayfields growing green, that bids were opened on the largest single contract ever let by the Bureau of Reclamation, the largest nondefense contract the United States had ever entered into. As defined by the Bureau planners in Denver, the prime contract for the Glen Canyon Project would call for construction of a concrete gravity-arch dam of no mean proportion—it would be the fourth highest in the world—a power plant structure at the toe of the dam, two tunnel-type spillways, a truck-negotiable access tunnel from the rim to the power plant, and all appurtenant penstocks, pipes, cables, valves, switches, and gates. Before it was finished, the project would require five million barrels of cement, ten million cubic yards of aggregate, three million board-feet of lumber, 130,000 tons of steel, 20,000 tons of aluminum, 5,000 tons of copper, and a peak work force of 2,500 men. The prime contract did not call for the installation of generators or the completion of the power plant or for a thousand and one finishing details, but it did demand that the enormous undertaking be completed in 2,500 days—a little less than seven years—and it required that the Colorado River be diverted, cared for, and kept out of the way for the duration of construction.

When the Bureau of Reclamation made its bid specifications public early in the year, you could have heard a hundred dramatic gasps throughout the heavy-construction industry. Even with Hoover, Shasta, and Grand Coulee already in place, this was going to be some kind of project. And even with the inflated dollars of the fifties, it was going to cost real money to get it built. Construction-company presidents around the country either shuddered once, then vowed to stay as far away from Glen Canyon as they possibly could—realizing that a hole in the ground like that had a terrible potential to swallow profits, careers, and whole companies if you encountered trouble—or immediately itched to get in on the action, to do something bold and brash, something with some challenge to it, to make some of the government's money by doing the exhilarating kind of work they'd be tempted to do for free.

Everybody who knew a little something about dam building was certain that the winning bid would be well over $100 million, maybe

$150 million when you factored in the remote location, and there weren't too many outfits around that could carry that kind of risk on their own. In order to secure the bonding the Bureau of Reclamation would require of its prime contractor, several companies—enormous operations by any other standard—would likely have to form partnerships, consortia, huge holding companies to pool the kind of financial clout the Glen Canyon job would demand.

The fellows at Boise's Morrison-Knudsen, one of Hoover Dam's Six Companies, did precisely that in order to become a player, the giant firm forging a bidding consortium made up of all manner of large and small outfits, each one with the expertise to fill a particular niche. A California-based partnership that dubbed itself Glen Canyon Contractors—and which included the firm now run by Dad Bechtel's boys—also got into the action, chartering its new corporation, securing its bonding, then frantically studying the more than 2,500 design blueprints prepared by the Bureau before anxiously drafting its bid.

The third player in this high-stakes game—the final hand dealt in a Mormon town where simple Saturday-night poker was firmly frowned upon—was Merritt-Chapman & Scott Corporation, a New York City conglomerate that had built the Columbia River dam at The Dalles for the Bureau of Reclamation, as well as having been involved in a very big way in ship building, marine salvage, smelting and steel fabrication, and the manufacture of paints, chemicals, fuels, and heavy equipment. With subsidiaries in Canada, the West Indies, Panama, and India, and with gross revenues in 1956 of $374 million and an after-tax profit of $24 million, Merritt-Chapman & Scott was one of the few companies in the country that didn't need to form any sort of cumbersome or compromising alliance before it went after a piece of work. From their Fifth Avenue offices, the men at the top of MCS simply decided which government contracts seemed particularly appealing, then set their subordinates to work, reminding them of the cutthroat truth that nobody ever made a dime off a contract he didn't get.

The old high school gymnasium was festooned with crepe paper on the day Lem Wylie was slated to open the bids. A crowd made up of curious townspeople and Bureau personnel in casual clothes, as well as the dark-suited, chain-smoking corporate men who had an awful lot riding on the afternoon, filled the room to overflowing; still photographers from newspapers around the region jammed together near the front, and a phalanx of newsreel photographers turned on their noisy, tripod-mounted cameras as Wylie, sitting at a long folding table in front of a blackboard that would be used to record the numbers, wearing a suit himself and appearing rather uncomfortable, his favorite sweat-stained brown fedora surrendered for this august occasion, got right down to business.

Glen Canyon Contractors' bid, dutifully scrawled on the blackboard for all to see, was for $118,336,476—no one rounding off so much as a single buck in this business. And although he sat calmly, his face inscrutable, Wylie was tickled to death by the bid: the Bureau's money men in Denver had estimated the cost of the specified work, plus profit, at almost $136 million. Here, on the very first bid, it looked like they could undercut that figure by as much as $17 million and save the citizens some money. Morrison-Knudsen's bid, the second one to be opened, was right in the same ballpark, but at $120,178,853, the Idaho outfit was out of the running. A round of applause rang out, and Lem Wylie couldn't help but shake his head in amazement a few moments later as he announced Merritt-Chapman & Scott's bid of $107,955,522. It was an astoundingly low figure, almost $30 million below the Bureau's estimate, but it was no bum bid, not coming from this company it wasn't. And if Merritt-Chapman & Scott wasn't worried about making a profit, well, Wylie sure as hell wasn't going to worry about it for them. They were big boys—as big as you could possibly get—and they had a decent reputation within the Bureau, so they were welcome to it. And if they were willing to leave almost $11 million on the table to get the job, well, maybe that was just the way these New York fellows played poker.

As the photographers were packing up their cameras and the festive crowd was filing outside, big Bill Denny, the head of Merritt-Chapman & Scott's construction division, an Irishman who was usually ebullient and full of bluster, approached Wylie looking ashen, looking as if he and his company had just suffered a great disaster. And perhaps they had. They had just underbid respected and shrewd competitors by more than $10 million, 10 percent of their own contract bid, and it looked as if they were going to start losing money before they dropped a single man with a three-dollar shovel into that canyon. It was one thing to submit a low-ball bid, but it was quite another to *pay* for the privilege of breaking the company's back.

But Wylie tried to reassure the executive, tried to tell him that he thought he'd come to town with the right price. You could build this thing for $107 million if you ran a tight operation, you bet you could, and you could make a little money in the process. Wylie knew for a fact, he told Denny, that Reclamation's own estimate was so high simply because Glen Canyon was a million miles from every-where. If MCS could compensate for that by keeping its labor costs down, and by pouring concrete faster than anyone had ever poured concrete before, well hell, the company could salvage a profit the same way it salvaged ships. Denny was buoyed a bit by the opinion of this pugnacious engineer he was going to encounter often in the coming years—this guy who knew dam building inside and out, after all—and just maybe he was right. Denny accepted Wylie's invitation to retreat to his office, where the project chief kept a bottle of scotch in a file drawer in his desk. And in private the two men lifted their glasses to toast the successful bidder, but they drank to try to settle his nerves.

Like it or not, Merritt-Chapman & Scott was officially awarded the prime contract for the Glen Canyon Project on April 29, 1957, but it was early June and the desert had begun to bake before the first MCS crews trekked into the area and got down to tasks, first building barracks, a mess hall, a general shop, and storage facilities on the east rim of the canyon, and beginning construction on the mesa above—up where this town of Page was supposed to stand one day—of a permanent twenty-five-bed hospital.

In the hole itself, the first order of business was transportation, devising means of sending men and materials from rim to rim, from rim to riverside, from one bank to the other down in the bottom. The cableway that Kiewit–Judson Pacific–Murphy had built for the erection of the bridge would be of no help on the dam itself—the bridge would span the canyon a few hundred feet downstream from the axis of the dam, far too far away. MCS would need a couple of giant cableways of its own directly above the dam site—massive four-inch-diameter cables weighing 38 pounds per foot and capable of delivering 50-ton payloads down the canyon floor every four minutes. The cables would be suspended from four towers, a tall one and a short one on either rim, each one traveling laterally on railroad tracks to provide drop-access to the entire area from the upstream face of the dam to the downstream wall of the power plant, the small-tower system capable of moving laterally beneath the cables tied to the tall towers.

Six monkey-slides, more formally known as personnel elevators, would have to be put in operation quickly, the big metal cages climbing the canyon walls on fixed guy cables and controlled by derrick hoists, dropping men down to the diversion-tunnel inlets, to the outlets, to the points on the walls where the giant keyways would be blasted. Two small pedestrian bridges, suspended from cables, would cross the river in the canyon bottom, and from the rims, a breathtaking, daredevil footbridge—nothing more than four feet of horizontal wire mesh and a shudderingly unobstructed view—would span the open gorge.

By Christmas of 1957, with snow draping distant Navajo Mountain and ice clogging the eddies of the ambling river, a gargantuan gopher hole now curled through the wall of the canyon, and the Mountain States crews who dug it had gone home to Colorado. High-scalers hung like spiders now from the rims of the canyon, prying loose rock from the walls, sending it crashing down to the river, drilling first hundreds, then thousands, of bolt holes throughout a half-mile stretch of the canyon, inserting rock bolts up to 30 feet long into each one of them, securing the sandstone face with a freckled mass of metal plates. And on either wall, the cutting of the keyways, the notches that very literally would prevent the dam from moving downstream, was under way—the ritual of drilling, blasting, and mucking debris beginning at the rims and ultimately descending 800 feet straight down.

Nearby, 135,000 cubic yards of sandstone already had been cut from the beehive—removed without any argument—the fat monolith flattened on its canyon side to make way for the tracks being laid for the traveling cableways. A thousand feet to the south, the luffing cableway that would lift the steel sections of the bridge into place stood ready to go to work. The bridge's skewbacks had been excavated 34 feet into the canyon walls and now were ready for concrete, and Fran Murphy's crews were readying a safety net that would be strung from rim to rim beneath the structure, designed to catch errant steelworkers who might otherwise plunge to their deaths.

The 25-mile highway up from Bitter Springs was completed by Christmastime, finished with a hard black coat of asphalt, and the

W. W. Clyde Company, the Utah outfit that had laid the highway in from Kanab, had just been awarded a second contract, this time to lay streets, curbs, gutters, sewer and water lines for the new town up on the Caterpillar-scraped, dust-drifted mesa. A total work force of 750 men was at ease in that icy outpost on the one-day holiday, lingering long over a bountiful Christmas dinner prepared by Willard Wood—ten times the number of workers that had been in Glen Canyon when work began a year before.

By the Fourth of July, 1958—the next time the job site strangely was idle—drillers and powdermen making $2.70 an hour had cut more than 200 feet down in the keyways, and on the rims upstream they had begun blasting the wide portals of the spillway tunnels that twenty years hence, maybe thirty, would receive the overflow of a reservoir filled to its brim. Ironworkers worth $3.25 an hour had begun to build the traveling cableway towers and had completed the tieback towers that would help hold the great arch of the bridge in midair as successive sections were moved into place. Section by weighty section, two arches were crawling out into the canyon, and by Independence Day, they appeared only a few feet from meeting in the middle of the gorge. That same day as well, an army of teamsters ($2.29 an hour) took a holiday from trucking seemingly endless loads of still more steel—the vertical columns that would rise up from the arch and the horizontal members of the roadbed—140 miles north from the Flagstaff railyard through the quiet and sweltering desert.

Then, on August 6, 1958, the final truss of the arch was lowered into place—missing its mate by just a quarter of an inch before tension on the tieback cables was adjusted minutely and the sections met and were pinned and the arch bore its own weight between the walls. In early August as well, the left diversion tunnel was completed, concrete lining of the two tunnels had begun, and excavation of the steep, 55-degree, 50-foot-wide spillway tunnels was proceeding at almost eight feet a day.

By the beginning of November, the keyways had been cut down 500 feet, their rubble bulldozed into the canyon bottom, where draglines and dozers slowly eased it out in the river channel, beginning the cofferdams that would send its water into the diversion tunnels and keep it away from construction. By the Thanksgiving weekend, when more than 200 Reclamation employees and their families made the move en masse from Kanab to the trailer and tin-quonset town on the mesa, MCS had more than 1,700 men under its employ—carpenters and crane operators, welders and mechanics and dozens of drillers, drivers of flatracks, drivers of graders and loaders and lowboys. The fat cables were in the slow process of being strung between the cableway towers; a concrete bed was being poured on the nearly completed bridge; the first permanent homes were being readied in Page, and inside a single tin barn located on what would become the town's main street, a grocery store, a service station, a beauty shop, and a bank were open for business. It had taken two years, but this was no longer nowhere. It had a name; it had roads and sewers and a brash and roughneck purpose. Five thousand people already were temporarily at home in this place where, as recently as 1955, the only inhabitants had been Manson Yazzie's sheep.

The dedication ceremony for the new bridge spanning Glen Canyon was scheduled for February 20, 1959, and it was going to be some kind of celebration. Thousands of people from Phoenix to Salt Lake City were expected to drive the spanking new highways to the site, to have a look at this mammoth project the *Saturday Evening Post* recently had said would "transform the face of the desolate Arizona-Utah frontier, with eventual benefit for millions of vacationing Americans," and to witness the biggest congregation of elected officials you could ever imagine. Every member of the legislatures of the two states was expected to attend. Arizona's Governor Paul Fannin would be there, as would George E. Clyde, the governor of Utah, himself a partner in the construction company currently at work in Page. Ole Larson, the Reclamation chief in Salt Lake City, would attend in the company of Wilbur Dexheimer, beleaguered by his jealous assistant Floyd Dominy but still the commissioner of the Bureau of Reclamation. Navajo Tribal Chairman Paul Jones would make the journey from the tribal headquarters at Window Rock, Arizona, joined by the members of the Navajo Tribal Band, who would entertain. The ceremonies, to be held in the middle of the bridge, with Utahns and Arizonans crowding in from either side, would be climaxed by Governors Fannin and Clyde cutting a chain strung across the roadway with a ceremonial blowtorch.

But before Lem Wylie could take any time to celebrate, he had a job to do down in the hole. The lining of the diversion tunnels finally had been completed; through the sections where raging water might otherwise have chewed away at the ragged rock, the tunnels were now smooth as babies' bottoms, ready to receive the river. The schedule called for the diversion of the river in February, in time to turn it laterally while it was still in its winter trickle, in time to build the upstream earth- and rock-fill cofferdam 200 feet high before the spring runoff bore down upon it. If they didn't get the river diverted now—within two or three weeks, at least—it would be eight months or more before they would have another opportunity, and the entire project would be stalled, crippled by the swollen river.

Wylie had spent long hours with Al Bacon, Merritt-Chapman & Scott's project manager, the company's head man on the local operation, coordinating with him what doubtless would be a dicey business. The river had dropped to less than 3,000 cubic feet per second in the days since the first of the year—as low as it likely would get—but work in the canyon bottom had already significantly narrowed the river's channel, speeding its flow, and even at very low water, it wasn't going to be easy to talk this river into taking a detour.

Beginning in October, growling D9 dozers had started pushing the spoil upstream from the west keyway excavation and tumbling it into the river. From the west bank, the low finger of a dam now reached halfway across the current, forcing the flow against the east bank, aggravating it into a feisty rapid. On Sunday, February 8, it was time to extend the cofferdam all the way across the channel, and no one doubted that the river now would get angry.

The first step on Sunday morning was for draglines and dozers to begin removing the earthen dike that protected that portal of the west tunnel, getting it out of the way so the river would have an alternative to fighting against the growing dam. The tunnel portal lay four feet below the river level, and Wylie and Bacon had hoped that simply by removing the dike as much as half the river's volume might immediately flow into the tunnel. As a dragline breached the dike and water spilled through it, the tunnel quickly received its first flow. The breach widened, then the river tore through the dike, churning up mud and tumbling boulders. Much of the river disappeared into the dark tunnel, but much of it remained in its channel.

Working round the clock—tower-mounted flood lamps turning night into a strange, otherworldly kind of day—dump trucks drove across the crest of the cofferdam, dropping their loads of sandstone and gravel into the bucking current. But by Monday afternoon, they had made negligible progress. The river was swift enough, strong enough that it carried boulders, muck, aggregate, everything that was thrown into its path, through the tight channel and on downstream. Wylie and Bacon tried speeding up the dumps, one truck on the hurried heels of another, in hopes that they could outrace the current. They gained ground, but in doing so, they sped the current up even more, squeezed it into standing waves, and the prospects for sealing it off didn't seem bright.

Exhausted, and growing dispirited now, the crews continued their fight, seemingly throwing every piece of rock that had come down from the west wall into the angry channel, before Wylie and Bacon tried a new tactic. The channel evidently was too smooth, too unobstructed for the boulders to gain some sort of purchase in the swift water. What they needed was something that would make a bit of a mess, tie things up a bit, catch the tumbling sandstone as it fell into the current. On Tuesday evening, the first of the steel tunnel-lining forms were dumped into the river, dozens of them, then steel ramps, re-bar, frayed cable, culvert pipe, and they would have thrown in old car bodies as well if they had had them down in the hole.

The driven, desperate littering of the riverbed finally began to work; large boulders began to take hold; gravel packed in behind them; even the sandy soil held after a few hours, tamping rollers packing it tight in the seconds after it was dumped. At 7:30 on Wednesday morning, seventy-two hours after the diversion attempt had begun, the cofferdam was sealed and all of the river was thenceforth required to take the tunnel route—Wylie and Bacon and the spent but elated crews at last riding up the monkey-slides, leaving the hole victorious.

The crest of the cofferdam continued to rise in the succeeding days, climbing a quarter of the way up the canyon before the big Cats topped it out. And the river slowed abruptly now as it neared the new impoundment. It seemed to stall, to puzzle a moment in utter astonishment, then for the first time in eons, in more years than you can imagine, it left the fine deep canyon of its own making and fell into a giant pipe.

5
THE SALVAGE SEASONS

This was marvelous country. Driving north in 1941 from the south rim of the Grand Canyon, through the strange spectrum of rock and sky and emptiness on the Navajo reservation, up through Cameron, Tuba City, Kayenta, and on into Monument Valley, Greg Crampton was enchanted by what he encountered. The thirty-year-old doctoral student from the University of California was en route to a conference in Laramie, Wyoming, and it had been a detour of only a thousand miles to deliver his professor and friend, Herbert Bolton, to Mexican Hat, Utah, a tiny settlement perched on the rim of the San Juan River, where an adventurer named Norman Nevills was waiting to ferry Bolton and a few other passengers down the San Juan and on through lower Glen Canyon.

Bolton was at work on a history of the 1776 Dominguez-Escalante expedition, and he wanted to see for himself the "Crossing of the Fathers," the place 26 miles downstream from the San Juan's confluence with the Colorado where the Spanish friars had forded the river, trekking back to Santa Fe after an unsuccessful attempt to reach the missions on the California coast. Crampton sorely wanted to climb aboard one of Nevills's "cataract boats," but he couldn't abandon his commitment in Wyoming, and the cluster of boats floated round the first bend without him.

Four years passed before Crampton got near the canyon country again. Bolstered with a doctorate in North- and Latin-American history, as well as a budding interest in what he termed "historical archaeology," he joined the Department of History at the University of Utah in 1945 and quickly began to make weekend and holiday forays south to the slickrock desert. But it wasn't until the early summer of 1949 that, at the invitation of Salt Lake City realtor O. Coleman Dunn, Crampton finally got to see a portion of the Utah canyons from the back of the big brown river. Embarking from Hite, Crampton and his companions floated the Colorado—so swollen with runoff it was close to flood stage—in the war-surplus rubber rafts that detractors such as Nevills (who used only hard-hulled wooden craft) liked to call "baloney boats." Though the rafts looked like outsized wading-pool toys, they were perfectly suited to desert rivers, and navigating them downstream, Crampton and company quickly were overwhelmed by the glories of Glen Canyon.

But in addition to the beauty of the serpentine river and the high, many-hued walls that contained it, in addition to the enchanting side canyons and legendary landforms like Rainbow Bridge, Gunsight Butte, and Sentinel Rock, Crampton was astonished to find such rich evidence of the human presence in the canyon. There were Fort Moqui, Defiance House, Wasp House, and dozens more easily visible architectural ruins of the Anasazi; there were the chimney remains of the cabin built by the first Anglo settler in the canyon and many, more recent domiciles scattered about in various states of disrepair; there were mining dumps and rusted hand tools; an ancient Model 30 Caterpillar filled with gasoline waited on Gretchen Bar as if it were still ready to go to work, and there in the middle of the current a couple of miles upstream from Bullfrog Creek was the listing wreck of a giant gold dredge. It was fascinating and somehow paradoxical from Crampton's perspective. This was wilderness, to be sure. The canyon was hundreds of miles from the nearest community of even a few thousand people. The river itself was its only avenue, its only accessible pathway. It seemed as wild as any place imaginable, yet it obviously held a whole sweep of human history, of long occupations and short attempts at exploitation. It was at once a naturalist's wonderland and a historian's treasure trove, a riverine world that was utterly empty of people when Greg Crampton floated through it in 1949, but one whose ghosts were almost palpable, a place that he was determined to visit again and again.

Although he managed a second sojourn in Glen Canyon in the summer of 1950, familial and professional demands kept Crampton from returning to it again before he heard in 1956 the crushing news that the canyon would be flooded. The river canyons of the Glen and the San Juan and their side canyons—more than a hundred of them—would begin to go under in only a few years, Crampton's wonderland lost.

Yet perhaps there was an opportunity amid the tragedy. Jesse Jennings, an archaeologist specializing in the prehistoric cultures of the desert Southwest and chairman of the University of Utah's anthropology department, was already at work trying to secure for the university the federal contract to conduct studies of the myriad archaeological resources soon to be inundated by the reservoir that would fill Glen Canyon. The Historic Sites Act of 1935 had charged the Department of the Interior, through the National Park Service, with the preservation and scientific dissemination of the nation's antiquities, and it had called for their "emergency salvage" in situations where major developments threatened to destroy them. The law had been passed too late to result in any sort of antiquities inventory prior to filling Lake Mead behind Hoover Dam, and only minimal salvage work had been undertaken in the intervening years at other Reclamation project sites. But because the Anasazi culture had had such an enormous impact on the Colorado Plateau, because the plateau's dry desert and semidesert environments so successfully preserved physical evidence of the culture, and because Glen Canyon itself had been heavily occupied by early peoples, Jennings was determined that this time the salvage should be comprehensive. It should include basic geological, paleontological, and ecological surveys, as well as extensive archaeological fieldwork, and the resulting data and their interpretations should be made readily available through a series of detailed, if necessarily voluminous, reports.

In concert with Jennings's efforts, Crampton urged the park service to authorize historical investigations in the reservoir area as a critical adjunct to the studies of the canyons' prehistory. And by early 1957 the park service was convinced—the agency entering into formal salvage agreements with the University of Utah and the Museum of Northern Arizona, a privately endowed research institution based in Flagstaff. The museum, with Alexander Lindsay, Jr., acting as field director, was charged with the archaeological investigation of the 74 miles of the San Juan River Canyon that the reservoir would inundate, plus its adjoining side canyons, as well as the 69 miles of the left bank of the Colorado between the confluence with the San Juan and the canyon's terminus at Lee's Ferry. The University of Utah's responsibilities included the archaeological studies of the 130 miles of the upper Glen and its tributaries, the right bank and side canyons below the San Juan's confluence, and the ecological and historical studies of the entire reservoir area. Jennings was named project director, and Crampton happily agreed to head up the historical studies.

There would be seven field seasons, 1957–63, during which time more than 800 miles of sheer-walled canyons would be investigated by boat and on foot, the looming deadline imposed by the dam allowing archaeological sites merely to be sampled rather than entirely excavated. Small teams of researchers—usually composed of ten or fewer professors, students, volunteers, and occasional hired hands—systematically would scour the reservoir area during the summer months, using baloney boats and hiking boots as their principal means of access, spending up to two weeks at a time ensconced in the canyon, roaming, digging, recording their findings before flushing themselves out at Lee's Ferry. Unlike traditional scientific fieldwork, normally undertaken as an inquiry into a specific issue or "problem," the mission of the salvage teams would be no more than to try—as completely as the limitations of time and related resources would allow—to document what was there in those secretive canyons, what once had been.

The first fundamental surprise encountered by the academicians and inveterate diggers was that, even in relatively recent times, Glen Canyon had been a different sort of place. In the late 1950s, the side-canyon streams the researchers observed coursed across bedrock en route to the river, but they cut through remnants of alluvial sediments that once had created broad, flat, and fertile plains in the canyon bottoms. They were astonished to discover that in some cases the silt beds had been as deep as 90 feet. What immediately seemed possible was that agriculture once had flourished in these side canyons, but that much of the evidence of the first farmers—their shelters, fields, ditches, and drainage systems—subsequently had been flushed from the canyons along with the alluvial soils. The ruins of masonry structures that might have been granaries, lookouts, or seasonal dwellings still clung to rocky outcroppings, some of them sheltered by overhanging cliffs, but none seemed to be accompanied by the large, multilayered trash heaps that archaeologists call middens, whose presence strongly would have suggested long-term habitation. The ruins of the first residential structures in Glen Canyon long since might have been washed into the Colorado and away, but there was another possibility. Perhaps these early inhabitants lived on the canyon rims and adjacent uplands and only ventured into the canyons to farm. To try to determine whether that indeed had been the case, the archaeologists would have to extend their investigations laterally onto the plateaus that bounded the canyons, and the salvage would have to become decidedly more complicated as it barely got under way.

The first nomadic inhabitants of the Colorado Plateau belonged to a culture that Southwestern archaeologists have labeled the Desert Archaic. Entering the region perhaps as early as the tenth century B.C., the people of the Desert Culture migrated seasonally, following game and the growing seasons of the plants they foraged. They shaped primitive tools out of stone and wove baskets, but left only scant evidence of themselves in Glen Canyon and elsewhere in the vast region they roamed.

By about 100 B.C., however, the people had begun to be more sedentary, building their first permanent dwellings and experimenting with the exotic business of agriculture. A cultural revolution—an elemental change in living groups, habitation, diet, and geography—subsequently was spawned by the successful cultivation of crops and it became widespread throughout the Southwest, beginning first with the people in the region that is now southern Arizona and New Mexico, then extending north to the peoples who inhabited a great arc of land reaching from the Rio Grande in central New Mexico to the Virgin River in southwestern Utah. It was roughly the same region the unrelated Navajos would begin to occupy 1,500 years later, and the Navajos referred to their predecessors as *Anasazi,* a word variously translated as "Ancient Enemies," "Old Strangers," "Those Who Were Here Before Us."

Anglo settlers in the Four Corners region of the Southwest (where the states of Colorado, New Mexico, Arizona, and Utah meet at a common point) tended first to refer to the Anasazi as *Aztecs,* mistakenly thinking that the physical evidence of them—their ruins were obvious and everywhere—linked them to the ancient culture of central Mexico. Later, the terms *Moqui,* a variant of *Hopi,* and *Pueblo* became current, and finally, archaeologists made a determined effort to establish Anasazi as the best of several possible appellations.

As archaeological investigations of the Anasazi grew intensive early in the twentieth century, it became clear that the culture had had a remarkable fluorescence, one that had lasted more than a millennium before it mysteriously collapsed. During the early years of what was the nascent Christian era on the banks of the Mediterranean, the Anasazi lifeway on this continent quickly began to grow complex, and the Anasazi population burgeoned. The people devoted far more of their time to growing corn, beans, and squash than to foraging and hunting; they wove intricate, excellent baskets, and they constructed pithouses—dwellings dug shallowly into the earth, covered with roofs built from posts and a latticework of branches, brush, grasses, and dirt.

An Anasazi move toward urbanization began between A.D. 700 and 900, at about the same time that technical abilities and aesthetic senses began to flourish. Pithouses began to be closely clustered, sometimes built with adjoining walls. Pottery became commonplace and was carefully and beautifully painted. By the year 1000 or so, houses were built almost exclusively above ground, with upright walls constructed of mortared sandstone, and the clustering of houses increased. For reasons that probably had much to do with their agricultural achievements—their increasing ability to support large numbers of people from carefully tended fields, and the corresponding demand for many field-workers—the Anasazi began to group themselves into larger and larger masonry villages, some several stories tall, built high on mesa tops and, later, in the arching caves in canyon walls. Concerns about defense could have precipitated the moves, but the migrations weren't made hastily. Architecture, in fact, reached its most sophisticated and aesthetic dimensions by about 1100.

At the height of their cultural expansion, the Anasazi were the Southwest's dominant culture. Trading within the Anasazi region and with neighboring cultures located even thousands of miles away became extensive. Elaborate road systems, some hundreds of miles long, were constructed throughout much of the Anasazi region, as well as what appear to have been line-of-sight communications towers. Irrigation, previously limited to the planting of crops in alluvial creekbeds or near intermittent springs, felt the rush of technology as well. Catch basins and dikes were built at the mouths of arroyos and washes to trap periodic runoff. Ditches, long canals, and reservoirs carried and stored the precious water for the fields and the people themselves.

It was evidence of the Anasazi's proficiency as farmers and irrigationists that was among the most important information salvaged in Glen Canyon. Jennings and his young, enthusiastic colleagues—many of whom would go on from the salvage project to establish prominent careers in southwestern archaeology for themselves—identified and excavated an otherwise unknown double-walled masonry dam at a place called Creeping Dune, a relatively sophisticated structure that had caught water which spilled to the surface of the ground where a block of Chinle shale was exposed. A water gate with a notched stone slab that likely was used to regulate flow was built into a corner of the U-shaped dam, and a stone-lined ditch system led from it to a terraced field. In Beaver Canyon, the archaeologists found an extensive ditch and field complex where an alluvial fan had been irrigated by water diverted from the intermittent creek and carried in carefully lined ditches. Near Castle Creek, an earthen aqueduct had carried water laterally across a shallow depression, making it possible to irrigate an otherwise inaccessible shelf. Terraced fields, often protected by mortared retaining walls and watered by subterranean seeps, were found on the large bars along the Colorado and in several side canyons; and in Lake Canyon, where piling sand had dammed the headwaters of a slender stream to form a lake, numerous sites testified to the fact that the Anasazi had been dependent on the canyons' water resources, water that wasn't available on the arid mesas above.

During their several summers of work, Jennings's and Lindsay's crews were able to identify a variety of trails—precarious toeholds chipped into sweeping slabs of sandstone—connecting the canyon bottoms to the rims. They found exquisite petroglyph panels—

pecked and incised figures at the bases of looming walls that depicted mountain sheep, birds, animal tracks, geometric designs, trapezoidal men and women. Collectively, the two institutions mapped and cataloged more than 2,500 canyon sites, and hundreds more on the peripheral tablelands. And before their emergency work was finished, they were able—to the satisfaction of most of their number—to make a basic conclusion about what first had puzzled them.

It seemed certain in the end that the side canyons of the Glen long had undergone a cycle of filling and flushing, sediments washing down and settling on the canyon bottoms for hundreds of years perhaps—perhaps only for decades—before the runoff from violent storms flushed them out and the filling began again. And although it was possible that large Anasazi villages had been constructed on these very tenuous and temporary plains, what seemed far more likely was that the Anasazi never lived in the canyons in great numbers. The trail systems, the ruins of large pueblos (with middens of corresponding size) that were found on adjacent mesa tops, and the Anasazi's long tenure in the Glen Canyon country (making it very likely that they would have known from the experiences and stories of their elders that the alluvial beds could be destroyed) all contributed to the archaeologists' conclusions that Glen Canyon served principally as a kind of garden, but a vitally important one. Water was prevalent in the canyons; the soil was fertile, the growing season long and dependable. But for some reason, the Anasazi never chose or were never forced to build the great fortress cities in Glen Canyon that they had constructed at Chaco Canyon, Canyon de Chelly, Tsegi Canyon, and Mesa Verde. It was hard to know why they did not, and the nature of the emergency salvage meant that these diggers into the dimly understood past, these examiners of bone and stone and painted ceramics, probably would never know.

It may have been the breaching of their fields by floods that led the Anasazi elsewhere; it may have been the depletion of myriad resources after centuries in a fragile landscape; almost certainly, a long and disastrous drought descended on the plateau region in the middle decades of the 1200s. But whatever their mix of reasons, the people began to go. There was no mass exodus from the cities and villages and the fields that had sustained them for so long, but everywhere the emigrations increased. People who for generations had known only intensely communal lives left their houses, their possessions, their buried kin, and wandered away in little bands—

drifting south, southwest, southeast, searching for those elements that had made life good in the slickrock north. The Anasazi abandonment reached its peak in the last quarter of the thirteenth century. By the year 1300, most of the Colorado Plateau was empty of human habitation, and no one was left in Glen Canyon.

The fundamental distinction between archaeology and history, historian Greg Crampton was quick to point out, was that while archaeologists were limited in their investigations by what they found on the ground and beneath it, historians could focus their attention on oral and written records of the past. What archaeologists knew, and could know, of an ancient culture was limited to the physical remains of that culture, to those few objects and effects that could survive the centuries. Historians, on the other hand, could work from legends, stories, eyewitness accounts, from official documents, diaries, letters, and books. The advantage to historians was enormous; much of what they paid attention to was how individuals and whole societies envisioned themselves, how they described themselves and their exploits. But if archaeologists were limited by the physical, in a very real sense historians had become bound by the walls of libraries, Crampton was convinced.

It was Herbert Bolton, Crampton's professor at Cal, who had instilled in him the notion that historians had to get out into the field, to acquaint themselves with landforms and the sometimes ineffable *feel* of a place, to come to know the focused spots where important and incidental events had taken place. Sixteen years before, Bolton had floated into Glen Canyon with Norman Nevills in order to see for himself the straight-walled gorge that had been such a barrier to the haggard members of the Dominguez-Escalante expedition. And now Crampton was following his mentor, not to see and study a single place and the things that once had happened there, but to try to draw together a kind of collective history of the entire riverine environment—how humans had experienced and exploited Glen Canyon in the nearly two hundred years since they had begun to record their observations.

Although Crampton had been charged with the historical salvage of the entire reservoir area, his resources for the job were rather limited. His share of the park service funding would only stretch far enough to pay the costs of acquiring resource materials, typing and editing manuscripts, and supplying himself and his crews with a sturdy baloney boat equipped with a three-horsepower motor, as well as enough food to sustain them on their sojourns. Crampton, like the archaeologists, would not be handsomely paid for his part in the project, and he would be able to recruit field assistants only with the promise that he would take them to extraordinary places that soon would disappear.

Beginning in the summer of 1957, Crampton and company made the first of what finally would total thirteen salvage trips into the canyon. Normally working as a foursome, the small "historical" crew embarked on each trip at Hite, but the focus of its work necessarily began at the dam site, then slowly moved upstream— those sites that would go under first obviously being the first to be studied. And although prior to each trip Crampton had pinpointed the canyon regions to be surveyed, the known sites that had to be investigated, he was never much of a taskmaster once afloat on the languid river. Mornings were usually given over to tramping around scattered industrial wrecks and ruined cabins, to hiking side canyons, to checking whether there was any evidence of activity at dozens of recorded mining claims. Then it was time to boat downstream a distance, and by midafternoon it was often time to stop. In the oppressive heat, the crew would camp in the shade of the canyon walls; they would swim, fish for the big-whiskered channel catfish that often were fare for supper, write up their field notes, and peruse the assorted documents and books they brought with them in rubber bags. It was work, yes, but it was also a splendid kind of break from the rote demands of academia, and a more glorious place than Glen Canyon was very hard to imagine. No wonder Professor Bolton had deemed it essential to his study of the circuitous trek of the Spanish friars to linger for a few days in the canyon that had caused them such trouble.

Ironically, the need for an overland route between Santa Fe and the Spanish missions near Monterey had been made all the more urgent by the realization of an unfortunate truth about the nature of the rivers in the vast region the Spaniards had begun to call New Mexico. There were many rivers, to be sure, but few of them were waterways. None was the calm, broad, perfectly navigable stream that Europeans long had hoped would connect the Atlantic Ocean with the Pacific. The rivers, instead, were a tumult of rapids; their courses twisted like corkscrews, and they ran at the bottoms of dark, inaccessible gorges. If a safe and direct passage to California could be found, it seemed late in the eighteenth century, it would have to be across the harsh and waterless wastelands.

With a small party of men, horses, and pack animals, thirty-five-year-old Fray Atanasio Dominguez and his friend and chronicler Fray Silvestre Vélez de Escalante, both belonging to the Order of Friars Minor, set out from Santa Fe at the end of August 1776. They traveled known trails northwest to the Dolores, Gunnison, and Grand rivers in western Colorado, then proceeded into the Uinta Basin in what is now Utah—into territory still unknown to Spain at that time. Reaching Utah Lake near present-day Provo in October, battered by the journey and the early onset of winter, and knowing that Monterey surely was still very far to the west, Dominguez decided that the party would turn south and return to Santa Fe by completing a great circle. They would not reach the Pacific, but they would make a reconnaissance of a wealth of unexplored territory, and they would bring the message of the Gospel to all the native inhabitants they encountered en route.

Since its abandonment by the Anasazi around 1300, the canyon country of southeastern Utah and northwestern Arizona—with Glen Canyon at its rough and rocky center—had been only scarcely settled. Paiutes, living much like the Desert Archaic peoples had a thousand years before, occupied the region intermittently until they were overwhelmed and pushed to its fringes by several more advanced Ute tribes. But the Utes were seminomadic as well, and largely preferred the highland areas of the Wasatch, Uinta, San Juan, and Elk mountains to the sere and severe landscapes of the canyons. The Navajos, although nearby, had yet to move west and north to claim much of the canyon country as their own.

It was members of the Kaibab band of the Paiutes who eventually told Dominguez and Escalante of the place where they successfully could cross the large river that would otherwise block their home-ward march. The young priests were well aware of the river that they variously called Rio Colorado, Rio Grande, or Rio de Tizón, and they were anxious to find the ford, cross the canyon barrier, and make their way expeditiously back to Santa Fe. But at the place at the mouth of the Paria River that one day would be known as Lee's Ferry, they encountered a brown river too deep to wade across, too

swift to swim, and their efforts to build rafts were equally unsuccessful. They had had to resort to slaughtering horses from their own
herd—accompanied by the broiled pads of prickly pear cactus—for
food, and they began to sense a kind of desperation.

With the weather growing steadily worse, the party found a way
up the Vermillion Cliffs and onto the plateau land that flanked the
river, slowly making its way north, traveling for four days before a
scout reported that he had found a place where the river was wide
and placid, as well as a steep but negotiable passage down to it
through a slender side canyon.

On November 7, the thirteen members of the expedition, despair-
ing that God had yet to show them a way across the river "perhaps
as a benign punishment for our sins," used their axes to hack steps
into the sandstone slope at the head of the canyon to give their
horses surer footing, then made their way down to a broad sandbar
along the river. The current was barely knee-deep there—40 miles
upstream from the ford the Paiutes had described to them—and by
nightfall the men and horses had crossed the Colorado without
incident and had celebrated by "praising God our Lord and firing off
a few muskets." Writing in his journal on the eastern bank of the
river that night, Fray Escalante noted that although the ford was a
good and probably dependable one, the canyon itself offered little
immediate solace and nothing more in the way of promise. "In all
that we saw around here," he meticulously penned, "no settlement
can be established along the banks, nor can one even go one good
day's march downstream or upstream along either side with the
hope of their waters being of service to the people and horse herd,
because, besides the terrain being very bad, the river flows through
a very deep gorge."

Four months after the signing of the Declaration of Independence
in Philadelphia, immigrants to the New World had entered Glen
Canyon for the first time, and Escalante's less than rapturous
account of it was the first time the canyon had been recorded. The
thirteen members of the failed expedition would return to Santa Fe
on January 2, the two priests thereafter dispatched to serve in
missions along the Rio Grande; and 400 miles away to the west, el
Vado de los Padres, "the Crossing of the Fathers," would become the
first of several fords and ferries in Glen Canyon, that strange and
secretive slit in the earth subsequently crossed repeatedly before at
last men traveled its length.

■■■■■■■■■■■■■■■■■■

John Wesley Powell often had heard of the place called the Crossing of the Fathers by the time he and his crew of eight men pulled their wooden boats ashore there on August 3, 1869. Powell had expected to find the ford somewhere below the mouth of the San Juan River, and he was soon certain that, in fact, this was the storied place. "A well-beaten Indian trail is seen here yet," he wrote in his journal. "Between the cliff and the river there is a little meadow. The ashes of many campfires are seen, and the bones of numbers of cattle are bleaching on the grass."

In the century since Dominguez and Escalante had pioneered this route across the river, it had become part of a system of trails linking Santa Fe with southern California. A trading party headed by Antonio Armijo had crossed the Colorado in Glen Canyon in December 1829, and completed the first successful Spanish trek to the Pacific when it reached the small settlement of Los Angeles early in the new year. For the next two decades, caravans of traders made annual marches between New Mexico and California, carrying woolens on the westbound journey, driving horses and mules—sometimes as many as a thousand head—on the return trip. But the Crossing of the Fathers was a poor one for large numbers of men and animals, and a longer but easier trail eventually began to be used, this route crossing the Colorado at present-day Moab, Utah, and fording the Green at a settlement called Green River Crossing, now Green River, Utah.

By the time Powell stopped to examine this one place in Glen Canyon that was well known, the Spanish priests' ford was seldom used except by Paiutes, Utes, and Navajos who periodically ventured into or across the canyon country. Powell, in fact, suspected that was why so many bones were scattered on the sandbar at the river's edge. "For several years," he wrote, "the Navajos have raided on the Mormons that dwell in the valleys to the west, and they doubtless cross frequently at this ford with their stolen cattle." But no one came down the cut descending from the western rim—which Powell named Padre Canyon—during the afternoon and evening he and his men rested there, and the canyon was still empty except for them when they pushed their boats into the sluggish current again the following morning.

Major John Wesley Powell had lost his right arm fighting for the

Union at the Battle of Shiloh seven years before, but it was a handicap to which he paid little heed. As a geology professor at Illinois State Normal University, he had made two summer explorations to the Colorado Rockies in 1867 and 1868, climbing mountains and overseeing a kind of all-purpose scientific reconnaissance. It was during these first western trips that his appetite was whetted for something truly adventurous. The great rivers that drained the western slope of the central Rockies had never been navigated, so far as anyone knew, and the country through which they coursed was still very largely unknown. There were plenty of tales of horrible canyons, terrible whirlpools, and waterfalls that could not be portaged along the twisting courses of the Green and the Colorado. Powell had heard them all, but he was suspicious and he wanted very much to head up an expedition that would prove the rivers could be successfully traveled, and, at the same time, to survey the Colorado all the way downstream to the point where Lieutenant Joseph Ives had ended his upstream expedition ten years before. Powell was obsessed with the idea of adventure, with a compelling personal need to go out and experience as much as he could of the western frontier and to discover its natural wonders. And, actually more adept at promotion and the art of politics than he was at science, he succeeded in organizing and funding a monumental, if possibly foolhardy, expedition in the spring of 1869.

At Green River City, Wyoming, on May 24, Powell, his brother Walter, and eight scrappy and game young men, most of them former soldiers, loaded themselves and an expected ten months' worth of supplies, equipment, clothing, and scientific instruments into three heavy, sweep-oared oak boats and a fourth pine pilot boat to which they had strapped a chair—a kind of seated sentinel post for Major Powell. With a wave of a little American flag he carried, and with alternating shouts of encouragement and derision from the shore, the boats moved into the Green's quick current and were waterborne.

For 60 miles, all was well. The river rolled and it occasionally roiled, but it was somehow playful, often serene. When the Green fell into the canyon that Powell called Flaming Gorge, the riffles grew bigger yet still were manageable, but by Red Canyon, the waves bucked high enough that the frightened adventurers took to the shore, portaging their heavy boats around the worst of the trouble. In the place Powell called the Canyon of the Lodore, the river became a nightmare of rock and boiling water. One of the three heavy boats, the *No Name*, was sucked into a narrow channel of racing water, then hit a boulder broadside and broke in two. Its passengers survived the wreck, and barometers, flour, clothing, and a smuggled-aboard barrel of whiskey were eventually recovered downstream, but the boat was lost. Now there were only three fragile crafts with which to continue the journey.

In Desolation and Gray canyons, the rapids were gentler, the boatmen now far more adept. In Labyrinth and Stillwater canyons, the river was placid, though its course was torturously twisted. Rations ran thin; one crew member had already abandoned the expedition—to discover that the entire crew long since had been reported killed in the national newspapers; and the one-armed Powell somehow had got himself stuck climbing a cliff in order to survey the surrounding country and had had to be rescued by his men. Now, after nearly 500 miles of hardships, the river, at its confluence with the Grand, was only just becoming the fabled malevolent Colorado.

Its volume now doubled, the river was pleasantly well behaved for a few miles, but on the morning of July 21, two months into the journey, the boats entered a maelstrom of menacing rapids, a new one beginning at the outwash of the one before it, rapids so furious that the Canyon of the Lodore now seemed peaceful in comparison. Powell's pilot boat, the *Emma Dean*, named after his wife, was swamped, then overturned, its oars lost. Downstream, the rapids still unrelenting, the men succeeded in portaging their heavy, battered boats over the difficult talus at the river's edge, at one stage traveling less than a mile in a long day's work of hauling, stumbling, and cursing this supposed adventure. The heat was becoming unbearable; the flour had molded by now; the bit of remaining bacon was rancid. The men were exhausted, anxious, openly afraid, and quarrelsome. Yet after eight days of alternately riding, lining, and portaging the awful succession of rapids—together with much talk about whether *this* would be where all those terrible stories came true—the weary party finally floated out of the 45-mile canyon Powell logically enough had called Cataract. The deep, craggy walls of Cataract Canyon dropped down; a dark stream the men named the Dirty Devil flowed in from the right, then quickly the canyon was very different, its walls smooth and mounded on top, colored red and orange and sometimes streaked with black. The waves somehow disappeared in this new and utterly different gorge, and Powell and his men now floated the flat back of the river with a swelling sense of relief, of euphoria.

After three days of inspecting the masonry ruins of some forgotten Indian civilization—one which Powell correctly speculated was akin to the contemporary Hopi culture—of floating past canyon walls "variegated by royal arches, mossy alcoves, deep, beautiful glens, and painted grottoes," the three boats reached the spot where the San Juan joined the river, virtually all of the Colorado Plateau's watershed now joined in a single channel, then made two more miles before they beached where the men spotted a particularly intriguing alcove in the left canyon wall. Passing a small grove of box elder and cottonwood trees at its entrance, they entered "a vast chamber, carved out of the rock. At the upper end there is a clear, deep pool of water, bordered with verdure. . . . Through the ceiling, and on through the rocks for a thousand feet above, there is a narrow, winding skylight; and this is all carved out by a little stream which runs only during the few showers that fall now and then in this arid country." It was too marvelous a place not to linger in, and the men were so enchanted by it that they brought their cooking and camp supplies up from the riverbank for the night. "When 'Old Shady' sings us a song at night," Powell wrote, "we are pleased to find that this hollow in the rock is filled with sweet sounds. It was doubtless made for an academy of music by its storm-born architect; so we name it Music Temple."

The travel-weary, hardship-worn men spent a second night in Music Temple, carving their names to record their stay, before Powell urged them on their way, hoping to find the Crossing of the Fathers not too far downstream. But although they had had their fill by now of this protracted river journey—and knowing that more rapids and hardships surely awaited them downstream—none of the group was eager to pass too quickly through this maze of rock and water, punctuated by dozens of inviting side canyons. "Sometimes we stop to explore these for a short distance," wrote Powell. "In some places their walls are much nearer each other above than below, so that they look somewhat like caves or chambers in the rocks. Usually, in going up such a gorge, we find beautiful vegetation; but our way is often cut off by deep basins, or 'potholes,' as they are called. On the walls, and back many miles into the country, numbers of monument-shaped buttes are observed. So we have a curious ensemble of wonderful features—carved walls, royal arches, glens, alcove gulches, mounds, and monuments. From which of these features shall we select a name? We decide to call it Glen Canyon."

But Powell was writing those words in 1874, five years after that first trip through the canyon, and three years following a *second* expedition down the Colorado, his riverside journals expanded and dramatized where it seemed appropriate, certain details (such as the fact that there *were* two trips) sometimes overlooked in the interests of self-congratulation and simplicity. On the first trip, Powell had been fascinated by "the mounds and cones and hills of solid sandstone," and in his attentively kept journal he labeled the stretch of the Colorado between the Dirty Devil and the San Juan *Mound* Canyon. Below the San Juan, he gave it the name *Monument* Canyon because of the predominance of "some variegated monument, now vertical, now terraced, now carved by time into grotesque shapes."

During the course of that first journey, Powell had done a remarkably descriptive job of naming canyons and other topographical features, and the names Mound and Monument too were appropriate enough. Yet somehow in retrospect, he wasn't satisfied. When he finally wrote a popular account of the expedition for *Scribners' Monthly*, in comparison with the other canyons of the Colorado he remembered these two as being very different, sublimely serene and inviting, and the two canyons really seemed to be one. From the Dirty Devil to the mouth of the Paria, for 155 winding miles, the river was a wonderland of gentle water, glorious rock, and in the side canyons there were more delightful mysteries than one could have imagined possible. And somehow, neither of the names he had bestowed on it captured the sense of the canyon's tranquil beauty. So it became *Glen Canyon* on further contemplation. If it wasn't a particularly parochial name—the Southwest was a place of creeks, arroyos, and gulches, not brooks, valleys, or glens—the old Scottish term sounded appropriate at least; it imparted a sense of serenity, an image of secluded loveliness. And if the words *glen* and *canyon* formed a sort of oxymoron when joined, that too was appropriate for this place where emerald-green grottoes were hidden amid bare and baking rock.

The nine men found the Crossing of the Fathers and camped there, then spent the night of August 4 at the mouth of the Paria

River, where Glen Canyon's walls dropped down and the surrounding country momentarily was accessible. But another canyon gaped open just downstream, and the sight of it immediately made them anxious. "We have learned to observe closely the texture of the rock," Powell wrote. "In softer strata we have a quiet river, in harder we find rapids and falls. Below us are the limestones and hard sandstones which we found in Cataract Canyon. This bodes toil and danger."

Twenty-six days later, six men in only two boats rounded a bend in the river beneath the Grand Wash Cliffs and utterly astonished three Mormon men and an Indian friend who had been camped beside the river for three weeks, hoping to salvage something from the wreck of Powell's doomed expedition. But by some miracle, not only were two boats intact, at least some of the party had *survived* the journey through the Grand Canyon.

As Powell described them to the incredulous Mormon scouts, the 280 miles from the foot of Glen Canyon had been a seemingly endless succession of long, slackwater stretches separated by rapids, by *falls* of awful and punishing power, rapids with towering, thundering waves, suck-holes like chasms, and terrible spinning whirlpools. But by stealing a few squashes from an Indian field they had encountered near the river, by portaging and lining and ceaselessly fighting the fierce water, they had made more than 200 miles by the time three members of the crew finally had had enough. Their only clothes were in tatters; their blankets were gone; their food supply now was little more than dried apples, and still there were more rapids to face.

Despite his imploring—his conviction that they weren't far from escaping the "prison" of the canyon, and his fears that the men might encounter even more danger if they tried to climb out and find their way to safety—Powell was unable to avert the mutiny. On the morning of August 28, Bill Dunn and the brothers Seneca and O. G. Howland had taken rifles and their share of the remaining rations, waved farewell, and begun to struggle up the cliffs.

Leaving the leaky and battered *Emma Dean* behind, Powell and the five men who remained steadfast successfully ran the fierce rapid that had been Dunn's and the Howland brothers' final straw, then another rapid a few miles downstream that was perhaps the meanest of all the hundreds they had encountered, then at last they floated into the open country at Grand Wash, "the first hour of convalescent freedom seeming rich recompense for all pain and gloom and terror." Three months and six days after their departure from Green River City, his crew and his boats depleted by half, himself and his companion adventurers starved and utterly drained, not only had Powell and his men survived, they had done what they set out to do. It would be weeks before Powell would hear the news that the three men who had left the expedition at the place he called Separation Canyon had been killed by Indians when they had reached the north rim of the canyon.

Although Powell had become a kind of folkloric hero by the time his second Green and Colorado expedition was mounted in May 1871, and although the subsequent journey was in every way better funded, better equipped, and better annotated than the first— virtually all the crew members kept journals this time, and hundreds of photographs were taken throughout the trip—Powell had lost his keen interest in the canyons as well as the personal commitment to proving himself a pathfinder that had spurred him on three years before. He had been uninterested enough in the expedition's progress, in fact, that he left it at the junction of the Uinta River to go overland to visit his wife in Salt Lake City, then had rejoined it at Green River Crossing before departing again at the Crossing of the Fathers in Glen Canyon. His crew had continued on to the mouth of the Paria, where, with the weather growing cold and wet and ominous at the end of October, they cached their boats.

Powell spent much of that winter in Washington, D.C., while his crew wintered in Kanab, Utah. And it wasn't until August 1872 that Powell and company returned to the river, where the crew readied the boats to continue the trip into the Grand Canyon, and where the peripatetic Powell renewed acquaintance with the storied John D. Lee, who lately had started a ferry service near the mouth of the Paria River, and who didn't seem much like the bloodthirsty murderer some tales claimed him to be. Ultimately, Powell floated only to Kanab Creek, halfway through the Grand Canyon, before he grew restless again and abandoned the river for the third and final time. Lee, on the other hand, who once had sworn that he "would want no greater punishment than to be sent on a mission to the Pahariere," decided that, under his current circumstances, it wasn't such a bad life after all, there where the little stream met the big brown Colorado.

John Doyle Lee had always been quick to defend his faith against the persecutions of the "Gentiles," and he did not suffer lightly his fellow practitioners of Mormonism who occasionally strayed from the church's prescribed path or showed signs of moral weakness. His loyalty to Mormon leader Brigham Young was unwavering, and he enthusiastically supported the doctrine of polygamy, marrying nineteen times. Before coming to the bank of the Colorado, Lee variously had been a farmer, miller, judge, legislator, and tavern keeper, and indeed he had played a role in the brutal killing of 120 men, women, and children.

As a member of the Nauvoo Legion, the Mormon militia in Young's self-proclaimed State of Deseret in the late 1850s, Lee had been called upon by church leaders to assist in making an example of a wagon train of emigrants into the region bound from Arkansas to California, a group that had had repeated altercations with the Mormons during its trek through Utah territory. With the help of a timely attack by allied Paiute Indians at a valley in southwestern Utah called Mountain Meadows, Lee had succeeded in convincing the travelers that if they would give up their arms and give some of their cattle to the Paiutes, they would be spared more trouble and could continue their journey. But as soon as the members of the wagon train had disarmed themselves on the morning of September 11, 1857, a cry of "Do your duty!" had rung out, and every male Arkansan had been quickly killed. The women and all but eighteen small children also were dead within a matter of minutes.

Years passed before knowledge of what had happened at Mountain Meadows became widespread, before federal officials began to try to determine who was responsible for the massacre, and before Mormon officials tried to assess blame of their own and to quiet the subject that was impeding progress toward statehood. Although Brigham Young considered Lee his "adopted son," there seemed to be little alternative but to make Lee out to be the principal culprit, then to make him scarce. Young's adviser and an early explorer in southern Utah, Jacob Hamblin, knew exactly where to send him.

Hamblin had pioneered a ferry route across the Colorado River at the mouth of the Paria, and he knew that the crossing was becoming increasingly more important as the Mormons expanded into Arizona. The place was incredibly remote; Lee would be safe there, and he would also be of service. So in December 1871, two months after Powell's crew had cached their boats there, Lee arrived with one of his sons and his seventeenth wife, Emma, at the spot at the foot of Glen Canyon that probably should have been called Hamblin's Ferry. After hearing his wife describe it when she saw it for the first time, Lee thereafter called the spot Lonely Dell, but it nonetheless became known as Lee's Ferry, although Lee himself carried travelers across the river, off and on, for only a little less than two years, first using one of Powell's boats for the service, then building a big ferryboat and a smaller skiff of his own.

In November 1874, while Lee was visiting his wife Caroline in the Utah town of Panguitch, he was arrested on an outstanding murder warrant stemming from Mountain Meadows. Although his first trial ended with a deadlocked jury, and although no one ever established how central Lee's role was in the decision to kill the emigrants, a second jury found him guilty of the mass murders. Lee was executed at Mountain Meadows on the morning of March 23, 1877. His faithful wife Emma continued to operate the ferry at the foot of Glen Canyon until May 1879, when the Mormon church officially purchased it from her for $3,000, the fee paid mostly in cattle.

On the morning of July 23, 1847, an advance group of members of Brigham Young's nascent Church of Jesus Christ of Latter-day Saints had unloaded their wagons in the Salt Lake Valley at the end of a historic migration from the plains of Illinois. By noon, a farmable plot had been staked and the first soil west of the Continental Divide had been plowed by Anglo settlers. At 2:00 P.M., work had begun on a dam and a series of ditches that would divert water from City Creek to the field. It wasn't until the second day of the Mormons' arrival, however, that a five-acre potato patch had been readied and water at last had been turned onto the land.

The early Mormons, buoyed by their faith and spurred by awesome determination, had been able to survive and to thrive in the Utah wilderness for thirty years now. Living and working communally, much as the Anasazi had done, they had established farming settlements in virtually every valley of their new Zion that held a trickle of precious water. In the 1850s and 1860s, Mormon colonies had spread into what Brigham Young called the "Outer

Cordon"—north into Idaho and south into "Dixie," the area that comprised much of southwestern Utah and the southern tip of Nevada. And from Dixie, Young had issued calls to his people to migrate east and south into the canyon country to help broaden and strengthen the Saints' domination of the whole of the western slope of the intermountain region.

In November 1879, 230 Mormon pioneers gathered at Forty Mile Spring southeast of the village of Escalante to begin a trek to a new outpost that was perhaps 200 miles away, at the place where tiny Montezuma Creek met the San Juan River near the border of the state of Colorado. An advance party had selected the site the previous summer, but rather than journey to it by known routes— one which crossed the Colorado at Lee's Ferry or another far upriver at Moab, either of which would have been 500 miles long—the settlers, led by elders Silas Smith and Platte D. Lyman, decided to try a shortcut. Calling themselves the San Juan Mission, pulling eighty-three wagons and driving a thousand head of livestock, the Mormon settlers headed southeast across country that had never been marked by wagon tracks, presuming that they could manage a way across the Colorado wherever they encountered it.

They first glimpsed the river from the high rim of Glen Canyon, about 50 miles upriver from the Crossing of the Fathers, between the mouths of the Escalante and San Juan rivers, but there was no way to lead the wagons and the stock down into the canyon, and now early snows effectively blocked their return to Dixie. Finally, Smith and Lyman and company found a narrow fault in the canyon rim—just a crack, a slit in the mounded sandstone—but it opened out into a short and very steep tributary canyon. It wasn't much, but with some work it could be made into a route. For six weeks, with the whole party camped near the rim, the settlers worked to make their way down to the river. They blasted the crack with dynamite until it was wide enough to accommodate the wagons—the crack they now called the Hole-in-the-Rock—then began to cut a road out of the rock below, chipping and blasting a route barely as wide as the wagons' axles. At one spot, where the canyon wall was too nearly vertical for a passageway to be blasted in, oak logs were set horizontally into holes drilled in the cliff face, stringers were laid across them, then sticks, brush, and stones were laid on top to form a kind of cantilevered road that hung precipitously in the air.

On January 28, 1880, the entire caravan began the descent to the river; all eighty-three wagons were braked, their wheels locked, and then skidded down off the rim, teams of men checking their speed with ropes as they proceeded. At last they reached the river, a descent of more than a thousand feet in less than three-quarters of a mile, but their work was just beginning. A ferryboat had to be built, then, two wagons at a time, the caravan had to cross the river. On the opposite side of the canyon, a roadway had to be chipped out of the rock to reach a bench 250 feet above the river. At the head of Cottonwood Canyon there was more work to be done in the rock, still more to allow them to climb The Chute, and finally, yet another series of "dugways" got them to the rim of Grey Mesa, where they encountered a foot of snow. But they continued on across the mesa to Castle Wash and up over Clay Hill Pass to the dense forest on Grand Flat. Pinon and juniper trees had to be cut continually now to make way for the wagons before they could begin to descend to Comb Wash, where they trudged through the sand to the San Juan River. But there were cliffs that blocked the route they had prayed for along the river, and there was no alternative but to cut another dugway up and over them. At last across Comb Ridge, they dug their way across Butler and Cottonwood washes, but only 18 miles from Montezuma Creek, they finally could go no farther.

It was April 6, 1880. The members of the San Juan Mission had traveled all winter across 200 miles of terrible terrain, "nothing in the world but rocks and holes." Three babies had been born en route, but almost miraculously no one had died. Just beyond Cottonwood Wash there was tillable land, there was water and wood, and the travelers had come far enough. That evening a town site was laid out for the community the Saints would call Bluff City, the name later shortened to Bluff. On the following day, April 7, Platte Lyman wrote in his diary: "We began laying off the lots and land and most of the brethren began work on the ditch."

Although its route was tortuous, the Hole-in-the-Rock trail joining the towns of Escalante and Bluff was used intermittently in the succeeding years. Charles Hall, one of the founders of the town of Escalante and the man who had fashioned the Mormon's Hole-in-the-Rock ferryboat, was optimistic enough about the trail's potential, in fact, that he had stayed behind in Glen Canyon, offering his services to the rare traveler for a full year before he determined that business in that spot probably never was going to be brisk. Thirty-five miles upstream, however, Hall found a ferry-crossing site that seemed far better than Hole-in-the-Rock. The creek that eventually was named after him provided relatively gentle canyon access from the west, and from the east the river was accessible down a series of descending benches. Hall built a 30-foot-long boat out of lumber hauled in from Escalante, reasoned that he would need to charge $5.00 per wagon and $.75 per horse for the ride across the river, then sat down to wait for customers, the first full-time resident in the interior of the canyon since the era of the Anasazi. Two years later, however, Hall had company in the canyon and his river crossing had competition.

Cass Hite was a prospector who had spent years searching the Navajo country for a fabled silver mine. A Navajo headman named Hoskininni, with whom Hite had become acquainted, informed him that although he wouldn't tell him the location of the silver mine, he'd show him where he could find gold, if that particular mineral was of any interest to him. Hite assured Hoskininni that gold would be just fine; the two made their way down White Canyon to the Colorado; and sure enough, Hite found gold in the sand and gravel terraces beside the river. Despite his jubilant mood, Hite also took time to notice that the spot where the two men stood would make a "Dandy Crossing," a name that stuck only a little while before it was replaced by Hite City and then simply by Hite.

But Cass Hite wasn't interested in anything as prosaic as operating a ferry, not when there plainly were fortunes to be made. He never built or brought any boats to the river, but he did do his best to publicize the crossing—certainly the most accessible one between Lee's Ferry and Moab, he claimed, probably correctly—and in letters to the *Salt Lake Tribune* and several newspapers in Colorado, he lauded the Glen Canyon gold discovery, speculating that the canyon would become the site of the nation's next great mining boom.

At first, however, there was anything but a rush to riches in Glen Canyon. Interest in the place was so slim, in fact, that Charles Hall abandoned his ferry business in 1884, aware that Hite, 45 miles upriver, was getting increasing cross-river traffic, convinced as well that this mining business would never amount to much. But soon after Hall had made his retreat from the river, reports of other strikes—big strikes—began to gain currency in the alpine mining camps of Utah's Wasatch and Uinta ranges and Colorado's San

Juans. Many a miner decided it wouldn't hurt to go have a look for himself, to see if those crazy canyons really were giant sluice boxes lined with gold.

Dandy Crossing Bar, Ticaboo Bar, Good Hope Bar, California Bar, Klondike Bar, Gretchen Bar, Oil Seep Bar, Boston Bar—before the eighties were out there were dozens of variously productive gold placers in operation along the river. Gold, washed down out of the Rockies for many millennia, so fine it was literally dust, was found along the river from Hite all the way to Lee's Ferry. And when William McKinley's election to the presidency in 1896 meant a victory as well for the gold standard, prospectors began to pour into Glen Canyon—entering at Hite, at Hall's Crossing, at Hole-in-the-Rock, and the Crossing of the Fathers, maniacal men getting into the canyon wherever and however they could, staking claims, establishing mining districts, hauling in sluicing and dredging and drilling equipment, building cabins and camps throughout its length. By 1889, Hite was so prosperous it had a post office, and an estimated 1,000 miners were at work in Glen Canyon, each one sure he soon would be a millionaire.

The fact that Glen Canyon was going to be another Klondike, so lots of fellows said, made a railroad through the canyon all the more attractive, all the more a certain financial success. A rail line already had reached Grand Junction, Colorado; if it could be continued on through the lower canyons of the Colorado River to southern California, what a boon it would be to the whole of the Southwest, what money it would make for its owners! Frank M. Brown, president of the newly incorporated Denver, Colorado Canyon and Pacific Railroad Company, was determined to see it accomplished, and he had hired the well-known railroad engineer Robert Brewster Stanton to design a suitable river-level route. Traveling through Glen Canyon by boat at the end of June 1889, Brown, Stanton, and a crew of sixteen men liked what they saw. Westwater Canyon on the Colorado-Utah border would be difficult to punch a railroad through; Cataract Canyon would be equally intractable; but Glen and the connecting canyons upstream seemed to have been built not by Powell's "storm-born architect," but rather by some sort of celestial engineer. They were perfect for such a plan. Then, having met Cass Hite at Ticaboo Bar—discussing their dream with him while they repaired their Cataract-battered boats—and having made a similar courtesy call on the Mormons operating Lee's Ferry, the men pushed on into Marble Canyon. There, within only a few miles, Frank Brown was caught in a whirlpool and drowned in Soap Creek Rapid. Neither the overweight entrepreneur nor any of his companion travelers had brought along the cork life-preservers that by then had become commonplace.

Stanton eventually was able to complete his survey all the way to the mouth of the Colorado, and although he remained adamant that a railroad through the canyons was both economically and technically feasible, the company never was able to raise enough capital to give the project a try. Daunted, but still convinced that his fortune awaited him along the river, Stanton returned to Glen Canyon in 1897, possessing substantial eastern capital this time and a grand plan indeed, though no longer one concerned with railroads.

Cass Hite and other miners in the Glen had convinced Stanton that the canyon *was* a gigantic serpentine sluice box. All that was required to exploit it, to extract its gold, was an efficient large-scale operation, it seemed. What Stanton wanted to do was to install a series of floating dredges throughout the canyon's length. He would build dams in several side canyons—and perhaps even across the Colorado itself—to generate electricity to operate the dredges. He would stake contiguous claims from Hite clear to Lee's Ferry, more than 150 miles of them, and he would become rich in the process of growing legendary.

Initial tests seemed to prove the abundance of the Glen Canyon gold; crews built primitive roads and improved trails—including the widening of the Hole-in-the-Rock access to the river—in advance of delivering the many dredges; and 145 claims were staked and duly recorded. In June 1900, Stanton's pilot dredge, the machine that would prove the viability of the grand scheme, was shipped in pieces by rail to Green River, Utah, hauled by wagons to Hanksville, then over the flank of the Henry Mountains, and finally down to a dredging site about four miles above Bullfrog Creek. Stanton's men had to cut a dugway in the canyon walls to get the dredge down to the river, but it was finally assembled, floated, and put in operation— five small gasoline engines operating the 46-bucket, 105-foot-long dredge. Yet nothing seemed to go as planned. The dredge tended to get stuck on submerged sandbars; the river's silt gummed up the amalgamating tables; and, somehow, the fine, powdery, almost ethereal gold dust kept escaping back to the river. Stanton's operation had already cost $100,000, and a return on that investment was growing problematic. After his first general cleanup, Stanton recorded in his diary a yield of $30.15 in gold. A second cleanup yielded a whooping $36.80 before Stanton abandoned his dredge in midriver in the summer of 1901 and left Glen Canyon for good.

The mining business wasn't going much better for anyone else at the turn of the century. There had been a brief boom along the banks of the San Juan, but it had played out quickly; Cass Hite was still at work, and so were scores of others, but the gold dust was so light, so flaky, that it simply floated over the amalgamators. No placer-mining technique known to humankind could deal with gold that tended to *float*. Yet before it all went bust, there would be one more flamboyant effort to capture the canyon's riches.

A prospector named Charlie Spencer had already failed dramatically in the San Juan Canyon when he went to the foot of Glen Canyon at Lee's Ferry in 1909 in possession of a pneumatic pipe dredge that surely would be the solution to the problem of the lightweight gold. His high-pressure hoses would force water to dissolve the gold-bearing Chinle shale in the area, and mercury-coated amalgamators then would latch on to the precious mineral and trap it. It was a system that was certain to work, but it would require a lot of coal to produce steam to power the operation.

Spencer and his men found a suitable coal bed on a branch of Warm Creek, 28 miles upstream; they mined the coal, hauled it to the mouth of the creek on ox-drawn wagons, then loaded it onto the *Charles H. Spencer*, the only steamer and the largest boat of any kind ever to float through the canyon. Although the *Spencer* occasionally would beach itself on sandbars as it came downstream heavily loaded with coal, its biggest problem was that it required almost a full load of coal to steam *upstream* to Warm Creek. Charlie Spencer finally surmounted that difficulty when he began to tow a barge behind his steamer, but by that point, it was becoming clear that all the coal in the world couldn't solve a fatal situation: something in the river water, the sand or the Chinle shale, was coating, clogging, gumming the mercury, rendering it unable to absorb the gold, and, as had happened countless times upstream during the preceding thirty years, the gold cavalierly slipped away and then returned to the river.

By 1913, Lee's Ferry was simply a ferry again, owned now by the

Arizona county of Coconino; Charlie Spencer had gone on to try to pursue greatness in other places. Cass Hite died at Ticaboo in 1914, and in 1915, the abandoned *Charles H. Spencer* sank. At the end of that same year, Bert Loper, a river man and occasional prospector, left the hermitage he had built for himself at the mouth of Red Canyon, near his diggings at Castle Butte Bar. Loper had been the last resident in the interior of the canyon. Now Glen Canyon was quiet, its flirtation with industry ended.

The physical evidence of the occupation of Glen Canyon from the time of the Anasazi until early in the twentieth century had been readily visible to Jesse Jennings, Lex Lindsay, Greg Crampton, and their baloney-boat crews. From Hite to Lee's Ferry, masonry ruins, small dams and ditches, astonishing dugways, mysterious petroglyphs and clumsily incised names, cabins and the sentinel chimneys of cabins long since burned, wrecked wagons, steel boilers and the remnants of sluicing equipment, a river-soaked dredge and a steamboat that now rested on the river bottom—all had been in easy evidence, certifying that people had known this place for two thousand years and more.

Yet this was a place that people simply had clambered across, a riverine world they had gone into to farm, to mine, to explore in the name of science and the marketplace. It had always been remote enough, harsh enough despite its myriad delights, that no one had ever gone there plainly and principally because it seemed like a good place to be. The Anasazi had never lived in the canyon in large numbers; the early Mormons, willing to endure the hardships thrown up by almost any other place, weren't even tempted to stay. The canyon had done little more than to get in the way of Utes, Paiutes, Navajos, and the long queues of horseback missionaries and traders. Powell and subsequent explorers at last had seen all of Glen Canyon, but it was only a place through which they slowly were in transit, a place to note in water-stained logs before they reached the next place. The miners, like miners everywhere—equally ignoring difficulty and splendor—had been interested only in rocky treasure.

In 1922, the year E. C. LaRue escorted federal and state water officials through Glen Canyon to show them what a fine environment it would be for a high dam and reservoir, the canyon had appeared much the way it did to the salvagers in 1962. The relics of its history already seemed as old. The trails that cut across it at the Crossing of the Fathers, Hole-in-the-Rock, and Hall's Crossing seemed equally archaic, equally impossible to believe.

At Lee's Ferry at the foot of the canyon, however, much had changed in that forty-year span. In 1929, Navajo Bridge had been built across Marble Canyon, six miles downstream from the ferry, by the Arizona Highway Department—at that time the only bridge across the Colorado in the more than 600 miles between Moab, Utah, and Needles, California. The fifty-year-old ferry subsequently—and very quickly—had gone out of business.

At Hite, at the head of the canyon, things were different as well. A cable-guided ferry finally had gone into operation in September 1946, when the state of Utah completed a serviceable gravel road connecting the communities of Blanding and Hanksville. Yet Hite had not gone modern; it was still a mere speck of a place set down in God's own outback; it was still an awfully slim excuse for a town, and its minuscule post office, barely bigger than a bread box—officially White Canyon, Utah, on the canceled stamps—was still open for occasional business. Historians and archaeologists, students and summer adventurers wearing T-shirts and sand-battered pith helmets, men like the grinning, impish-faced Greg Crampton, often were wont to mail wish-you-were-here postcards from Hite before they embarked in their rubber boats, headed into splendid Glen Canyon to look again at the old days.

6

THE LAST FRONTIER

The *Page Signal*, a biweekly barely six months old, broke the story. Reclamation Commissioner Wilbur Dexheimer was resigning and would be replaced by Floyd Elgin Dominy, currently the associate commissioner but the man who had been the Bureau's de facto boss for some time now. The *Signal* had gotten the scoop because the commissioner-to-be readily had offered the information when he was in town in April 1959, traveling west—as he did on almost every opportunity—to check on the progress of this, the Bureau's one hundred and ninety-fourth dam.

Yet Glen Canyon Dam was more than simply Reclamation's current focus; numerous other dams also were then under construction, and dozens more were in the planning stages. Glen Canyon was special because of its size and its deep-canyon location—it was a dam builder's dam and a grand kind of centerpiece. It would be the dam that would drive the Colorado River Storage Project, the dam that would settle decades-old water squabbles and provide 27 million acre-feet of insurance against a catastrophic drought. It would be a monumental edifice blocking the flow of an altogether mythic river, and it was precisely the kind of project that the man-who-would-be-commissioner was enchanted by.

Back in the Depression, Floyd Dominy had discovered that he loved to build dams. Fresh out of the University of Wyoming in 1933, Dominy had gone to work as an agricultural agent in Campbell County, Wyoming, a long rectangle in the northeastern corner of the state that was about to blow away. During his four years in that outpost, Dominy oversaw the construction of more than 300 dams across intermittent creeks and washes, tiny catchments designed to do little more than keep cattle from dying of thirst. And it was his flurry of bar-the-door dam building that first had caught the attention of people in the Roosevelt administration's agriculture department. After several attempts were made to lure him away from Campbell County, Dominy finally accepted a job as a field agent for the Agricultural Adjustment Administration, later transferring during the war years to the Inter-American Affairs Bureau, where he worked under Nelson Rockefeller developing farms in Latin America and the South Pacific to feed fighting men, miners, and construction workers.

Back in the United States in 1946, handsome, brawny, hail-fellow Floyd Dominy had been put out of work by the end of the war. Deciding that dam building was the one enterprise he knew something about that wouldn't quickly bore him silly, he placed a call to the Bureau of Reclamation's Washington headquarters and talked himself back into government service in a matter of minutes. In his thirteen years at Reclamation, Dominy had risen like a rocket, moving from the land development office to Allocation and Repayment, then on to Operation and Maintenance, to Irrigation, to assistant commissioner under Wilbur Dexheimer, and finally to associate commissioner—a post that was created specifically for Dominy, one that would allow him to *run* Reclamation if it didn't quite give him the biggest desk in the building.

But the Nebraska-born Dominy, a man with a water-impounder's soul and an autocrat's convictions, was worried that prior to his wresting control from Dexheimer—the mild-mannered engineer who had succeeded Mike Straus in 1953—the Bureau very nearly had gone to hell. Its administrators had almost lost the war for the CRSP on Capitol Hill—looking like ill-informed, uncommitted *bureaucrats*, for Christ's sake, in their testimony before the congressional subcommittees—and Dexheimer himself was happily acquiescing to the Republicans' desires to apply the spending brakes to the Bureau. If it hadn't been for his own skill and tenacity, as well as the virtually assured support of Colorado Congressman Wayne Aspinall, now head of the House Interior Committee, and Senator Carl Hayden of Arizona, the longest-tenured member of the Senate and chairman of its powerful Appropriations Committee, Dominy was convinced that the Bureau of Reclamation might have been mortally wounded.

In the spring of 1959, Dominy finally could abide his subordinate position no longer. He informed Dexheimer that if he didn't turn Reclamation's reins over to him, he would quit. Then Dominy likewise informed his several friends in Congress that either Dexheimer walked or he did. It was a gamble that the gin rummy master confidently took. As he toured the progress at Glen Canyon in April 1959—down in the hole and up on the mesa where they were trying to build a town that the wind couldn't cover with sand—members of Congress in Washington were calling for an investigation into Dexheimer's moonlighting on private dam-building jobs. The commissioner was about to resign, Dominy told Janet Cutler, editor of the *Page Signal*, and on May 1 he did so.

What the new commissioner had seen during his short trip to Arizona was "the most beautiful bridge in America," a shiny steel web spanning the canyon, its roadbed laid and painted with road stripes, the bridge now open for traffic. Seven hundred feet below, two pyramidal cofferdams built out of rock and sand and impermeable clay now kept the river at bay—the upstream dam diverting it into the tunnel in the western wall, the downstream dam preventing the river from backing up into the dam site once it was free again in its channel. Giant diesel-powered sump pumps had drained the water from between the two dams, and now dozers, draglines, earthmovers, and dump trucks were working round the clock to excavate the river's silty, unstable bed, to peel it back to the bedrock that would form the dam's foundation.

Thirteen hundred men now were at work on the project, digging out the riverbed, completing the keyways, erecting the cableways, mining aggregate on Wahweap Creek, and constructing the concrete mixing plant—the largest one ever built. In January, an arbiter in Los Angeles had ruled that Merritt-Chapman & Scott no longer was required to offer its workers free room and board—a provision that had been in effect since the job began under the terms of a master labor agreement between the Association of General Contractors of Arizona and unions representing the five basic construction crafts—carpenters, teamsters, laborers, cement masons, and operating engineers. Adequate residential and municipal facilities now existed near the job site, the arbiter had declared, and the workers no longer were entitled to "remote projects" subsistence pay.

Angry that room and board in Page now would begin to cost them more than $100 a month—that amount subtracted from take-home wages that averaged less than $500 a month—800 workers had begun a wildcat walkout on January 23, just two weeks before the upper cofferdam was scheduled to be closed and the river diverted. Al Bacon, representing MCS, had refused so much as to discuss the issue with the men who, he claimed, were striking illegally, and the several unions' own officials had been unsuccessful in convincing their members to go back to work so that formal negotiations could begin. Finally, after a twelve-day stalemate, the workers had agreed to go back into the hole until June, when the master labor agreement was scheduled to expire. If the subsistence pay issue wasn't settled to their satisfaction under the terms of a new agreement, they vowed, they would walk out again, and next time they would be prepared to stay out.

When Floyd Dominy visited in April, a major strike was beginning to loom as a real possibility. The new master labor agreement in the works called for a $6-a-day "subsistence premium" to be paid to union members working 30 miles or more from one of twelve cities and towns in Arizona designated as "nonremote centers," and Flagstaff, 135 miles away, was the nearest such center to Page. MCS's Al Bacon—who with his flat-top haircut, plaid workshirts, and steel-toed boots looked like he wasn't too far removed from the union halls himself—had announced that his corporation could abide nineteen of the twenty provisions in the new agreement, including wage and overtime pay increases, the institution of a health plan, and a variety of changes in job schedules and conditions. But *under no circumstances*, said Bacon, would MCS agree to any form of the subsistence pay that the arbiter fairly had ruled no longer was required. What Bacon didn't say was that Merritt-Chapman &

Scott was losing enough money on this goddamn job as it was, and it was not about to add to its woes by spending six extra dollars a day on every oiler and bellboy and sandblaster on the job for the next five years. The company already had spent $2.5 *million* on housing and on municipal, recreational, and health facilities in Page—$700 for every man, woman, and child who resided there at the moment—and that plainly was plenty.

For his part, Floyd Dominy was optimistic, telling the *Signal* that although a strike surely would put the project a bit behind schedule, he was certain that the pay impasse would be resolved before too much time was lost, telling the paper parenthetically that Page didn't look like too remote a spot to him, coming as he did from Hastings, Nebraska, and Campbell County, Wyoming, up there at the back of the neck of the world. Page, in comparison, was a bona fide city already, said the commissioner-to-be. The construction workers had virtually every amenity they could hope for, and he looked forward to the not-too-distant day when the Bureau of Reclamation could relinquish control of the town, allowing it to be incorporated under the laws of the great state of Arizona—a fine independent and thriving community. Page would be a construction camp only for another five years or so, at which point its future would be defined by the hundreds of thousands of tourists who would pass through it en route to the shimmering reservoir that would fill Glen Canyon—a reservoir, by the way, that Reclamation had decided to call Lake Powell.

There were those who lately had urged that the dam and its reservoir be named for Dwight D. Eisenhower, the war-hero president who would leave office in a year and a half. But the western Democrats who long had championed the reclamation cause weren't too keen on the idea, claiming that if the Republicans had had their way, the Colorado's water forever would have flowed to California. Those Democrats were willing to honor former Reclamation Commissioner Mike Straus, or the retired Harry Truman, or better still, crusty old Carl Hayden, who had fought Arizona's federal water fights since 1912, but this was a Republican administration, after all, one that probably wouldn't agree to giving Glen Canyon a socialist cast by naming either the dam or the reservoir after a Democrat. The solution to this ticklish dilemma had come to Dominy in a clear-eyed moment: the dam officially would be called what it had always been called, Glen Canyon Dam. And the reservoir would be named for

John Wesley Powell—a patriot, certainly, and a reclamationist, a man who had explored and named the canyon, a man dead so long that no one would object. *Powell Reservoir*. No, that was rather pedestrian. *Powell Lake? Lake Powell?*—perfect. The reservoir that would fill Glen Canyon would be called Lake Powell, Floyd Dominy told the *Signal*, and not too far into the future three million people a year would visit this jewellike lake in the desert and the spanking new city at its shore, he proudly, if foolishly, forecasted.

The sand had begun to blow—really blast across Manson Mesa—at the end of 1957, just as soon as bulldozers ripped away the thin cover of blackbrush, saltbush, shad scale, and snakeweed, exposing the buff, beige, pink-hued sand, freeing it to be whipped and spun by the constant winter winds, freeing it to pelt everything in its path. A rough airstrip had been bladed in early, then dozers had scraped the street rights-of-way, carving the same kind of curving, casual thoroughfares that suburban planners across the country currently were enamored of—everywhere shunning the grid. Trenches for waterlines, for sewers and telephone cables were next to snake across the mesa top, their sandy spoil piled beside them until it was returned quickly to the ditch, or until it blew away.

Designed by engineers at the Bureau's Denver offices, the town of Page would have a park, a school complex, a warehouse district on its north rim, a curving row of churches on the south. A long and only slightly winding commercial boulevard, Seventh Avenue, would connect the industrial with the spiritual, and would separate the 300 permanent homes to be built for Reclamation personnel, MCS officials, and their families on the west side of town from the thousand or so trailers that would fill the eastside Page Trailer Court—the sea of temporary housing serving as residences for workers and their families for six months, for a year, some of them for six years.

In the beginning, water would have to be hauled from a deep well near the Navajo community of Coppermine, then stored in a 75,000-gallon tower; electricity would be provided by a bank of noisy diesel generators. Merritt-Chapman & Scott's mess hall would have to serve as a multipurpose restaurant, supply store, community center, and post office—each day's mail simply spread out on the sandy dining tables to await retrieval. There wouldn't be any phone service for a few months; water would be in short supply, its quality leaving a lot to be desired; and electrical brownouts were virtually ensured. But given time and perseverance and the constant presence of sand, Page Government Camp, *Page, Arizona,* certainly had potential.

By the time the Bureau of Reclamation moved its offices and its personnel to town—dozens of families and filing cabinets making the move from Kanab, Utah, during the long Thanksgiving weekend in 1958—there really was a kind of a town on top of the mesa. Most of the trenches were closed now; powerlines cross-hatched the scarred scrubland, and a few of the arterial streets were actually paved. The Mountain States Telephone & Telegraph Company had hauled the first tin building into Page near the end of 1957, later replacing its temporary phone line from Kanab to the construction area on the west rim with a permanent line on poles, then continuing it on across the canyon and establishing service. Dialing MIdway 5 at the close of 1958, residents could reach the First National Bank, a grocery store called Babbitt Brothers Trading Company, Page Barber Shop, Page Beauty Salon, and Page Service, a gas station—all housed in a large sheet-metal barn built alongside Seventh Avenue. The post office (still without phone service) had moved in for a while as well, before it acquired its own cramped but adequate trailer; the Bureau of Reclamation had erected warehouse and office structures; Merritt-Chapman & Scott had built separate facilities for itself, and was about to finish the twenty-five-bed hospital called for in the prime contract. At Page Trailer Court on the east side of town, more than 500 pink, portable "Transa Homes" and shiny aluminum-sided trailers were perched side by side on cinder blocks and hooked up to the tenuous municipal services. And on the opposite of Seventh, in the "Bureau area," where little broomstick trees already had been planted and where residents were trying to anchor the sand in place with bluegrass, ten lap-sided and brick-veneered bungalows by now were occupied, with many more nearing completion on winding streets prosaically named Elm, Birch, Fir, and Cottonwood.

Utah's W. W. Clyde & Company had gotten the big $1.6 million contract to construct the town's streets and utilities, plus a second, smaller job building the sewer treatment plant, and its work was well under way. Southern Engineering and Construction was at work on the permanent water system that would pump water 1,200 feet up from the bed of the Colorado, filter it and treat it, then store it in the massive new tank being supplied by the Pittsburgh–Des Moines Steel Company. Security Construction had built the Bureau's laboratory, offices, and warehouse; Sierra Construction was at work on the police and fire stations; and Mobilhome Corporation was supplying $3 million worth of homes designed to stay put. They were straightforward contracts, straightforward jobs, and Lem Wylie administered them with no more special concern than he gave to the tunneling and keyway cutting going on down in the hole. But somehow, turning this thing into a *town*, running a town, making the place livable, even likable, was just a little daunting. He was an engineer, after all; he built things. He could get streets and sewers and buildings built, and he could construct them in a hurry, if he had to. But in this case, he was also going to have to be the de facto mayor, police chief, housing and welfare administrator, and recreation director of this government burg—jobs that did not sound nearly as simple as raising a ten-million-ton dam. When Wylie had gotten in touch in 1957 with Elmer Urban—the thin, balding, friendly, facts-and-figures fellow who had been his administrative assistant on the Eklutna Project in Alaska—asking him to come to Arizona to help him run a town, Urban had protested that towns weren't his area of expertise. "Well, I don't know a goddamn thing about it either," Wylie had responded, "but I guess we'll learn how to do it together."

Right off the bat, Wylie and Urban had decided to advertise, to let the wide world know that enterprising, pioneering businesspeople could get in on a good deal down in Arizona. They could come to this last frontier in the lower forty-eight, and with fortitude, faith, and eight days a week of work, they could build businesses, secure incomes, and raise their families amid the splendor of the desert. There was only a single catch: a permit system would regulate the number of particular kinds of businesses that could be started—only so many service stations; a single jewelry store would do; they'd need a drugstore and a store that sold appliances and another one or two that traded in clothes. They would need doctors and dentists and a lawyer or two, repairmen and mechanics and teachers. Nonunion construction workers needn't bother to come to this union camp, however; it was a few years too early for businesses selling boats; and no, there wouldn't be bars or liquor stores, not on government land bartered from the dry Navajo Nation.

When 200 Bureau of Reclamation employees and their families arrived in Page from Kanab at the end of 1958, George Koury, manager of Babbitt's market, was already there to greet them, daily dusting the sand from the tops of his tin cans, futilely trying to keep it from infiltrating the produce. Johnny Keisling was selling Chevron gas from two pumps sprouting out of the sand alongside Babbitt's, his office his pickup truck, his cash register the pockets of his pants. Royce and Dora Knight, who had flown their former home in Cedar City, Utah, already had built a hangar out at the airstrip and were flying mail and occasional passengers up from Flagstaff, servicing the Bureau planes that landed and took off almost daily, keeping their own planes earthbound in the midst of sandstorms with guy lines tied to rocks. Katherine Pulsifer, somehow always elegant in this hardscrabble kind of outpost, was sorting small mountains of general delivery mail; and over on H Street, young Dr. Ivan Kazan— transplanted from Hoover Dam's companion town of Boulder City, Nevada—was suturing cuts, treating chest congestion, and practicing primitive pediatrics out of an eight-foot-wide trailer, next door to an identical trailer that he and his family called home.

The amazing thing was that the circulars touting the town brought results. People *wanted* to come to Page, it appeared. Like the Mormon pioneers before them, they wanted to go to a place that was virtually empty, entirely unformed, a town unfettered by convention. The stories of chronic power shortages and water as red as the river didn't scare them; they didn't care if there wasn't any television, no churches yet, precious little to occupy the kids. It didn't matter in the beginning. They would build what they needed when they needed it, organize whatever had to be organized. Page was a place on the make, and there weren't too many of those around anymore.

By the time Floyd Dominy came to town in April 1959, Earl Brothers and Bill Warner had opened a Firestone store across the street from Babbitt's chaotic barn, selling the tubeless tires that were the coming thing, selling refrigerators, toasters, and coffeepots. There was the Style Center (featuring work clothes instead of fashions), Ernie Severino's Page Jewelers, Grant and Flora Jones's shoe shop, pharmacist Mack Ward's Rexall drugstore (where sales representatives were required to arrive bearing cases of beer)—all in identical steel Butler buildings, separated from each other only by the several proprietors' trailers. Nearby, Marie and Roger Golliard had opened the Glen Canyon Trading Post, catering to the Navajos who had begun to frequent this new supply center, as well as to the occasional tourists who were beginning to filter through en route to look at the big dam doings they had heard about; "Chinaman" Bill Lee, a former fighter-pilot instructor, had opened the Page Restaurant, the mess halls' first true competition; and over at the newly completed hospital, Dr. Lavon Gifford had started a dental practice.

Lavon and Adrienne Gifford first heard about this coming "Giant of the Colorado" in a story in the *Saturday Evening Post*, and decided it wouldn't hurt to go have a look at the canyon, the project, and the new town the government was said to be building. What they discovered a few miles south of the Arizona line captivated them. Despite its raw and roughneck feel, despite its remoteness from the bosom of the Mormon church, from all of Utah and almost as much of Arizona, Page looked like a place they could help create, a town where that old "ground floor" expression actually meant something.

Dr. Gifford immediately applied to the Bureau of Reclamation to come to Page as the community's dentist, but Elmer Urban cautioned him not to get his hopes up. Only one dentist would be granted a permit to practice in town, and that person would be selected by drawing from among many applicants. When, after weeks of waiting, a letter arrived in Ogden, Utah, informing Gifford that he had been selected, the news seemed miraculous, absolutely wonderful. And it also seemed utterly clear that Page was where the Giffords were meant to be.

A month before Christmas, 1958, the Giffords arrived with their three children on cold and windy Manson Mesa to discover that there weren't any motel rooms available in Page—there weren't any *motels*—and every trailer, Transa house, and station wagon already was occupied. They spent their first night in Page on beds in the soon-to-be completed hospital, and it wasn't until the next day, when he was examining the fine new equipment in the dental clinic he would operate, that Gifford learned that his had been the only dentist's name in Elmer Urban's hat.

Life in a trailer town on tableland that appeared to be blowing away may have seemed adventurous to some, but often it was downright depressing. Workers' wives tended to spend their first few weeks in Page in tears, realizing only after it was too late that *this* was where they would spend the next year, the next half-dozen years if their sentences were particularly harsh—in a boiling, sand-filled trailer inches from the trailers that surrounded it on every side. Once they were settled in, the tears usually disappeared, but for many people, the situation seemed to improve very little. The trailers did have evaporative coolers mounted on their roofs, so the heat had a hard time *killing you*, but the coolers needed water, which sometimes wasn't available, and they needed electricity, which was always in short supply. On the hottest days, the power would be spread so thin that the fans would barely spin, temperatures climbing so high inside the aluminum boxes that it was actually better to go outside and face the sun and the transient sand.

Socks tied around kitchen faucets, shower heads, and the inlets to the washing machines down at the community laundry did little to trap the water's silt, and everything hung outdoors to dry took on a decidedly sandy hue. Kids seemed to delight in unplugging the freezers that stood in the sand behind most of the trailers, leaving the food inside them to thaw and spoil; more than a few mothers seemed to sustain themselves on late-morning cocktails and afternoon liaisons with fellows who worked the graveyard shifts; fathers were prone to disappear after supper, drinking bootlegged beer with their buddies, getting rowdy and more than a little western, until it was time to work in the morning. On sultry summer nights, you could hear your neighbors making love, hear them making war; and if the lingering strife between you and the son of a bitch next door with the barking dog and the bad attitude ever got out of hand, Lem Wylie would warn you once, then the second time he'd have his men jerk both trailers off their blocks and haul them to the edge of town, the message awfully explicit.

But it wasn't *all* bad, not all the time. Page Trailer Court did have its cookouts and catfish fries; the kids had packs of playmates, and as long as they stayed out of the way of the bulldozers and far from the rim of the canyon, it was hard for them to end up in too much trouble. If you were lucky, you liked the people two trailers away; the people at E-11 were from Pinetop, too, and the guy at E-14 would tune your carburetor for a fifth of something fermented. Four nights a week, you could pack up the kids and the lawn chairs and go down to the cement slab by Johnny Keisling's station to watch the movies MCS supplied. The gals had sewing circles and baby showers and hand-me-down parties; guys would lean on the hoods of their Ramblers, their Impalas, and yeoman Willys pickups late into the night, swapping stories about the hole, telling tales about the whores at the beer joint over at Fourteen Mile. The youngsters racing through the amber-lighted, sandy streets after dark, playing tag and incessantly squealing, seemed to think this was some sort of paradise; and the teenagers, sitting in their jalopies out beyond the airport, listening to the rock and roll that somehow reached them all the way from KOMA in Oklahoma City, trading the hickies that they would have to hide from their parents, tried to convince themselves that they weren't as far away from the rest of America as the radio sometimes made it seem.

Yet if you lived on the other side of Seventh, if you weren't there yet but your house would be finished soon, if you worked for the Bureau, if your husband or maybe your dad did, your prospects in Page were rather different. You too might have cried when you first saw what there was to see—hearing stories beforehand about a "city on a hill" and imagining San Francisco, then discovering that the reality was even grimmer than Tuba City. But by now, three months or even a year into your frontier life, you were probably volunteering for the church construction projects on weekends, working with Father Sullivan to build the Catholic church if your allegiance wasn't otherwise fixed—there always being a beer or two available for those who labored for Rome.

Bureau people who had purchased one or two of the rangy, hammer-headed reservation ponies that roamed the region, and who kept them in a big collective corral down off the edge of the mesa, seemed to devote all their idle time to reliving the wild West. There were gun clubs and motorcycle clubs for those who preferred to approach the desert with a little more loutish commotion; there was a club for people who liked to jump out of airplanes, and there were loose confederations of fishermen and arrowhead hunters, hikers and jeepers. A Navajo Ceremonial Club had been formed to promote the appreciation of Navajo culture among the newcomers, and now, added to a virtual laundry list of organizations, were the Lions and Masons, Knights of Columbus and Daughters of the Eastern Star.

Over on the south side of town, where the expansive new school would stand one day—where the tin-sided Butler buildings with the terrible acoustics still served as interim classrooms—a rough athletic field was beginning to take shape. Teenagers were running track, playing sand-lot baseball in a truly sandy lot, and come fall, the mighty Sand Devils (never did a school adopt a more appropriate mascot) would host the football team from Williams High in the first homecoming game in the community's history, the festive Saturday afternoon lacking only old grads who could come back home.

The kids who lived on the permanent side of town made their own sexual discoveries in backseats on Saturday nights; their siblings too tore through the streets making mischief, imagining they were Apaches—or Navajos who were somehow more sinister than the shy ones they actually knew—imagining they drove the big dumps and haulers down in the hole, emulating with sticks, picks, and cherry bombs the drillers and powdermen who tore into the rock, idolizing the high-scalers who appeared as agile as Superman.

On occasional Thursdays, their long-suffering moms—the wives of Bureau men, wives of the MCS brass, and even a specially invited storekeeper or two—put on their best dresses, their heels, pillbox hats, and white gloves, then trudged through the sand to the ladies' bridge luncheon at the eastside mess hall. The hall was noisy, and it looked like an army camp; most of them were new at this cultured game of cards, and by the end of the afternoon, the sand had always powdered the tables and soiled their delicate gloves, but still it was something special, a bit of refinement and social edification, an escape from the sometimes awful isolation.

Their husbands, on the other hand, saw little reason to dress up just to play cards. Nights down at the barber shop, at crowded kitchen tables, or, by special invitation, over at the MCS guest house, where the company kept a cook on duty and where at least the liquor wasn't laden with silt, they wore chinos and Ban-Lon shirts, basket-weave summer shoes and shiny nylon socks, playing rummy and quarter-ante poker, playing high-stakes, balls-out blackjack till the sun started to climb the sky, agreeing without having to discuss the matter that there were things about Page a fellow damn near could enjoy.

In the midst of all the strike talk, Merritt-Chapman & Scott had argued against the need for remote-area pay by pointing out that with its current population of about 3,500, Page was already the tenth largest city in the state of Arizona. Page had become a *place*, and Manson Mesa had spawned a city on a hill of sorts in less than a year and a half. Sure, the burlap bags draped across snow fencing to help keep the sand settled attested to the truth that work remained to be done. Yes, you could still get stuck, buried in sand up past your axles, driving from the trailer court to the service station. The town needed a bona fide school and surely something more to keep the kids active and out of their cars; Page needed something to capture the interest—and the dollars—of all those tourists who, according to Floyd Dominy, would begin barreling across the Kaibito Plateau, and it needed some vegetation, anything *green*, in the very worst way.

But if Rome wasn't built in a day, you couldn't expect Page Government Camp to turn into a showplace in thirty minutes. It would take time and truckloads of almost everything to make this the kind of place you'd see in *Sunset* magazine, yet already there was much you could point to with a parochial kind of pride: Page now had stores you actually could spend your money in; *twelve* churches were under construction; the park had been planted and seeded with some sort of Sahara-resistant grass, and there was a brand-new swimming pool in which the kids could splash. The town had *two* doctors now; the swank new Glen Canyon Motel was rising at the edge of town; Bill Lee was planning the adjacent Glen Canyon Steakhouse, and Ted Selna was going to build a bowling alley.

In January, the newly incorporated Glen Canyon Golf and Country Club had secured a long-term lease from the Bureau of Reclamation on a piece of rock and blowsand down near the rim of the canyon, hard by the sewage-treatment plant and the portal of the access tunnel that was currently being dug from the rim to the riverbed. A clubhouse already was under construction—big enough to rent out to a variety of local organizations as a much-needed meeting place—volunteers were working two-hour shifts through the dead of night watering the recently planted fairways, and just the other day, Johnny Bulla, a golf pro from Phoenix who was supervising the construction of the new nine-hole course, had come to town and lied through his teeth, proclaiming that this soon would become one of the finest links in America. People would be playing *golf* in Page, of all places, in only a couple more months. It was kind of hard to believe.

No one had had the nerve yet to try to talk Lem Wylie into helping with the all-night watering down at the golf course, but Elmer Urban gladly had signed up for a regular 2:00 A.M. shift, deciding that making a town out of nothing out in the middle of nowhere really hadn't been any big trick as he watched the cool spray from the sprinklers shoot into the hot, hushed nights.

■■■■■■■■■■■■■■■■■

The work force that peaked at 1,700 back in November as the keyways were falling away, the tunnels were receiving their concrete lining, and the bridge was being paved, had dropped sharply by May 1959. The river was swirling through the west diversion tunnel now; and with the spring runoff, it would rise high enough against the giant cofferdam for a few weeks that it would spill into the east tunnel as well. The bed between the cofferdams was dry and was disappearing—the hole where the dam would stand now looking a lot like an open-pit mine, haul roads curling down to the lowest point in the excavation, 30 feet already chewed from the sandy bottom, 100 feet still to consume. The access tunnel down from the rim still was under construction; the towering batch plant wasn't yet operational, but the payroll had dropped to 622, the lowest level it was expected to reach before it shot up again as concrete began to rain into the canyon in the fall.

At the trailer court, the vacancies were becoming obvious. Laborers, drillers and powdermen, dump drivers and pipefitters were hitching their trailers to their trucks and heading up to the Curecanti Project in Colorado, to Flaming Gorge on the Utah-Wyoming line, to California, where, so people said, good work went begging. Few men were wedded to this particular job in this particular canyon. Sure, some would be back when things got cracking again, and there were worse places to make a buck, but Glen Canyon meant nothing more than a paycheck to most of them, and, for now at least, the pay window was shut.

During the second week of May, however, with the layoffs complete and the foundation excavation proceeding on schedule, men with jobs—jobs that weren't due to end any time soon—also began to go, giving a day or two notice, then hightailing it out of town, hoping to beat the rush of hardhats who would be hustling for work on down the road if the unions kept their word and called for a Glen Canyon strike. The Association of General Contractors of Arizona had come to terms with the five unions on a new master labor agreement, the details of wage increases, benefits, *and* subsistence premiums successfully hammered out prior to the June 1 expiration of the old agreement. But Merritt-Chapman & Scott had never joined the associated contractors—a bone of contention in itself—and the unions had been forced to negotiate directly with the three-piece suits from New York City who, according to the union men, were prone to talk through the side of their mouths. In the months since the January walkout, MCS had refused to budge; the company still would agree to nineteen of the twenty provisions in the new pact signed between the Arizona contractors and the five unions, and it recently had announced that, beginning June 1, it voluntarily would begin complying with those nineteen provisions as if there were an agreement in place. But it would not—and would *never*—capitulate to paying that $6 per worker per day.

In a statement distributed to news outlets throughout Arizona, MCS contended that under the terms of its initial agreement with the unions at the beginning of the Glen Canyon project, all parties were aware that as soon as adequate residential and municipal facilities were in place at Page, subsistence pay would be dropped. In the event that the company and the unions couldn't agree on whether the town was sufficiently developed to have reached that point, the question would be resolved by arbitration. The press release noted that when an arbitrator first had ruled in May 1958 that the Glen Canyon area should still retain its remote status, the company dutifully had continued to make subsistence payments. Eight months later, however, the arbitrator had ruled that Page now was sufficiently operational for the payments to be suspended; and now it was the unions' turn, according to the company, to act in good faith and accept the arbitrator's edict. MCS plainly had played by the rules, and as far as it was concerned, there was nothing more to negotiate. The company would shut down the job in a minute, if it came to that, rather than spend an additional $14 million over the next four years just to meet the workers' capricious demands.

What the statement to the media didn't make clear—didn't so much as mention—was that, as MCS officials saw it, they simply had to bide their time. Under the terms of its prime contract with MCS, the Bureau of Reclamation had agreed to assume 85 percent of the contractor's labor-cost increases during the span of the project—85 percent of all wages and benefits *except* subsistence payments, those being specifically exempted back when the ink still was wet in 1957. Had it now been possible for MCS to pass along 85 percent of the cost of the subsistence pay the unions still were demanding—$5.10 of that $6 a day—to the Bureau, MCS might well have agreed to the ransom, letting the government pick up all but about $2 million of its total cost. But since that clearly wasn't going to happen—at least not for the moment—the MCS strategy was simple, and it was classic hardball: if the unions wanted to go the tough-guy route and walk, they were welcome to. It was true that most of their people could find work elsewhere; the company would have continuing administrative expenses, and a lot of valuable machinery would sit idle down in the hole, but sooner or later, Reclamation was going to decide it needed its dam built, and Reclamation would have to do the dirty work, either somehow muscling the unions into submission or, ultimately, going dutch with the company on the cost of subsistence pay.

For their part, the fellows at Reclamation—Floyd E. Dominy at the helm now—were not about to panic. As a policy, the Bureau had always tried to stay out of disputes between its contractors and their workers, encouraging open lines of communication, good-faith bargaining, and all the rest, but otherwise keeping clear of the gunfire. In this case, the Bureau wanted it known for public consumption, if a strike were to ensue, and if it appeared it would become protracted, well, the government simply would monitor the situation and would take whatever measures it deemed necessary whenever they seemed necessary in order to protect its interests and its ends—namely the completion of a dam in Glen Canyon—Commissioner Dominy expressing optimism all the while.

The unions themselves were finding it hard to be hopeful. "Looks like a stalemate to me," said Lew Reddick, business agent for Laborer's Local 383 and the only union official living in Page. "If there's no settlement, a strike is inevitable, and if we have a strike this time, it'll be a union-sponsored strike. We'll put up pickets, but most of the men will just clear out. They won't be compelled to stay here during a strike."

On Wednesday evening, May 27, four days before the labor agreement deadline, the movie playing down at the cement slab was *The Unguarded Moment*, starring Esther Williams and John Saxon, and over at MCS's offices, Al Bacon was momentarily unguarded as well, telling a group of concerned townspeople that the unions were blatantly trying to circumvent the arbitration ruling, telling the anxious shopkeepers that the unions' claims about the town's remoteness were bogus as almighty hell. "The facilities in Page," he said, "are now at least equal to, and in some cases superior to, those available in some of their so-called nonremote centers." The shopkeepers proudly agreed with Bacon on that point, but the company man never spoke the kinds of conciliatory words they had come hoping to hear.

The dozers and haulers and draglines droned all day down in the hole on Monday, June 1, then on into the night on the swing shift. The men were at work on Tuesday, then at work again on Wednesday, talking about little else besides what the hell was going to happen, when word spread that meetings had just been set for Sunday and Monday in Page, then continuing in Phoenix on the following days if necessary, between the international vice-presidents of the unions and a clutch of MCS brass who were going to fly out from New York City. Two hundred and fifty men down in the hole that afternoon and 3,000 people up on the mesa top were elated; both sides were finally going to bargain. At the last yawning moment, they were going to avert this ridiculous mess.

During the negotiating sessions the following week, the two sides did succeed in keeping their deliberations secret and they agreed to a three-week "study period," during which time each promised to give careful consideration to the other's demands, but there was no real progress toward an accord. The study period quickly degenerated into a war of waiting, a war of nerves, and a trailer or two a day continued the slow roll out of town. On Tuesday, July 1, the day the study period expired, there was a flurry of phone calls, followed in the next days by another round of charges and accusations, then a weekend of awful silence before, in the wee hours of Monday morning, July 6, the unions ordered their men not to report to work.

Down at the guard station near the rim of the canyon, the unions posted four men carrying placards during the day shift on Monday; only two manned the picket line during the swing shift, and the graveyard hours passed unprotested. Seven hundred and fifty union members—employees of MCS and its current subcontractors— stayed away from the canyon, where all was silent except for the white noise of the river as it tumbled out of the tunnel, except for the drone of the sump pumps sucking the construction site dry. Lew Reddick reported that all was peaceful in Page, and the union official sounded surprisingly cordial as he noted that MCS had "gone all out" to cooperate with the picketing. Sounding in turn as if he expected this thing to last a while, Al Bacon said the company would immediately begin to curtail its office and administrative staffs. Lem

Wylie publicly would say no more than to hope the two sides would be able to resolve their disagreement quickly. Privately, the project construction engineer was every bit as disgusted with the unions for refusing to back off from their asinine demand as he was with MCS for its obvious stonewalling. Privately, Wylie didn't give a damn about wages or premium pay or the complexities of this remote-status business; all he was concerned about was the fact that he had a canyon to plug, a canyon that was awfully quiet at the moment.

The union men who had been at work on the school complex up in town joined the strike as well, shutting down a job that already had been riddled with delays; and every other heavy-construction job in the state of Arizona was quiet as well that Monday morning, members of the five unions striking elsewhere solely in solidarity with the Merritt-Chapman & Scott workers, their own labor agreements already negotiated and in force. If all of Arizona was shut down, so the thinking went, it wouldn't be long before the Association of Contractors would pressure maverick MCS into negotiating seriously, into settling, into spending some money on premium pay. It was a logical enough strategy, and it might well have worked except for the fact that the *last* thing multinational MCS was concerned about was the opinion of the piddling little Association of General Contractors of Arizona. What Merritt-Chapman & Scott did care about was nothing more than getting this contract completed, losing the least amount of money possible in the process, then getting the hell out of the wild West. Besides, the company continued to contend, the arbitrator already had settled everything there was to decide.

Seventy trailers rolled down the sandy streets of the trailer court on Monday; by Wednesday, the 617 trailers that had packed the east side of town back in February, when the bridge was completed and the cofferdam closed, had dwindled to 448, and half of those were in the process of being tied on to the bumpers of trucks. The unions had told their men to expect a long strike, and few of them appeared to be waiting around just in case the unions were wrong. There were other fish to fry—if not on jobs in Arizona, then up in Utah, over in Colorado, better yet, out in California, where the wages were great and the living was awfully easy.

When the hot weather hit, the afternoon temperatures soaring to 105 early in August, half of Page was a ghost town. The only people living in the trailer court were those Reclamation and MCS personnel who had not yet been able to secure a house over on one of the Bureau streets. Hundreds of trailer pads were bare except for the piling sand, their hookups hanging on poles, abandoned. By now, much of the MCS office staff had been laid off, and dozens of the 200 Bureau people stationed in Page were away on "details"—temporary jobs on active Reclamation projects. The two sides still weren't talking, weren't talking about talking as far as anyone knew, and from his office in Washington, Floyd Dominy was beginning to sound impatient, the commissioner regularly mentioning how critical the early completion of Glen Canyon was to the whole of the CRSP.

At the end of September, when rumors ran rampant that the Bureau had initiated secret talks between the unions and MCS, several Page businesses already had locked their doors, their proprietors—most of whom had staked every penny they owned on fledgling businesses in this erstwhile boom camp—having gone to Flagstaff to pump gas, to wait tables until something finally broke. There had been many days when Marie Golliard at the Glen Canyon Trading Post had sold nothing more than a single roll of film, days when she sold nothing at all. Mack Ward long since had begun filling prescriptions on credit, hoping that somehow, someday he'd get paid. George Koury at Babbitt's market had taken to overlooking a few items, somehow not noticing quarts of milk and cartons of eggs at his checkout stand. Lately, he had been stocking up heavily on cans of pork and beans, his meat counter now offering little more than hamburger. And the *Signal* had had to suspend regular publication, the community news now printed on the back page of the Tuesday edition of the Flagstaff paper. A town of 3,500 just three months before, Page had dwindled to 800 residents, their collective anger now a kind of numbness. People seldom smiled, seldom stopped to greet each other at the once-chaotic post office, where now no one ever had to wait in line.

On into the fall, nothing came of those secret sessions everyone had heard about. Either the two sides weren't actually talking or the talks had come to nothing. Why wasn't the Bureau—the *government*, for God's sake—imposing some sort of settlement? No one expected much of the unions; if this dam were ever built it would be built by union men, and the union bosses, still getting *their* weekly paychecks, would simply wait till they got what they wanted. But didn't MCS need to get on with the project? Get it done and get on with other work? None of it made any sense to the sand-weary, strike-weary townspeople of Page, but then it hadn't made sense from the very beginning.

It was going to be the grimmest of Christmases. For more than five months now, the only work done down in the hole had been the sucking of seep water by big diesel-driven pumps. Up on the mesa, the little town that had been bursting at its seams had shrunk to derelict status. The trailer court looked as though it might soon erode back into the desert; buildings downtown were boarded shut; the school still sat empty, its windows uninstalled, its construction never resumed; and the few people that remained in Page were taunted by the sand and relentless winter winds. During the week before the holiday, church groups distributed food and hand-me-down toys to the folks who were in the most desperate straits. George Koury's generosity in the midst of trouble meant that people continued to go home from the market carrying more food than they had purchased; Mack Ward had begun giving away baby formula, saying it was just his way of keeping the population up; Earl Brothers was still in business, but he wasn't selling toasters or teapots as presents, nor was he selling tires in advance of holiday trips; Grant Jones wasn't even selling shoes. Royce Knight had flown Santa Claus into town for a brief visit to try to lift everyone's spirits a couple of days ago, and Merritt-Chapman & Scott was supposedly talking to the unions again—there were rumors that representatives from the two sides lately had been ensconced in a resort outside Phoenix—but no one dared to hold out much hope. And whatever it might have been doing behind the scenes, the Bureau of Reclamation publicly was still keeping its hands clean. Who knew? Maybe the Bureau was reconsidering the entire project; maybe this goddamn dam would never be built. Perhaps this would be the last Christmas anyone ever spent on miserable Manson Mesa.

But Christmas came early in 1959. Just before dark on the afternoon of December 22, Al Bacon took a call from company headquarters in New York City and heard the astonishing news that MCS just had reached an agreement with the five construction unions. In return for the unions' withdrawal of their premium pay demand, MCS would agree to an across-the-board wage increase of fifty cents per hour, one that would affect only the Glen Canyon job;

workers at construction sites elsewhere in the state would return to their jobs at the same pay scales that were in effect when the strike began. The wage increase would amount to $4 more per worker per day instead of the $6 the unions had held out for—roughly $9.5 million instead of $14 million for MCS to cough up—but the beauty of the deal was that, as wages, 85 percent of that additional $9.5 million would have to be borne by Reclamation. It said so right there in the prime contract.

The unions had gotten better than half a loaf; it was a great agreement for them. MCS had done what it had always hoped to do—get the government to pick up most of the settlement tab. And at Reclamation, Floyd Dominy and an entire floor of accountants were willing to look the other way for the time being; that was a small price to pay for getting Glen Canyon going again. But Dominy wasn't the hayseed he sometimes pretended to be, despite what the men at MCS might have suspected. He would play appropriately dumb for a week or two, wait for the workers to return to town and the hole to fill with a frenzy of activity again, then he'd submit the matter of this premium pay disguised as a wage increase to the comptroller general for a ruling, certain that the comptroller would smell the obvious rat. Merritt-Chapman & Scott would rue the day that it tried to hoodwink the commissioner of Reclamation.

In his office in Page, Al Bacon was ecstatic. Everyone else in the MCS office heard the glorious good news within seconds; everybody in town, in turn, was in jubilant spirits within a matter of minutes. This was the best Christmas present anyone could have hoped for; the strike was history—it was *over*—and on Monday, January 4, the company would begin rehiring men and would return them to the hole. The trailer court quickly would bulge out into the desert again; the post office and the bank soon would be overrun with people; and the stores—what a blessing it was—at long last would take in a dollar or two.

The air was cold and there was wind, of course, but soon it seemed as though all 800 stalwarts left in this forlorn-looking camp were out in the streets joining an impromptu parade, screaming the good news, shouting for absolute joy, singing Christmas carols and "Auld Lang Syne," embracing, kissing friends with the warmest and most grateful kinds of emotion, kissing bare acquaintances with something akin to passion. Suddenly, somehow, there was a lot of liquor in officially sober Page; the justice of the peace got out his old musket and began to shoot it into the air; the volunteer firemen turned on the loud emergency siren mounted high on poles above the fire station and pulled their trucks into the streets to join the merrymaking. Bureau people, MCS personnel, shopkeepers and teachers and cops, Navajos and newcomers saw no distinctions between themselves that night, no classes or ranks or special status. From house to trailer to house to house, they traveled en masse, still caroling, still shouting the truth that all was right with the world.

Sometime after midnight, the party shifted to Al Bacon's house, the MCS boss an appropriate host at the end of this long ordeal. Lem Wylie was there, feeling no pain at that late hour and assuring Bacon that all was forgiven, Bacon in turn assuring the government man that he'd build him a hell of a dam. Elmer Urban, seated at the piano, playing every Christmas song he could remember, felt for the first time in half a year as though he'd again have a town to manage. George Koury, Mack Ward, Earl Brothers, and their wives, the Golliards, the Warners, the Joneses, the Keislings, the Knights, Drs. Kazan and Washburn, the teetotaling but high-spirited and grateful Giffords, people who had bet the pot on Page and lost, it seemed, now joyously were game to play another hand. It would take time, they told themselves as it finally began to grow light, time for the town to swell again, time for the paychecks to fill up the pockets, time for the shopkeepers to recoup their losses. But the bottom line was that the boys *were* going to build Glen Canyon Dam; Page Government Camp *was* going to become a city someday, and their separate and varied treks to this last frontier no longer seemed in vain.

It was breakfast time on Wednesday morning before the party at Al and Wanda Bacon's broke up, and by noon Page had gone into a kind of quiet stupor lasting all the way through Christmas and into the following week. Many townspeople, most of them it seemed, spent what little money they had left to get out of town briefly now that relief was in sight, and unexpectedly celebrated in motels with television in Flagstaff, in summery Phoenix, where the grass was wondrously green and the shopping centers dazzled, in wintry Salt Lake, where the Saints warmly shared their fellowship and the Tabernacle Choir sang the joys of the season.

But within a week after the announcement of the settlement, a pilgrimage back to Page had commenced. The boards came off the windows, the OPEN signs went up, crews went to work sweeping the sand from the pads in the trailer park, and whole caravans of sedans and station wagons, panel trucks and pickup trucks, and all manner of vehicles towing trailers filled the highways heading for Manson Mesa, heading back to the job in Glen Canyon.

The work would be slow at first, Lem Wylie was well aware. Many of the men would be new, and even those who had been on the job before the strike would need a little time to get squared away after that strange six-month hiatus. There would be quite a few wrinkles for Al Bacon to iron out, and Wylie was willing to give him some time to get his ironing done, but the commissioner of Reclamation, "the Kmish," as everyone was calling Floyd Dominy already, was not going to be endlessly patient, and neither was Wylie. The core sampling in the riverbed had indicated that somewhere in the neighborhood of 125 feet of sand and silty muck would have to be excavated before the foundation rock was reached, and the keyways, which had almost reached the riverbed when work was halted, would have to be cut on down that additional distance as well. Before the strike, MCS crews had begun stockpiling washed material out at the aggregate plant, and now that job would have to begin again in earnest. The taller of the traveling cableways was operational, but the shorter of the two had yet to be strung and tested; the batch plant that ultimately would mix ten million tons of mud and the refrigeration plant that somehow would keep it cool were still months away from doing business. Before the strike, it had looked as if MCS might beat the prime contract's March 1964 deadline by several months—might save some real money in the process and avoid taking the bath that the company had seemed destined to take that day back in 1957—but now, having sacrificed half a year to the premium pay issue, it appeared to Wylie that if the company was going to finish in four years, these New York City boys would have to kick some ass to see it through.

By April 1, 1960—the day the comptroller general ruled that Reclamation did *not* have to pay 85 percent of the fifty-cent wage hike because it was a subsistence payment in poor disguise, the day that MCS promptly appealed—Merritt-Chapman & Scott had 800 workers on the job again, down in the excavation pit, at the batch and refrigeration plants perched on the west canyon wall, at the aggregate operations six miles out at Wahweap. Up in town, the trailer court was a sea of aluminum again, every pad occupied, kids and dogs and cars and constant commotion the unmistakable signs that Page had come back to life. The spanking new school complex finally was nearing completion; the post office, the bank, and Babbitt's again were cacophonous, always crowded; Ted Selna's ten-lane Canyon Bowl was booming, as was the new drive-in theater, which recently had put the cement slab out of the movie business. The Navajo Cafe—"The Best Food by a Dam Site"—had opened its doors, as had the Windy Mesa Lounge, The Cove, and Bottle Stop Liquors within days after the *Signal*'s Janet Cutler had convinced Floyd Dominy to lift the ban on alcohol sales in an effort to keep drunks from driving the far-flung highways. KOMA in Oklahoma City still came in loud and strong after dark, and you could even get cloudy reception from a couple of Phoenix television stations now. But strangely, another kind of evening entertainment was beginning to captivate both young and old. Nights after supper, families, teenagers, single workers who had been down in the hole all day themselves, but who now had nothing better to do, would drive down off the edge of the mesa, park near the rim of the canyon, and walk out onto the bridge. The work lights down in the foundation pit seemed to illuminate that whole stretch of the canyon, to give it a strange and otherworldly quality. For an hour, sometimes more, people simply would watch, observing the slow movements of the seemingly minuscule machines, watching as if to make sure they were really at work, visually convincing themselves that all still was okay.

Townspeople jammed the bridge, jockeying for vantage points against the cyclone fencing that kept them safe, on the June morning when Interior Secretary Fred Seaton—the man who had been at President Eisenhower's side for the first blast almost four years before—joined the governors of Arizona, Utah, and Colorado, and Reclamation and Merritt VIPs for a ride down the monkey-slide and a climb down a series of ramps and ladders to the bare Navajo sandstone 127 feet below the riverbed, where forms outlined what would become the base of Block 10-B. On a radio command from Lem Wylie, who dutifully had put on a white shirt and tie for the occasion, a 12-yard bucket of concrete swung away from the loading dock far above, then seemed to drop out of the sky, falling 500 feet before it stopped only inches above the shiny, ceremonial hardhats worn by the cluster of dignitaries. Seaton pulled the lanyard on the bucket, and 2,000 pounds of concrete slumped onto the sandstone. "I didn't help you very much, did I?" Seaton asked, turning to Wylie. The men in ties and shirtsleeves laughed, then stepped out of the way as a vibrator crew moved in to settle the mud.

Up on the bridge, the people who hadn't been able to discern anything when the gates of the bucket were opened cheered as it rose up through the void in the canyon again. And they remained there, fixated, some almost motionless, through the rest of the morning and into the sweltering afternoon, watching each of the twelve big buckets begin its hundreds of thousands of trips into Glen Canyon, the hopeful townspeople peering 800 feet into the earth, trying to see a dam rise.

7

THE WAY THINGS WERE WHEN
THE WORLD WAS YOUNG

Near the close of the 1950s, nine decades
had passed since Powell first plunged
through Cataract Canyon, then washed
down the languid waters of the Glen;
five decades had come and gone since an
Ohio industrialist named Julius Stone
had become the first—and for a long
time the only—man to float through the
canyon not in pursuit of science or notoriety, of pelts or illusive pieces of gold,
but solely for the pleasure of doing so. It
wasn't until a great dam had begun to
rise across the southern end of the canyon that large numbers of people began
to do what Stone had done at the turn of
the century—to go into the Colorado's
canyons in hopes that they harbored delight.

Stone had been president of the
company that planned to make millions
from Robert Brewster Stanton's gold
dredge, and he had traveled west in
1900 to oversee the dredge's initial
operations. But if his pocketbook had
taken a beating in the ill-fated effort to
capture the canyon's gold, Stone did,
however, acquire an abiding interest in
the Colorado River during the months

he spent at the dredging site near the mouth of Bullfrog Creek. It was in Glen Canyon, too, that one of his employees, a trapper, hunter, and itinerant river man named Nathaniel Galloway, began to regale him with tales of his two-man trapping expedition down the Green and Colorado four years before, a trip, as he recounted it, full of high adventure with little hint of hardship.

With the dredge abandoned and Galloway economically adrift in 1907, Stone wrote to the river man at his base in Vernal, Utah, with an intriguing idea. If Galloway would design and build four river-worthy boats and agree to head up a similar trip two years hence, Stone would pay for every expense. Galloway gamely agreed, and on September 15, 1909, Galloway, Stone, and three more men pushed their four boats into the Green to begin the thousand-mile journey. Whether it was Galloway's pioneer luck or his skill as a boatman (he developed a stern-first technique for running rapids that allowed the oarsman actually to *see* where he was headed) that accounted for the group's good fortune, they reached Needles in only two months, having overturned just twice, otherwise having successfully run or lined every one of the hundreds of rapids along the way. And although the fifty-one-year-old Stone was incapacitated by painful pleurisy for much of the trip, it was nonetheless the holiday of a lifetime. At the foot of Glen Canyon, Stone wrote in his diary, "Here where the world is shut out, the spirit of the wilderness still abides and welcomes one into the full freedom and magic of the night and morning; uplifting and swaying the beholder with a sense of being that is delightful past compare." It was the first time—Powell's own journal entries not excluded—that anyone had asserted that instead of offering riches or fame or the advance of technology, the Colorado's canyons might be most valuable in terms of what they offered in the way of wonder. Prior to Stone's 1909 trip, and for two decades thereafter, the very reason anyone else *bothered* to float down the Colorado seemed to be because of a kind of P. T. Barnum belief that scores of people quickly would pay to hear about it.

Powell had combined his two Colorado trips into a single narrative for his article in *Scribners' Monthly* in 1874 as well as in a later book, *Canyons of the Colorado*, the two gaining him wide notoriety. And Powell's second expedition had produced a series of articles written by his cousin Clem Powell, hundreds of photographs taken by a former teamster named Jack Hillers, plus lectures, essays, and two turn-of-the-century books—*A Canyon Voyage* and *The*

Romance of the Colorado River—by a young fellow named Frederick Dellenbaugh, who had been only seventeen years old at the time of the 1871–72 trip. That fledgling tradition of risking your life in the deep canyons in hopes that you would survive possessing exploitable, profitable adventures continued in 1911 when two Pennsylvania-born brothers, Ellsworth and Emery Kolb, set out from Green River City with an eight-by-ten-inch plate camera and a newfangled hand-cranked motion picture machine. After 101 days on the river, 76 different camps, and five flips of their two little boats, the Kolb brothers completed the long run on January 18, 1912, a trip that subsequently spawned a book by Emery, an estimated 30,000 lectures by Ellsworth, plus a companion film that was shown at the brothers' small studio on the south rim of the Grand Canyon for the following fifty years.

Then, as the public controversy over where and whether to build a great dam on the lower Colorado crescendoed in the late twenties, moviemakers discovered the drama and the profit-making potential of the river's desert canyons. During the summer and fall of 1927, in fact, two separate filmmaking expeditions set out from Green River, Utah, traversing Labyrinth, Stillwater, Cataract, Glen, and Grand canyons—their wooden boats bobbing through waves and bouncing off rocks while cameramen recorded the daring deeds from the shore. The first of the two expeditions, mounted by a drug company executive named Clyde Eddy and funded in part by the makers of Mercurochrome (who presumably had in mind some compelling magazine ads featuring river runners doctoring their cuts with the antiseptic), was meant to be a documentary, one that would be shot by the International Newsreel Company, then later distributed by Metro-Goldwyn-Mayer as a short that would precede its feature films in theaters around the country. The second project was the sole enterprise of the Pathé-Bray Company of Hollywood, which had in mind either a fuzzy history of human daring on the river entitled *The Pride of the Colorado* or a romance called *The Bride of the Colorado*, depending on how the raw footage looked at the end of the expedition.

But Eddy, attempting above all to be secretive about his endeavors, made the mistake of assuming that the potential profit and notoriety should come solely from the film itself, and when his expedition collapsed in a series of minor traumas—including the rather more serious mutiny and defection of the International

Newsreel cameraman at Lee's Ferry—he literally was left with nothing to show for his trouble. The Pathé-Bray people ended up with plenty of river footage, but for a variety of reasons—not the least of which was some truly terrible acting—their picture was never released. Nevertheless, the struggling film company was able to garner valuable publicity by vigorously promoting the filming expedition itself—making room in its six boats for a studio publicist, a *New York Times* reporter, and a radio operator, whose job it was to send the reporter's daily updates to a receiving station the company had constructed on the plateau above Lee's Ferry, where they were subsequently relayed to the *Times* and to radio stations around the country.

Much of the nation was terrified, in fact, when the Pathé-Bray party radioed on November 14, 1927, that it was *lost*, somewhere in darkest Glen Canyon. The filmmakers' lives were at stake, claimed myriad national broadcasts, and a rescue mission would have to be mounted immediately. No one seemed to consider that in order to escape the canyon to safety, all the group would have to do would be to travel in the direction of the rather formidably flowing river, and the bald-faced publicity stunt was deemed remarkably successful when the six boats bobbed around the bend at Lee's Ferry to meet their "rescuers"—including Clyde Eddy—on November 30, the *Times* man reporting to the nation that thankfully everyone had been spared.

Glen Canyon did claim a life in 1934, that of a twenty-year-old artist, poet, and desert mystic named Everett Ruess. A Californian whose parents had encouraged his vagabonding, Ruess had spent much of the previous four years traveling the Southwest on foot and burro-back, getting to know Navajos, Hopis, traders, park rangers, and archaeologists, but seemingly always preferring solitude amid the rocks to prolonged companionship. Although he sometimes exhibited a kind of silly self-indulgence and a pretentiousness befitting his years, Ruess often seemed remarkably mature, and no one who met him disagreed that he possessed a strange and rather profound sensitivity to the beauty and wonder of the natural world. His woodcuts and watercolors, his poems, and the hundreds of letters he wrote to his family and friends—keeping them abreast of his wanderings and waxing purple in his descriptions of the desert—

underscored his determination to live life fully in a landscape of "such utter and overpowering beauty as nearly kills a sensitive person by its piercing glory."

Writing to his brother on November 11, 1934, a few days before the last time he was ever seen by anyone, Ruess averred that he was sure he could never settle down. "I have known too much of the depths of life already," the young man wrote, "and I would prefer anything to an anticlimax." He was writing from the rim of Glen Canyon, near the place where the small, mazelike Escalante River emptied into the Colorado. He had told people in the town of Escalante, as well as two sheepherders he had encountered, that he and his two burros were headed deeper into the Glen Canyon country, toward Hole-in-the-Rock, and ultimately across the Colorado toward Rainbow Bridge and Navajo Mountain. And he had asked his parents to forward letters to the post office at Marble Canyon, near Lee's Ferry, where he expected to retrieve them sometime in January.

But Ruess never arrived at Marble Canyon, and the postmaster finally returned his mail to his parents' address. Understandably alarmed, they made inquiries, which led to months of organized searching. The burros, fat and healthy, were found grazing in Davis Gulch, a tributary of the Escalante, together with a bridle, halter, and rope believed to belong to Ruess. Searchers subsequently found his footprints, a site where he undoubtedly had camped, and, at a small Anasazi ruin perched above the canyon floor, as well as on a pictograph panel a distance away, the strange inscription *NEMO 1934* carved into the rock.

Ruess's parents first speculated that their son remembered that Odysseus once called himself Nemo, "Nobody," and since he often chose nicknames for himself instead of the embarrassing "Everett," this Nemo almost certainly had been their son. Later, to the satisfaction of a growing number of people who were fascinated by Ruess's disappearance, his parents suggested that perhaps their son was referring to Jules Verne's Captain Nemo in *Twenty Thousand Leagues Under the Sea*, who had tried to escape from civilization. Had Everett left these cryptic clues before intentionally vanishing into the Glen Canyon wilderness? Had he fallen from a cliff, or fallen prey to cattle rustlers or nomadic Navajos who might have resented his intrusion? His body was never found, nor were his bedroll, his wallet, his art supplies, or his journal. No one reported seeing him or any more sign of him for more than twenty years until, in the summer of 1957, Robert Lister, an archaeologist at work on the Glen Canyon salvage project, found a rusty cup, a fork, kettles and pans, and a canteen in Cottonwood Canyon, ten miles southwest of Davis Gulch. Lister also found a box of razor blades purchased from the Owl Drug Company in Los Angeles, where Ruess's parents lived. But two decades after the fact, Ruess's mother could not remember whether she had ever sent her son razor blades.

Aficionados of the swelling legend of Everett Ruess were pleased that his mother could not positively identify the objects Lister had found. If she had been able to, she would have lent credence to the speculation that her son had been robbed, then murdered and buried. As the Glen Canyon country was becoming better known, its profound beauty and beguiling mystery better understood, many people—young people who had also experienced something of the passion of the wilderness Ruess had tried to describe—preferred instead to think that the artist, who once had written that "finality does not appall me, and I seem always to enjoy things more intensely because of the certainty that they will not last," had very consciously and reverently chosen to surrender himself to the most beautiful place on the planet, somehow simply disappearing into its miraculous loveliness, choosing Glen Canyon for eternity instead of an anticlimax.

Like Everett Ruess, Norman Nevills had been born in California in the buoyant, burgeoning years before World War I. Like him, he had discovered the canyonlands of the desert Southwest while he was still a youth, and similarly had explored them with a kind of indefatigable passion. Much like Ruess's, Nevills's life was shaped by the sere wilderness, and he too met an early death.

The only child of a prospector and petroleum man who had come to southern Utah in the midst of a brief oil boom, Nevills had completed two years at the College of the Pacific in Stockton before he joined his parents late in the twenties at their isolated lodge and trading post perched on a sandstone ledge 30 feet above the roiling San Juan River in a settlement with next to no population but the memorable name of Mexican Hat—one borrowed from a rust-colored, sombrero-shaped pinnacle a couple of miles upriver. Immediately enraptured by the strange surroundings, Nevills decided to stay, lending his parents a hand at the lodge, doing odd jobs for the U.S. Geological Survey crews at work in the area, learning to climb, learning to fly an airplane, eager too to learn something about the river that relentlessly swept past the lodge and into the deepening canyon.

In October 1933, twenty-four-year-old Nevills persuaded a young woman he had met only recently at a dance in the nearby town of Monticello to marry him, and early the following spring the two took a honeymoon trip down the San Juan in a boat Norm had built out of lumber requisitioned from an outhouse and a horse trough, with oars fashioned from the rods of an abandoned oil pump. The 60-mile trip to the mouth of Copper Canyon helped forge an enduring bond between Norm, his bride, Doris, still only nineteen, and that desert landscape, and it convinced the two of them that the river would be their future and their fortune.

By the summer of 1938, Norm had designed and built three wide-beamed boats out of stout marine plywood, calling them "cataract boats" in allusion to the punishment he was sure they could take (although local skeptics began to call them "sadirons" since they resembled the old clothes irons that were heated atop wood-burning stoves), and he had rounded up several paying members of a major expedition as well. By now, Norm and Doris had gotten to know the San Juan intimately, traveling from Mexican Hat to Lee's Ferry several times each spring and summer. But if their river-running avocation was ever going to amount to much in the way of a livelihood, they knew, they were going to have to mount longer, more daring trips that would attract adventurers from around the country.

There were only four fares on the Nevills's first commercial expedition—a run from Green River, Utah, to Lake Mead—and although they had taken in only $1,000 in payment for many weeks' work and worry, Norm pronounced the long trip highly successful. Passengers Elzada Clover and Lois Jotter, two University of Michigan scientists, had become the first women to ride the river through all of its major canyons; the cataract boats had performed well; and a bona fide company had been born, Nevills Expedition, the first recreational river business in the desert Southwest.

By the summer of 1940, another major trip had been mounted, this one beginning at Green River, Wyoming, exceeding the length of the earlier expedition by more than 300 miles. In addition to

Nevills, his wife, Doris, two boatmen, and three customers from back east, this summer's adventurers included Barry Goldwater, the thirty-one-year-old head of his family's Phoenix-based dry-goods chain. It was July in Glen Canyon; its walls seemed to hold the heat like an oven, and according to Goldwater, "you picked up a good idea of what the inside of a weenie must feel like." Yet despite temperatures that soared to 110 degrees, and a chronic knee injury that limited his ability to hike side canyons, Goldwater was enthralled by the canyon's grandeur. And although he was still nine years away from beginning his political career as a member of the Phoenix city council, his fierce chauvinism for his native state was evidenced when he scratched this inscription into a sheer wall just downstream of Warm Creek Canyon:

ARIZONA

WELCOMES | UTAH

YOU

Writing in his journal at a camp at that border crossing that night, Goldwater noted facetiously that now there was "a more bracing quality in the air; a clearer, bluer sky; a more buoyant note in the song of the birds; a snap and sparkle in the air that only Arizona has, and I said to myself, without reference to a map, that we were now *home.*"

It was August 23, 1940, when the Nevills Expedition boats finally floated into the upper reaches of Lake Mead—sixty-four days since the travelers first had pushed into the stream in far-away Wyoming. This time around, Norm and Doris had collected $3,900 for their efforts, and a rough guess was that Doris and the six customers had become numbers 69 through 75 on the list of people who by now had run the great canyons of the Colorado in the seventy years since Powell first had done so.

Wallace Stegner, a Stanford University English professor, was at work on a biography of John Wesley Powell in 1947 when he and his wife, Mary, joined ten other landlubber tourists for a week's sojourn in four of Norm Nevills's cataract boats, a kind of research trip during which Stegner hoped to get a feel for the still-little-known chasms that Powell and his men first had explored almost eighty years before, to gain some sense of the hardships they endured and the delights they encountered.

Near the end of the trip, sitting at the edge of the river following a 13-mile trek to Rainbow Bridge and back, with the setting sun turning the bald Navajo domes to rust, to red, to crimson, and with the catfish virtually begging to be yanked from the big brown river, Stegner opined that he just might stay behind when the rest of the group pushed on, to write his biography of Powell in this place both he and the major had taken very kindly to, it seeming to him then that "this was the way things were when the world was young." Norm Nevills assured the writer that he understood exactly what he meant, and he enthusiastically offered to resupply him from time to time if he took up residence beside the river, but by the next morning Stegner reluctantly had decided to climb back in a boat and float out of the storied and spectacular Glen.

By the time Wally and Mary Stegner joined Nevills and his ten-year-old daughter Joan (Doris was at home in Mexican Hat with their second daughter, Sandy) in San Juan and Glen canyons, Nevills had made the complete run of the Colorado seven times—more than anyone else had ever done. Nevills Expedition had become a going, growing concern, and Norm himself was becoming a kind of legend, his ebullient love for the canyons and the rivers helping to make him the most renowned river man in North America, if not the world. Unlike Powell, for whom the canyons had been an avenue of science and exploration, or men like Stanton, for whom they were a path to possible riches, Nevills was more emotionally akin to the wealthy Julius Stone, who had been captivated by the canyons simply because of the "spirit of the wilderness" they imparted, the charismatic man from Mexican Hat a kind of gregarious Everett Ruess delightfully alive in a desert landscape.

But on September 19, 1949, as Norm and Doris took off in their small plane from the airstrip at Mexican Hat, bound for a short holiday—their two daughters standing beside the runway, waving good-bye—the couple's Piper Cub crashed. Their bodies were cremated, their ashes spread across the canyons.

As the decade of the fifties opened, river running—still a strange and suspect enterprise to the vast majority of Americans—was steadily gaining interest among a small but fiercely committed group of boatpeople. Army-surplus assault rafts made of rayon or neoprene were readily available for fifty dollars or less, and the tough-skinned, stable, and forgiving boats were perfect craft for the canyon-bound desert rivers. The mountain rivers of Idaho, Oregon, and California were undeniably challenging, maelstroms of granite and growling whitewater; they were lovely to look at and delightful to travel down, but somehow, it was the desert rivers that captured the essence of this addictive new outdoor pursuit. The Southwestern rivers were cut into astonishing gorges; they alternately languished so slowly they seemed to hate to be on their way, or roared like express trains down rapids so steep they were properly waterfalls. These rivers—the Green and the river once called the Grand, the crooked Colorado itself, tributary streams like the Yampa, Dolores, and San Juan, the Salt and the Gila in southern Arizona, the great Rio Grande bisecting New Mexico north to south—all were oxymoronic, wonderlands of water winding through country as dry as a sun-bleached bone, navigable highways in terrain so rough you otherwise couldn't be sure you could get there from here. The worlds of the desert rivers were an unending delight, and once you had floated a hundred miles or more you were hooked.

For the young men who had been protégés of Norm Nevills—his boatmen and river companions during the forties—there was no question that they would find ways to stay on the water once Nevills was gone. Some formed river running companies of their own, outfits that variously prospered or went bust, operated with great financial resolve as well as the most casual kind of attention. Others simply made the rivers their lives, unencumbered by the details of livelihood, happily able to avoid turning their passion for rivers into efforts to turn a profit.

There were Bus Hatch, Malcolm "Moki Mac" Ellingson, and Al Quist and his clan—boating from scattered bases in northern Utah; there were Frank Wright and Jim Rigg, who had taken over Nevills's company, renaming it Mexican Hat Expeditions; Kenny Ross was 30 miles upstream at Bluff; Kent Frost operated from a base in Monticello; Georgie White, a Los Angeleno dubbed "The Woman of the River," fashioned gigantic boats out of Army bridge-building pontoons, lashed assault rafts side by side, and began taking large groups of people down the Glen and the Grand on what she called "Share-the-Expense Plan River Trips"; and Art Greene had already abandoned downstream travel in favor of ferrying his passengers *upstream* in a screaming airboat so loud he had to issue cotton as ear stopples and notepads for conversation.

Greene was eighteen in 1913 when, traveling down from his home high in the San Juan Mountains in Telluride, Colorado, he first floated the San Juan River—getting a taste of the desert that he held as a kind of hunger until 1943, when he and wife, Ethyl, and their four children moved to the rim of Marble Canyon near Lee's Ferry to operate a small lodge, restaurant, and trading post. Rainbow Bridge, 70 miles upstream and accessible only by a six-mile hike from the river or a difficult 13-mile trail leading down off the slopes of Navajo Mountain, had become a popular tourist destination, and Greene was soon packing passengers, boats, and supplies overland to Hite, then floating Glen Canyon—with a visit to Rainbow Bridge en route—back to the lodge near Lee's Ferry. Experiments with a skiff powered by a 25-horsepower outboard convinced him that traveling upriver as well as down would make his logistical life a lot easier, and in 1948 he built the *Tseh Na-ni-ah-go Atin*, Navajo for "Trail to the Rock That Goes Over," a 22-foot flat-bottomed scow powered by a deafening deck-mounted 450-horsepower aircraft engine, its big airplane prop spinning fast enough through the air to scoot the boat swiftly over the surface of the water, and to lurch it across the occasional river-level sandbar as well when the water was low. With his infectious, toothless grin and his boundless enthusiasm for the Glen Canyon country, Greene could communicate successfully with his passengers despite the terrible noise on the upriver runs; and in the starkly different silence of Forbidding and Bridge canyons, as well as during the slow, motorless float back to Lee's Ferry, Greene would regale everyone with stories about the Navajos, about the discovery of the wondrous natural bridge, about what it was *like* to live and raise a family in the redrock back of beyond. Pulling the big boat ashore on the willow bar at the mouth of Music Temple, he would suggest that this might be a place he kind of remembered, suggest that the several passengers from Denver or Dayton or Dubuque wander up past the cottonwood trees to see if there was anything worth looking at, then bemusedly smoke a cigarette back at the boat until he heard their echoing shouts of delight.

If you limited your count to the lower Glen, Art Greene certainly traversed it more times than anyone else, but if you tallied instead the number of trips through all of Glen Canyon, from Hite to Lee's Ferry at first, then Hite to the Kane Creek landing once the Bureau of Reclamation blocked travel farther downstream, the award for the most recurrent Glen Canyon traveler surely would have gone to one Kenneth Sleight, native of cosmopolitan Paris, Idaho, hard by the Utah border, a Mormon town of 500 or so hearty Saints, a farming community where ditches and dikes and dams were fundamental facts of life. When Ken Sleight ventured south to the University of Utah in 1947, he had known he well could end up an irrigator himself, but it was the notion that he just might become a park ranger that propelled him to study geology and to start making weekend trips south to the geologist's paradise in the southern tier of the state.

Before he left Utah for two years in 1951—on a mission to Korea for the U.S. Army instead of the Mormon church—Sleight joined friends for a trip down the Canyon of the Lodore on the Green, a brief journey that was nothing less than a revelation to him of the sublime world of the rivers. With the army waiting, and with a new wife in tow, Sleight quickly mounted his own river expedition down in the Arizona border country he had become enamored of, floating in a yellow vinyl raft of decidedly dubious construction from Hite to Lee's Ferry, seeing Glen Canyon for the first time, and forswearing while he was in its depths every ambition other than to figure out a way somehow to stay on the river.

Back in Utah in 1953, Sleight bought a few surplus assault barges for forty-five dollars apiece, painted "Ken Sleight Expeditions" on their fat black tubes, got back to work on a bachelor's degree (this time in business), and became an entrepreneur. For the first few summers, Sleight's principal river passengers were kids, members of Mormon youth groups, willing to be fed weenies and beans for a week in exchange for water fights, mud fights, sunburns, rock scrapes, and more fun than they had ever had. Glen Canyon was a perfect place for youngsters, a straight-cliffed amusement park where there was no entrance fee and all the rides were free, and Sleight might have decided to specialize in the youth-group market except for the fact that, despite his Spartan menus, he figured he was losing $10,000 a year.

By the time he had moved his operation south to the town of

Richfield on the western fringe of the canyonlands, then on to the town of Escalante in its redrock heart, Sleight was surviving, if not necessarily prospering, by running the river in Glen Canyon from April to September and heading up horse-packing trips into the canyons in the months before and after. Trailering his boats behind him, using the beds of pickups to transport dust-choked passengers, Sleight and his itinerant boatmen would begin each week with a bone-jarring drive down North Wash, a journey sometimes punctuated by burying an axle or two, then would spend six rather unstructured days guiding their group down the river, needing to average only 25 miles a day, dawdling, relaxing carefree on the current between hikes into Hidden Passage and Music Temple, to Rainbow Bridge, Gregory Arch, and the Cathedral in the Desert, poking into dozens of canyons on dozens of trips, before disembarking at Lee's Ferry, where the trucks would await them and ultimately bounce them back to a resupply and passenger pickup in Richfield or Escalante, Torrey or Hanskville, and on down the rutted road in North Wash to begin the cycle again.

Still fascinated with the place despite the demands of his business, and even after more trips through the canyon than he could count, Sleight liked to camp on sandbars where he had never camped, to search for Anasazi sites still unknown to him, to wander each trip into side canyons that somehow he had never seen before. Short, mischievous, gregarious, and sweet-tempered until you crossed him—an epic beer drinker in the great jack-Mormon tradition—Sleight was suited to his chosen field, suited to its pace and unpredictability, always at home away from home, always at home on the moving water, and at first it had seemed utterly impossible when he heard that plans were afoot to build a dam near the mouth of the canyon. Glen Canyon was a natural wonderland, a national treasure, and no one in his right mind—even the beaver-minded boys at the Bureau—would ever decide to destroy it, Sleight assured himself.

As a youngster in southern Idaho back in the Depression days, Ken Sleight had liked dams, he'd liked the *idea* of dams, but this Glen Canyon business didn't make any sense. It was one thing to dam a stream to send its water into fields, to conserve the water—that was what conservation was all about, as far as he had been concerned. But it was absolute insanity to drown nearly 200 miles of canyon just to generate a little electricity, just to make sure you could send water on downstream, a direction in which it so far had exhibited no trouble in traveling. It was such a ludicrous notion that surely it would never happen, and for a while he figured the rumors weren't worth too much in the way of worry.

Sleight, the canyonlands businessman—president now of the Escalante Chamber of Commerce—Sleight, the peripatetic desert rat, had joined his competitors early in the 1950s in forming the Western River Guides Association, a loose confederation of laissez-faire river men and women designed to promote their recreational offerings and to protect their commercial interests. As the battle for Echo Park had begun to mount in 1953, the Western River Guides had been approached with a plea for their support by a variety of conservationists active in the struggle to prevent the construction of a high dam in Dinosaur, and the small, motley, geographically scattered group—most of them sartorially complete in sand-caked sneakers, grease-stained shorts, and straw cowboy hats that had long since begun to crumble—had wanted to offer what limited help it could. Yet there was something that was increasingly troubling WRGA members such as Sleight: for some reason, the conservationists weren't trying to upset the whole stinkpot of the Colorado River Storage Project. Their opposition was aimed solely at potential dams inside Dinosaur. They had expressed mild concerns about how a dam in Glen Canyon might affect little Rainbow Bridge National Monument, but aside from those, the conservationists seemed strangely willing to let Glen Canyon go under. Sleight knew how important Dinosaur was; he first had discovered the wondrous desert water for himself in the Canyon of the Lodore, after all, but good Christ, was surrendering Glen Canyon the only way to save it?

In an effort to focus some attention on Glen Canyon, Sleight, a few of his fellow outfitters, a contingent of individuals who variously had gotten to know the canyon, and a smattering of outdoor activists at the University of Utah formed the Friends of Glen Canyon, a group with a name and a clearly understood objective but next to no idea of how to begin to pursue it. From the Friends' perspective, most of their potential allies already had been hoodwinked by the Bureau. Virtually every politician in the western United States already thought Glen Canyon belonged under water; so did the chambers of commerce, civic clubs, and church groups; the wildlife organizations—the goddamned hook and bullet boys—were thrilled by the prospect of a shiny, slackwater lake; and the conservation groups were trying to appear reasonable in the midst of demonstrating their surprising clout, supporting so-called *responsible* reclamation even when it was aimed at destroying a unique and wondrous place.

Working out of Salt Lake City with little money and less in the way of political acumen, the Friends did their neophyte best to remind the conservationists and everyone else that Harold Ickes and FDR had come within an eyelash of making the Glen Canyon country the largest national monument in the nation little more than a decade earlier; they tried to make a case for the canyon's important history—from the Anasazi to Dominguez and Escalante, from the Hole-in-the-Rock pioneers to Powell and the gold-crazed prospectors—as well as its stunning beauty; they even argued that the upper basin should have a variety of smaller water-storage facilities located in the upstream valleys where the fields and farms and towns were, instead of a massive reservoir deep in the empty desert, little more than a giant evaporation tank just 15 miles from the boundary of the water-gulping lower basin. But their arguments were unpersuasive, their voices of protest too distant and too soft, the bizarrely combined opposition of the dam builders and the conservationists too great. Glen Canyon got its death sentence in April 1956, just as a new season on the water was getting started.

Ken Sleight, Al Quist, Moki Mac, and Georgie White floated friends and paying acquaintances through Glen Canyon that summer just as they had done in the summers before, but now the canyon's beauty somehow seemed heightened by the certainty that it was fleeting, about to be destroyed. All of Glen Canyon was ephemeral now, and each trip was tinged with tragedy. It would be eight years, perhaps a decade, before the waters would rise, the willows and cottonwoods would go under, the deer and bobcats and coyotes would scurry high into the side canyons, but it would happen; and at night on the sweeping sandbars by the light of a succession of campfires, instead of telling the story he loved to tell about the time in a drenching afternoon storm he had seen literally hundreds of waterfalls cascade off the slickrock rims into the canyon below, Ken Sleight would speak with an increasingly bitter voice, his words increasingly angry as he described to his tourists what inexorably was on its way.

Sleight and his companion river rats—the several mainstays of the Western River Guides Association—were on the river again in 1957, floating through the canyon that now was a terminal case, but perhaps fittingly, it was twenty-year-old Joan Nevills Stavely, elder daughter of Norm and Doris Nevills, who—in the company of her husband, Gaylord Stavely, and Jesse Jennings, the archaeologist heading the Glen Canyon salvage project—made the last trip through all of the Glen. With her husband, Joan recently had returned to Mexican Hat and, as her parents had done two decades earlier, started a little river-running company, this one called Canyoneers. The salvage investigations were getting under way in earnest that summer, and with the Bureau of Reclamation's blessing, Jennings was making a survey of the dam site—where spoil from the west diversion tunnel was already narrowing the river's channel—as the Canyoneers' boats (newfangled neoprene things now instead of Joan's father's fabled cataracts) rounded the bend at the mouth of Wahweap Creek and flowed through that high, straight-walled section of the canyon for the final time on June 4, 1957.

Thereafter, the trip downriver from Hite would run 118 miles instead of the 155 miles it previously had, the float from Mexican Hat just 144 miles instead of 180 or so. Never again would sun-scorched, sand-caked boaters disembark at Lee's Ferry at the end of a trip through the length of the Glen, and for the next six years, commercial river runners piloting rubber surplus boats and Boy Scouts paddling inner tubes would encounter a large rectangular billboard standing in rocks on the river's eastern shore near the narrow mouth of Face Canyon. The first traffic sign in Glen Canyon, the first time in almost a century of canyon sojourns that travelers had been admonished not to follow the river's flow, it read:

ATTENTION

YOU ARE APPROACHING GLEN CANYON

DAM SITE   ALL BOATS MUST LEAVE

RIVER AT KANE CREEK LANDING ONE

MILE AHEAD ON RIGHT   ABSOLUTELY

NO BOATS ALLOWED IN

CONSTRUCTION ZONE

VIOLATORS WILL BE PROSECUTED

U.S. BUREAU OF RECLAMATION

Like Everett Ruess, who had always experienced things more profoundly when he knew they would not last, whose own life was short but intensely passionate, Glen Canyon began to take on an *importance,* a kind of terrible and tender beauty that it had never seemed to have before—even to those who knew it intimately—once its days were numbered. And the knowledge that Glen Canyon's grottoes, its sinuous side canyons and dazzling sandstone walls stained dark with desert varnish wouldn't last for another generation, would go under in less than a decade, made it a place many people now wanted to see—people who lived in nearby Flagstaff or St. George or Salt Lake City, who had always wanted to float through the canyon and who knew it was now or never, folks from farther away who increasingly had heard that river trips were something special and that one through Glen Canyon gave you good odds of coming home alive, who knew the name now because wasn't the government building a dam there? The commercial river men and women grew busier than they ever had been before, escorting through the canyon conservationists who more and more were hearing that this was the place that had been sacrificed on the altar of reclamation, history buffs who wanted to see the Moki ruins and the listing dredge and the dramatic steps cut into the Hole-in-the-Rock before they were gone, assorted mourners who simply wanted to say good-bye.

There were others as well—scores of them in the several years characterized by the aging Ike, by *Sputnik* and the pelvis of Elvis Presley—who went into Glen Canyon on their own, knowing little about it except that it was about to be taken from them, paddling craft that might have got them killed in any other canyon, delightedly on the bum in that awesome slickrock country—people like Ed Abbey, a thirty-two-year-old philosophy graduate from the University of New Mexico in Albuquerque and the author of two novels, a fellow who had seen the spires and hoodoos and crevice canyons before and had become mesmerized by them, a native of the Allegheny Mountains of Pennsylvania who now was determined to lay personal claim to the canyonlands. "We were desert mystics, my few friends and I," Abbey explained, "the kind who read maps as others read their holy books. . . . Anything small and insignificant on the map drew us with irresistible magnetism. Especially if it had a name like Dead Horse Point, or Wolf Hole, or Recapture Canyon, or Black Box, or Old Paria (abandoned), or Hole-in-the-Rock, or Paradox, or Cahone (Pinto Bean Capital of the World), or Mollie's Nipple, or Dirty Devil, or Pucker Pass, or Pete's Mesa. Or Dandy Crossing."

Abbey and friends had first crossed the Colorado at Dandy Crossing (Hite)—their derelict pickup truck bobbing across the current atop old Art Chaffin's ferry—on an outing early in the fifties. But by June of 1959, with Glen Canyon Dam under construction and with a million canyons still to be explored, Abbey and fellow New Mexican Ralph Newcomb decided it was time actually to get down into the labyrinth of the Glen, to get lost if they could manage it, to see the canyons from river level, to ride the red, runoff-swollen Colorado into the doomed canyon. Not since he was ten, and he and his brother had ridden a wooden box down Crooked Creek back in the Alleghenies, had Abbey been afloat, but in two tub-sized vinyl rafts (purchased from a drugstore en route from Albuquerque, almost as a kind of afterthought), the rafts lashed together to facilitate steering and conversation, Abbey and Newcomb paddled away from the landing at Hite purposely knowing little about what they would encounter downstream. Minus life jackets, but in cautious possession of a gas-station map for navigation, and with two weeks' worth of tinned food, they suddenly seemed wonderfully free. "I am fulfilling at last a dream of childhood," Abbey wrote, "and one as powerful as the erotic dreams of adolescence—*floating down the river.* Mark Twain, Major Powell, every man that has ever put forth on flowing water knows what I mean. . . . Cutting the bloody cord, that's what we feel, the delirious exhilaration of independence, a rebirth backward in time and into primeval liberty, into freedom in the most simple, literal, primitive meaning of the word, the only meaning that really counts."

For twelve days they were adrift in the twisting canyon, their boats taking on water in the river's riffles and sliding by side canyons that seemed to beg to be explored. Game-legged Newcomb caught schools of catfish, suckering the stupid fish with rancid salami, while Abbey explored miners' camps and the vaulted, cathedrallike canyons of the Escalante and hiked a succession of creek bottoms by the light of a slivered moon. Abbey climbed up to the rim at Hole-in-the-Rock to survey the surrounding wonderland; he hiked to Rainbow Bridge and beyond beneath the slopes of looming Navajo Mountain; he accidentally set fire to a willow thicket in a slot canyon alcove (when he too carelessly burned a bit of toilet paper); and together the two men drifted hour after hour in their banana-yellow boats down the obliging river before they finally saw the billboard they knew they would have to see, the declaration that they could proceed no farther, the government's announcement that it was about to destroy this place.

Although he never floated down Glen Canyon again, years later Abbey, by then the most outrageous and influential voice in the defense of the wild canyon country, would remember that trip at the close of a too-innocent decade as "a voyage like no other in my life, a continuous dream of marvels, wonders, splendors, that has haunted me ever since."

That Abbey and Newcomb had not taken a camera into the glories of the Glen with them in 1959 had made a simple kind of sense: how in the world could anyone reproduce in two dimensions the visual symphonies, the plays of light on stone, the splendid skies, the magic movement of water? Better not to try. Yet two renowned landscape photographers, a Californian named Philip Hyde and Eliot Porter, an eastern physician transplanted to New Mexico, were brave enough before long to try to fix on film a series of suggestions, of intimations, of how it looked, how it *was* in actuality. Hyde, a student of Ansel Adams, had learned his craft in the sylvan landscapes beside the Pacific, photographing first in black and white, and now increasingly in color, the coastal ranges, the Sierras and high Cascades, and when he first had set eyes and apertures on the Glen Canyon country in 1955 as a member of a Sierra Club float trip, he had been shocked yet still intrigued by what he saw. "I had to learn how to cope, both physically and photographically, with the heat, haze, and dryness," he explained. "I needed more time to digest what I saw in the arid lands. . . . In my memory of the river trip, nights on rocks radiating too much heat for sleeping are mingled with days of growing awe of the strange forms of this stone country. My awareness of water as a miracle was born in the shining trickles in canyon bottoms and the sudden springs that gushed out of rock as though piped through the waterbearing Navajo sandstone. These imprints went deep. This landscape took hold of me."

Hyde returned to Glen Canyon in 1961 and again in three more summers thereafter, floating the friendly, now familiar river, some-times hiking from the dry mesa lands down into the canyon's fringes, each time packing his tripods and box cameras in a kind of homage to the place, discovering that its abstract stone, its strange and brutal geometry, its bizarre juxtapositions of barrenness with water made it a photographer's dream, a place where you could encounter a whole lifetime of images from the vantage of a camera's viewfinder.

Living outside Santa Fe, New Mexico, Eliot Porter had heard for years that this Glen Canyon was nothing much—just rock walls and a languorous river. But when, in the autumn of 1960, he accepted an invitation to join friends for a week-long trip from Hite to Lee's Ferry, he was, as he put it, immediately "overwhelmed by the scenery—both in prospect and in description grossly underrated. The monumental structure of the towering walls in variety and color defied comprehension. So powerful was the impression, I didn't know where to look, what to focus on; and in my confusion, photographic opportunities slipped by. I always seemed to be photographing the wrong things—details of iodized rock while ignoring convoluted cliffs, or dominant canyon vistas when a sheet of water-stained wall was staring me in the face. There was so much and such variety; all so new and so young that I could not escape the impression I was witnessing a dynamic remolding of the earth itself, as indeed I was."

Dissatisfied with the photographs he had taken during that first fall visit to the canyon, Porter was determined to return and did so twice in the summer of 1961, both times accompanied by painter Georgia O'Keeffe, a friend and distant neighbor who lived in the New Mexico village of Abiquiu. Porter had exhibited photographs at the renowned 291 Gallery in New York City, operated by O'Keeffe's husband, Alfred Stieglitz, and both painter and photographer long had shared an interest in the visual drama of desert landscapes. When Porter described Glen Canyon to O'Keeffe during the winter following his first visit, she quickly demanded to join him on his return, the fact that she was nearly seventy-four deterring her not at all.

In August 1961, Porter, himself almost sixty, organized a week-long trip through the Glen that included ten friends and acquaintances, O'Keeffe among them, the *grande dame* of American painting wearing sneakers to help her negotiate the riverside rocks and boulders, an ocher sundress specially chosen to match the sandstone, and her trademark black, flat-brimmed fedora. O'Keeffe seemed to love the strange new experience of sleeping in a sleeping bag, choosing a spot for it each night that would give her a pleasing view of the dark shapes of the cliffs and the slivers of starry sky; she greatly enjoyed floating on the broad brown river and entering into the dimly lit, labyrinthine worlds of the side canyons; but she was perturbed that the pace of the trip allowed her no time to paint. The tall, bespectacled, and rather patrician Porter was able to take photographs en route, but painting demanded hours of time that the group wasn't able to spare, so O'Keeffe simply chagrined Porter into taking her back into the canyon as soon as her first trip was ended.

On the second sojourn, the two traveled by car to the Kane Creek landing—where all river runners now were required to disembark—and with the help of an outboard motor boated upstream to Music Temple, camping there and at Hidden Passage across the river, spending almost a week watching the river roll by, photographing, sketching, painting, idly observing the narrow sky through openings in the soaring rocks. But although O'Keeffe made a series of Glen Canyon paintings once back home in Abiquiu—abstractions of smooth water torn by the blade of an oar, of flat water fenced by vertical rock— and although she made half a dozen more trips to the canyon in subsequent summers, the only work derived from her experiences in Glen Canyon that she was ever well enough satisfied with to exhibit was a few paintings of the smooth and brilliantly black river rocks she collected along the shore.

For his part, Porter believed he was slowly learning how to photograph the Glen, increasingly shunning vistas in favor of the visual delights of detail—branches of redbud trees, reflecting pools, lichens and mosses on slabs of rock, even the charred willow twigs that were the legacy of Ed Abbey's earlier journey. And he was becoming convinced that the photographs he initially had made solely as aesthetic statements also possessed a didactic dimension; they sadly were instructive of the incredible beauty humankind was willing to destroy; they were terrible lessons of folly.

It was ironic—and personally dismaying—that it was Porter's brother-in-law, Mike Straus, commissioner of Reclamation in the forties and early fifties, who had set in motion the complex of planning and legislation that now had resulted in the raising of a dam downstream from Music Temple, downstream from the phenomenal side canyons of the Escalante, downstream from the Cathedral in the Desert at the head of Clear Creek Canyon—the single most spectacular place in all the canyons as far as Porter was concerned—a dam that would inundate them all. Visiting Glen Canyon twice in 1962, and four more times during the years when the water inexorably began to rise, Porter became increasingly outraged. With each return, he hatefully observed the climbing, spreading reservoir that was "inundating the sparkling river, swallowing its luminous cliffs and tapestried walls, and extinguishing far into the long, dim, distant future everything that gave it life. As the waters creep into the side canyons," he wrote, "enveloping one by one their mirroring pools, drowning their bright flowers, backing up their clear sweet springs with stale flood water, a fine opaque silt settles over all, covering rocks and trees alike with a gray slimy ooze. Darkness pervades the canyons. Death and the thickening, umbrageous gloom take over where life and shimmering light were the glory of the river."

Porter wanted his photographs of the Glen to be appreciated as art, but in addition to that, he wondered if they might not also serve as memorials, as elegies to that wondrous place. A book of Porter's intimate, somehow sensuous photographs of the hardwood forests of New England was about to be published by the Sierra Club, accompanied by excerpts from the writings of Henry David Thoreau and titled *In Wildness Is the Preservation of the World*, part of a new "exhibit format" series of books edited by David Brower—large, lavish books, this one featuring Porter's photographs beautifully printed on lacquered Kromekote paper. Never before had the technology existed to reproduce such color, and never before had a publisher gambled that people would be willing to spend as much as twenty-five dollars on books filled with pictures of nothing but wild country.

Porter decided he wanted Brower to have a look at his Glen Canyon work as well; it was starkly different from the work in *In Wildness*, pushing the possibilities of using detail and color to create a kind of abstraction even farther than the New England work had done, and unlike the four current books in the exhibit-format series, the Glen Canyon photographs couldn't be published as a kind of celebration. A portfolio of photographs from the Glen would be nothing less than a kind of tombstone, Porter knew, but nonetheless, he thought Brower should see them.

Way back in the spring of 1956, with dams in Dinosaur National Monument successfully deleted from the final Colorado River Storage Project legislation, Dave Brower desperately had wanted to keep up the fight. The dam now planned for Glen Canyon was more than simply suspect; it would be wasteful of water and ruinous of wilderness, he recently had come to believe; it epitomized all that was wrong with the CRSP, yet many of his colleagues still were certain that it was only the alternative offered by Glen Canyon that had allowed the dam builders and their brethren to abandon Echo Park.

Several senators and congressmen on Capitol Hill who had always been adamantly opposed to the whole of the CRSP had urged the conservationists to stay in the fray once agreement had been reached on deleting Echo Park. They knew support for the entire project, particularly in the House, was tenuous and potentially subject to collapse, particularly if the conservationists could sustain their remarkably successful lobbying in opposition. But as far as the National Parks Association was concerned, it had won a complete victory and now it wanted no further involvement with the machinations of Congress; members of the Wilderness Society, who always had been leery of opposing all of the CRSP, were bent on continuing to appear responsible by accepting the compromise, their leader, Howard Zahniser, now wanting specifically to move on to efforts in support of proposed legislation that would create a variety of federally protected wilderness areas; and the board of directors of the Sierra Club, still uncertain of their organization's new national role, one foisted on them—like it or not—by Brower, had ordered their executive director to come home to California, where work awaited him in getting ready for that summer's high trips into the Sierras.

Brower's instincts told him to go home immediately but, once there, to call an emergency meeting of the board to discuss what giving up the CRSP fight would mean, to try to convince his colleagues that even with just the Sierra Club leading the scattered opposition, the CRSP actually could be defeated. But Brower was apprehensive; he was not yet sure of his own role within his organization, uncertain about how much national fighting in support of conservation causes he personally wanted to undertake, how high a profile the little Sierra Club should dare to assume. He was deflated by the board's instructions, but he accepted them, and as the Green flowed unfettered through Echo Park that summer, and as construction began in Glen Canyon in the fall, David Brower turned his attention away from the desert canyons and back to the high Sierras.

By 1959, the Sierra Club's board of directors had grown cautious enough about the organization's role in future struggles (despite the fact Brower and company had been impressive indeed during the club's first major fracas in almost forty years) that it passed a resolution stipulating that henceforth the club would "offer cooperation with public officers in development of plans and policies," "criticize . . . in an objective and constructive manner," and never question "the motives, integrity, or competence of an official or bureau." In other words, as Brower viewed the resolution, the club now wanted, now *demanded* to muzzle itself. But if the Sierra Club no longer could openly argue with the Bureau of Reclamation or the Forest Service or Congress itself, if Brower no longer could testify or give speeches unless they were somehow "constructive," there was at least one other avenue available for getting the club's conservation message out.

Brower had worked for the University of California Press before assuming the job of executive director of the Sierra Club in 1952, and during that era also had served as editor of the *Sierra Club Bulletin*, the organization's bimonthly journal. His background in publishing convinced him that there was a growing market for books about the natural world—big, beautiful books, long on eloquent photography and necessarily short on dry discussion of conservation issues. What Brower wanted to do was to publish books that effectively brought the wilderness to readers, showed them images of the wonderful wild country still extant in North America, then somehow made them fervently want to take part in its defense. *This Is Dinosaur* had been extremely successful in alerting people to what would be lost if a dam were built in Echo Park; surely other books—better books—could prove as potent.

The hierarchy of the club was supportive when Brower outlined plans for the first book in his imagined series, a collection of photographs by board member Ansel Adams accompanied by a text by Nancy Newhall asserting that the current generations of Americans were only brief tenants of their lands and that they should preserve them carefully for their children and grandchildren. Co-published—and substantially financed—by McGraw-Edison early in 1960, *This Is the American Earth* was a critical and commercial success. Brower quickly followed with *Words of the Earth*, text and photos by Cedric Wright, then *These We Inherit: The Parklands of America*, another book based on Adams's photographs. By the time the club's executive director was making plans for Eliot Porter's *In

Wildness Is the Preservation of the World, however, its board of directors was growing wary again.

Since its inception seventy years before, the conservation-minded Sierra Club had prided itself on its *conservatism*. Members, each one joining the club only when sponsored by another member, tended to be male, Republican, in business or the professions, and to believe that the best way to influence public policy was to have a quiet chat with a congressman over an after-dinner drink. Sierra Club members long had tended to belong to cloistered, elite organizations such as San Francisco's Bohemian Club, and as late as 1945, when the southern California chapter of the Sierra Club refused to accept a black man as a member, no one in the parent organization had raised so much as an eyebrow.

Initially, Dave Brower had comfortably entered into this milieu. He had been a twenty-one-year-old anti-FDR Republican when he joined the club in 1933, encouraged by Ansel Adams, whom he had met on a trail in the central Sierras, and sponsored for membership by fellow climber Richard Leonard. Brower had looked on dams and the like as part of the responsible stewardship of the earth, and surrounded now by people who shared his enthusiasm for the mountains, he had found a kind of community, a home. A dropout from the University of California and an increasingly dissatisfied member of the Presbyterian church, Brower had been something of a loner earlier in his youth. As a child, he had been taunted by schoolmates, and he grew up leery of groups, spending many solitary hours collecting butterflies in the Berkeley hills, then traveling throughout the Sierras as a teenager, teaching himself to climb.

Once in the company of people who shared his devotion to alpine country, to climbing and skiing and waxing theological about the meaning of nature, Brower grew steadily more self-confident, more outgoing, and after service with the army's elite Tenth Mountain Division in World War II—earning three battle stars and a Bronze Star in the process—he returned to California with a renewed and decidedly more activist notion of how the club should go about protecting the mountains from which it derived its name. With the support of Adams, and of his friend and fellow climber Dick Leonard, Brower was elected to the club's board of directors, and the three then worked hard to steer the club's focus away from the socializing that was epitomized by the southern California chapter (the "Booze Brothers," Brower called the Los Angeles arm of the organization) toward a decidedly more sober defense of the mountains against the many pressures posed by California's growth. When Brower left his job as an editor at the University of California Press in 1952 to become the club's first full-time paid staff member, club membership—now open to anyone—was about 7,000 people. Eight years later, turned into something of a *club célèbre* by its role in the battle for Echo Park, and with the increasing notoriety Brower's exhibit-format books currently were bringing it, the Sierra Club listed more than 30,000 members.

Yet despite the renown and the several victories Brower had brought to the club, many board members openly worried that he was too much of a maverick. Brower was too radical, too unwilling to compromise when it came to opposing developments that would encroach upon pristine places; and now that he appeared to be trying to turn the club into a nonprofit publishing company, some were worried that he literally would bankrupt the organization. In order to keep him in check, the board of directors established a publications committee, a ten-person oversight body whose members included Adams, Wallace Stegner, and August Frugé, Brower's former boss at the university press, a committee designed to keep him mindful of what the board viewed as the club's rather more narrowly defined interests.

The first three books in the exhibit-format series—all of them produced, designed, and edited by Brower—had been successes, although to varying degrees, and with their black-and-white photography, and the copublishing arrangements Brower had made with commercial publishing houses, each one had cost the club a fairly modest amount. But when in 1961 Brower proposed a monumental new book containing state-of-the-art *color* reproductions, a book that would cost a comparative fortune to produce, the debate among members of the publishing committee grew heated. Yes, Eliot Porter was probably the best photographer currently working in color, and certainly, the wisdom of Thoreau was enduring, but did the club really have any business publishing a book about *New England*? Could the club really afford to pour such a substantial portion of its resources into a single book?

With the help of a slim majority of sympathetic members of the publications committee and with a $20,000 grant and a $30,000 interest-free loan from Kenneth Bechtel—who with his brother Stephen headed the now giant heavy-construction firm that had helped build Hoover Dam—Brower finally got the go-ahead to publish *In Wildness Is the Preservation of the World*. And it was in the midst of producing that book that the ebullient administrator-publisher-conservationist, now turned fifty, first saw a separate portfolio of Porter's photographs—desert photographs which *had* to become part of a book of their own, Brower knew at once, photographs that left him feeling as though he'd been hit in the solar plexus by a very powerful fist.

By the time David Brower wrote the foreword to *The Place No One Knew: Glen Canyon on the Colorado River* in March 1963, he at long last personally had visited the canyon, had gotten to know it a little during the course of three separate trips. He had hiked to the Cathedral in the Desert and to Rainbow Bridge like a latter-day Everett Ruess, waded the creeks into Mystery and Moki canyons, sat in awe as Powell and Stone, Nevills and Stegner and Sleight had done in Music Temple; three times he had floated the slow and silty river, seen the canyon the way it was when the world was young, and he had become certain that *acquiescing,* allowing a dam to be built in that place, was the most terrible thing he had ever done.

More than anyone else, he now believed in retrospect, he had had an opportunity to save Glen Canyon, but because he had been willing to sacrifice it, to give it away before he had even seen what was being surrendered, it soon would begin to go under. Before long the dam would be done and the awful waters would rise. Brower's failure haunted him; it broke his heart, and some of his friends actually wondered whether in the midst of his depression he might consider suicide. But those who knew him better correctly assumed that he would do what he did instead—throw himself headlong into a new book, fully supported this time by the Sierra Club's publications committee, a book he hoped would be "as ageless as the canyon should have been," a book containing eighty of Porter's intimate images, photographs that this time didn't celebrate what was best in the remaining American wilderness, but rather that starkly and tragically exhibited what was about to drown.

8
TEN MILLION TONS OF MUD

The first film featured Alvin Tsi'najinii, headman of the outfit based in the dune-mounded country beyond Leche-E Rock, driving his single-horse buckboard out amid the saltbush. "One fine morning, the old Navajo tribesman left his hogan to jog across the desert to the big digging he'd been hearing about," intoned the narrator of *Canyon Conquest*, produced in 1959 for the Bureau of Reclamation—the kind of film you used to see in the sixth grade on days when you had a substitute teacher, the color footage purporting to follow the senior Navajo during a day spent watching the wondrous work going on in the canyon of Old Age River.

At the guard station near the rim, a young Navajo guide with a hard hat in hand took time to "fit the old fellow with the proper headgear," then proceeded to show him the wire-mesh footbridge that spanned the canyon, the swift monkey-slide that carried men into its depths, the wide mouths of the diversion tunnels, the keyways being cut to hold the high dam—which Tsi'-najinii still couldn't quite imagine—

and most impressively for him, high-scalers hanging from ropes, trimming loose rock away from the walls with long steel pry-bars. At the end of the government movie, the old man was able to walk out onto the newly completed highway bridge that linked the opposing rims and pronounce it good. Bridges he understood and approved of, and if the dam was still a mystery to Tsi'najinii, the narrator offered the assurance that at least he knew it would mean "energy—pulsing, flashing power . . . life for the swarming population as yet unborn."

These public-relations films weren't the sort of thing the Bureau ordered up on every piddling irrigation project it had under way; even the Reclamation budget had its limits, after all. But in the case of Glen Canyon, Commissioner Floyd Dominy was convinced that a publicity barrage was absolutely essential. Eastern congressmen had groused about the Colorado River Storage Project from the very beginning and had whined even after it won approval; newspapers had gotten cocky, calling it pork-barrel, calling it reckless, claiming Glen Canyon couldn't hold water; and now the conservationists— who had *won* the Echo Park fight, for Christ's sake—were beginning to make noise about what a lousy thing it was to plug Glen Canyon and fill it with a reservoir's pool. Yet Dominy was certain that if he got the truth out, if school kids and Kiwanis clubs, Rotarians and Lions and chambers of commerce across the country got the straight stuff about what a boon, what a *blessing* this dam would be, the carping would quiet quickly and Reclamation could get back to the business of building things.

By the time the Glen Canyon strike finally was settled at the beginning of 1960, Dominy had made sure that Lem Wylie was keeping the lines of public communication open, the project engineer writing rote articles for trade journals like *Western Construction* and glossy tourist magazines like *Arizona Highways,* as well as meeting with every elected official who ever wandered within a hundred miles of the Kaibito Plateau. Special-services officer Bud Rusho, the lanky young man from Montrose, Colorado, who had come on board to coordinate such showcasing, was keeping dozens of VIP hardhats at the ready, demonstrating to engineers from India, water men from Mexico, car dealers from Phoenix—anyone and everyone who could muster a group and a preparatory phone call—what was under way down in the hole. Al Turner was taking thousands of black-and-white Bureau photographs, sometimes *hundreds* during the span of a single day, each one catalogued and filed for future reference. Jean Duffy, the pretty, petite Page reporter whom, like Janet Cutler, Dominy liked to shower with occasional scoops, was stringing for the Salt Lake and Phoenix papers. Harry Gilleland— nominally the office-services supervisor—an effervescent Arkansan who had never met a stranger, was making regular treks to the Flagstaff train station to meet and escort Louis Puls, the dam's chief designer, back to Page. A Denver concrete engineer who had nothing to do with public relations per se, Puls nonetheless was a crucial component in Dominy's efforts to see that the project played well to the public, the commissioner understanding that a dam that wouldn't stand, that couldn't hold water, would make the worst kind of news.

Floyd Dominy never had expressed any real concern about the structural integrity of the dam that would block Glen Canyon, nor did the porous sandstone cause him any loss of sleep. He knew that despite the broadsides it had begun to take, the Bureau still employed the best and brightest engineering minds in the nation. It was true that a few young electrical engineers were defecting to NASA and to the several private aerospace companies that were assisting that agency in its efforts to get an American into space, but the structural fellows were staying steadfast; they were men for whom nothing seemed so exciting, so challenging as stopping a river and turning it forever into a lake, and Dominy had the best boys in the business working for him.

At the top of the list was Louis Puls, part of the old school, a man born before the turn of the century and reared in a time when engineers seemed as heroic as those seven Mercury astronauts did, a man still given at the dawn of the 1960s to wearing jodhpurs, polished Wellingtons, and Pendleton plaid shirts, to wearing a rakish fedora when the situation in the field didn't demand a hard hat, to smoking expensive cigars or sometimes a corncob pipe that put you in mind of MacArthur.

Puls had been assigned the job of designing Glen Canyon Dam long before it was authorized, before anyone at Reclamation could be sure that the canyon would call for a concrete dam. But concrete seemed a likely enough possibility that Puls and his underlings—a whole cadre of fellows in the concrete design division—were put to work drafting preliminary plans back in the early fifties. And what a smashing prospect it was: the width of the canyon and the projected storage demands of the reservoir meant that this dam could well contain half again the mass of Hoover Dam. If it reached over 700 feet in height from bedrock foundation to crest—which it certainly appeared it would—it would be at least the third highest dam in the world, perhaps the second, depending on whether it inched over Hoover's height. And given the low plasticity and correspondingly low strength of the Navajo sandstone, the dam's foundation and abutment pressures would have to be minimal and remarkably consistent throughout the structure to keep it stable. Yes, you could build a dam in Glen Canyon, but you couldn't build just any dam.

The first order of business was to determine whether the dam should be a gravity-type structure—a conventional pyramidal shape, wide at the bottom and narrow at the top—which would direct the waterload vertically down to the dam's foundation, or an arch-type structure—thin top to bottom, arcing concavely into the waterload, a design that would transfer the load into the abutments in the canyon walls. Although a gravity dam would demand far more mass than an arch dam—and mass meant money—it would be much more straightforward to design and build, and it would allow for an overflow spillway down its downstream face, eliminating the need to tunnel spillways through the abutments. But an overflow spillway would require that the power plant—this was going to be a *big* power-producer, after all—be located somewhere underground or within the wall on one side of the canyon, a location that would not be simple or inexpensive given the sandstone.

An arch dam would require far less in the way of concrete, and that was in its favor. It would require tunnel spillways through the abutments, but they in turn would allow the power plant to be built atop the riverbed at the toe of the dam. The only real difficulty with an arch dam in the context of Glen Canyon seemed to be that the arch would have to be awfully thick. Contemporary arch dams were often called eggshells; the action of even the thinnest arch was able to transfer the force of water pressing against its concave face into thrust carried by the abutments. But at the points where the arch and abutments joined, the stresses were tremendous, far greater than the brittle sandstone could endure. If an arch dam were to be built at Glen Canyon, it would have to be a very thick arch indeed, thick enough to spread the stresses over a wide enough area to reduce them to safe and sustainable levels.

Based on the twenty-one preliminary designs that preceded it, based on hundreds of test holes drilled at the Mile 15 site, on a battery of investigations into the nature of this Navajo sandstone, on grouting and concrete compression tests, arch and cantilever stress tests, arch and cantilever shear tests, trial load analyses and a hundred separate considerations, the ultimate choice of Puls and his people was Dam Study A-22, completed in June 1957, a hybrid of the two basic dam types, a design not dissimilar from Hoover Dam, planned nearly forty years before.

Glen Canyon Dam would be a gravity dam, its base 300 feet thick, its crest only 25 feet; but it would be an arch dam as well, curving 1,560 feet—more than a quarter of a mile—from keyway to keyway, its radius of curvature 900 feet, a sharp enough arch to further lessen the stress at the abutments, but shallow enough still to deliver the stress laterally into the walls. At the abutments, the dam and its full reservoir would create stresses no greater than 750 pounds per square inch, and in the interior of the dam itself, they wouldn't exceed 1,000 pounds per inch—the dam wouldn't suffer from its own weight, and the canyon hardly would know it was there. The downstream face of the dam would rise at a steep and continuous pitch of roughly 3 to 1, vertical to horizontal, while the upstream face would rise vertically for half its 710-feet height, then curve slightly downstream for the rest of the distance to the crest in order to reduce cantilever stresses on the downstream face. On blueprints, the top view of the dam made it look like the rind of a slice of watermelon; the side view resembled a traffic cone.

Like Hoover Dam before it, Glen Canyon Dam would be built out of blocks—in this case two adjoining rows of twenty-four blocks each, plus four wedge blocks meeting the sandstone in the keyways at either end—the biggest blocks 60 by 210 feet, all of them rising in lifts 7½ feet high, their joints ultimately sealed with grout to make the structure monolithic. Like Hoover Dam, this one would be laced with cooling pipes to control the temperature of the concrete as it cured; like Hoover, spillway tunnels would plunge down and around the dam through the canyon walls. But unlike its companion dam 355 miles downstream, where intake towers were mounted on the canyon walls and the penstocks were tunneled through them, Glen Canyon Dam's eight penstocks would be integral to the dam itself, their intakes mounted midway up the upstream face of the dam, the 15-foot-diameter penstocks falling 330 feet through the concrete core of the dam to reach the downstream power plant.

It wasn't going to be a revolutionary dam; its technology already was well established. And its status as the third highest dam in the world would last only until the Swiss finished a high dam they were building contemporaneously, but Louis Puls was well aware that you didn't get to design and build a dam of this magnitude every Monday morning; one, maybe two of these in a *career* really was all you could hope for, and Puls savored every minute—the long hours bent over drafting tables and huddled around conference tables at the Bureau's dilapidated offices in an abandoned army depot outside Denver, the regular trips down into the redrock country to see the site, then finally to see the dam rise. Afraid to fly, unwilling to take advantage of Reclamation's ready airplane, Puls would board a Pullman car at Denver's Union Station at 6:30 in the evening, arriving in Flagstaff the following morning. "Sonny," he would announce to Harry Gilleland, whose car was always parked beside the track at precisely the spot where the Denver Pullman would come to rest, "I believe we need some breakfast," and the two would dine on linen tablecloths at the Hotel Monte Vista before beginning the two-hour drive to Page.

Like all the high-level Bureau personnel, like the senators and congressmen, and Merritt-Chapman & Scott's own administrators, Puls would stay at the MCS guest house while he was in Page—a big, rambling, ranch-style structure built near the rim of the mesa on broad Navajo Drive, the guest house hosting some legendary evenings of entertainment and a well-lubricated conference or two in its time. But prior to the evening's conviviality, Puls would have to meet with Lem Wylie to get a feel for the flow of things. Vaud Larson, Wylie's assistant project construction engineer, a man involved with Glen Canyon since the Bureau began its first geological surveys back in the forties, would sit down with the two of them to discuss the latest change orders; and Norm Keefer, the chief field engineer, the man charged with making sure MCS's work was up to specs or better, would assess for Puls the job done to date. Louis Puls wasn't anyone's boss; he didn't have the kind of clout that could get you into hierarchical hot water, but he demanded deference and an unusual kind of courtesy because, well, he was part of the old school, and that dam that was climbing up out of the hole somehow belonged to him.

■■■■■■■■■■■■■■■■■■■■

As far as Merritt-Chapman & Scott Corporation was concerned, the work was proceeding apace now. By August 1961, little more than a year after they finally had begun to pour some mud, 2,200 men were working on the project, punching in on three shifts a day, five days a week. The highest block on the dam now reached 250 feet above the foundation, and, on average, crews were placing 300 cubic yards of concrete every hour—600 tons of it, enough to fill four boxcars. Yet because of the six-month strike, MCS now really would have to highball to get the dam completed on time, and Al Bacon, the project manager who had seen the company through the awful quiet of the strike, just hadn't been able to boost production to levels the company brass could live with. In late July, Bacon had been ousted, told he was through, and ushered out the door in the same day, replaced by Jim Irwin, a burly, blue-eyed heavy-construction veteran with a grin that reached each ear. Irwin, lately the project manager on Oregon's Cougar Dam, hadn't won the job as Mr. Congeniality, however. His single responsibility at Glen Canyon was to get still more concrete into the hole in a shorter span of time. "Our job is to get her up fast," Irwin had told his troops as he introduced himself. "We simply must place concrete faster than anyone ever poured concrete before." If all went well, the prime contract could be completed by March 1964, two and a half years hence, on schedule despite the strike, and Irwin and 2,000 other workers and their families could bail out of booming, brawling, hardscrabble, sandy Page and hightail it back to civilization.

At the moment, Katherine Pulsifer at the post office was estimating that there were 6,000 people in Page—about 120 employees of the Bureau of Reclamation plus their families, roughly as many merchants and teachers and assorted service personnel, the rest the transient construction people who would be gone the minute the work gave out. But for now, this was a town in a kind of controlled chaos. You never could find a place to park within three blocks of the post office, and you couldn't rent a box at any price. You simply stood in line at the general delivery window like everyone else, in just the same way as you queued up at the bank and the spanking new Babbitt's store—a true supermarket at last. The water supply wasn't much improved, largely because of the enormous demand on the system, but the school—the finest facility in Coconino County, everyone said—was a going concern now, serving 1,500 students. Two hundred homes owned by the Bureau of Reclamation were occupied on the west side of Seventh; a hundred more belonged to MCS and fifty were owned by individuals—all bungalows with bright green lawns now and globe willows shading the sidewalks. With more than a thousand units, the trailer court was a little city unto itself, though no one on that side of Seventh made any pretense about permanence, and the rows of barracks on the north edge of town were crowded with single men catching a few hours' sleep between their shifts in the hole and their shifts in the bars and bowling alley.

Despite the fact it was still in the awkward process of anchoring itself to the mesa top, you couldn't pretend that Page was some sort of outpost anymore, not with its expansive new motels sporting architecture straight out of the space age, not with its lively bars and a nearby bordello, a golf course and an all-comers country club, Antennavision for TV and the new indoor Mesa Theater for movies (*Sex Kittens Go to College* with Mamie Van Doren and Tuesday Weld lately had sold out every showing), an AM radio station—KPGE— playing a lot of Andre Kostelanetz and a little Johnny Cash, a modern hospital and a street with nothing but churches, a "Teen Canteen" and a spacious swimming pool to keep the kids out of mischief, ringing cash registers that sounded like music and biweekly paychecks that looked like easy street. Page had so thoroughly become a *place,* a real live town that even had made the Rand McNally maps, that the producers of the hit television program "Route 66" had come to town to shoot two episodes, Martin Milner and George Maharis and a busload of young actresses bringing the whole community to a kind of screeching, gawking halt. And nobody had seemed to mind too much when Page was described on national television soon thereafter as "nothing to see, just a town on the mesa for the construction gang and their families. There's churches and schools, a shopping center, a mess of trailers." Nobody minded because, although it missed the spirit, the excitement, the *life* of the place, it was, nonetheless, a pretty accurate description.

Although no concrete was poured until June 1960, the aggregate processing operations—the mining and moving of the gravel that would comprise the bulk of the concrete's mass—had been throttled into high gear just as soon as the strike was settled. The Mile 15 dam site had been selected principally because of the configuration and geology of the canyon at that particular point; yet regardless of how suited to a dam the canyon seemed there, if Reclamation geologists and concrete specialists hadn't been able to find a suitable aggregate source in the neighborhood, Mile 15 would have remained un-touched. While you could haul cement hundreds of miles if you had to (and at Glen Canyon, you had to), you couldn't deliver nearly five million cubic yards of fine sand, coarse sand, pebbles and cobbles and rocks over truly long distances if you ever wanted to get your dam built.

But although the aggregate deposit in the canyon of Wahweap Creek was indeed nearby, as well as being enormous—enough to build a matched set of dams if there had been a reason to—it was a deposit of only passing quality. Still and all, in a landscape where seemingly every stone was otherwise soft sandstone or crumbly shale, the presence of a sizable bed of decent, hard conglomerate gravel—washed down from the slopes of the Kaiparowits Plateau by eons of thunderstorms—was a thing to be prized and put to use. Located about nine miles from the dam site, the Wahweap pits contained too little 30-mesh sand and far too few 3- to 6-inch cobbles (while having more 100-mesh sand than anyone knew what to do with), but knowing how bad the local materials *could* have been, no one at MCS complained too loudly about what the company had to contend with a ways upriver.

Wahweap Creek flowed so shallowly and intermittently that a simple dike was all that was needed to keep it out of the way of the digging; the amount of soil and overburden lying above the gravel deposit usually was less than 3 feet, and it took only two big diesel shovels—equipped with 6-yard dragline buckets and clawing round the clock—to provide a steady supply of raw aggregate. Twelve big Euclid bottom-dump trucks growled in head-to-tail rotation, hauling their heavy loads a little over three miles from the pits to the company's aggregate plant, which stood near the mouth of the creek. At the plant, the Euclids dropped their damp loads of sand and gravel and muck into a huge 150-ton hopper that funneled down to a vibrating "grizzly" capable of separating and immediately discard-ing materials over 6 inches in diameter. A broad conveyor belt carried the mix that survived that first test to the top of a screening plant, where the spray from hoses washed the aggregate clean and a succession of progressively smaller screens caught and separated it into six different sizes and grades of gravel plus two types of sand. Additional hoppers descending to more funnels leading to still more conveyors finally deposited the materials in separate bins equipped with bottom-opening gates and built above a 575-foot tunnel.

One after another around the clock, a fleet of 30-ton haulers would enter the north end of the tunnel. The driver of each rig would proceed till a red light, signaled from a control room, aligned him beneath the proper and predetermined gates, then, standing on the running board outside his cab, the driver would pull a lanyard releasing the load, filling his Cook Brothers hauler in as few as fifteen seconds, then exiting the tunnel and highballing south down the haul road to the stockpiles that were adjacent to the batch plant at the dam site, the 11-mile round-trip taking precisely twenty-six minutes—two dozen truckloads an hour, a total of 350,000 such deliveries required before enough rock and sand were mixed with enough cement and water, pozzolan and ice to make ten million tons of mud.

Under the terms of the prime contract, Merritt-Chapman & Scott had no responsibility for the mining, manufacture, or transportation of the three million barrels of cement or the 200,000 tons of pozzolanic material the dam's concrete mix would require. Reclamation was understandably picky about the quality of the concrete that was supposed to plug the canyon for a thousand years or so, and a decision was made early on to initiate separate contracts to purchase the cement and pozzolan—crushed volcanic material, pumice in this case, that would be mixed with the cement and aggregate to improve the concrete's "set" as well as reduce its cost. If the cement and pozzolan contractors were working directly for Reclamation instead of MCS, the theory went, a variety of quality-control headaches could be minimized, and for its part, MCS wouldn't have to so much as think about the cementing materials until giant screw conveyors transported them from storage silos adjacent to the beehive to the twenty-story batch plant built on a bench blasted into the west wall of the canyon.

Limestone was the principal raw material used in the man-ufacture of cement, but the Utah-Arizona border country cached precious little limestone, a fact that the Bureau's designers and cost controllers had known from the project's inception. There had been no option other than to have cement delivered to the dam site from some distant mining and milling location, and Reclamation hadn't cared where it came from, so long as it met its exacting specifications and so long as its price was right. In April 1958, the American Cement Company had successfully bid just under $10 million for the privilege of hauling to Glen Canyon more cement than had gone into any single structure since the glory days up at Grand Coulee. American Cement owned a high-quality white limestone quarry near the little Arizona town of Clarkdale, southwest of Flagstaff, and it had whipped a cement mill capable of producing 5,500 barrels a day into operation by the middle of 1959, just in time for the Glen Canyon strike to shut it down for half a year. A Washington company, J. G. Shotwell, had won the pozzolan contract for a piddling $2.5 million, latching on to a pumice deposit 30 miles north of Flagstaff and constructing a series of crushing mills nearby, both outfits raring to start delivering their goods just as soon as the strike was settled.

Beginning early in 1960, and for almost four years thereafter, a truck hauling pozzolan delivered a load to the dam site every two hours on average. Every *forty minutes*, day and night, seemingly endlessly, in the heat and sometimes the heavy snow, one of a constantly moving fleet of White Autocar diesels pulling tandem trailers filled with 28 tons of bulk cement from the Clarkdale plant pulled onto the government scales near the project's storage silos, each load a mere 142 barrels, meaning that 21,500 round-trips totaling 376 miles each ultimately were required to bring enough cement to Glen Canyon to make the mud.

Lem Wylie never had to worry about his cement supply; it always arrived like clockwork. The pozzolan supply, rather the lack thereof, made him nervous the few times when the Shotwell mills were down for renovations and the pumice in the local silos dwindled down to less than a day's supply, but by and large the raw materials were always at the ready. If you wanted to know the truth, despite the strike that had come close to crippling the project; despite having to manage 2,000 men and keep abreast of more than 3,000 technical drawings, hundreds of contracts, change orders, tests, reports, and assorted headaches; despite the fact that if this wasn't the best dam ever built, and if it wasn't built on time, he alone would be responsible, Wylie still considered the business of running the town and placating the politicians and keeping the parasitic con men at bay to be the toughest part of his job.

It seemed as though he always was wasting a morning showing the governor of Arizona or one of those gung-ho western congressmen how you placed mud in these modern times. He constantly was threatening to kick out of town some ne'er-do-well with a penchant for picking fights and regularly arbitrating the squabbles between businesspeople and the Bureau men who served as town administrators. After five and a half years, he still was battling the shysters who were endeavoring to sell the government's land for lakeside homesites—occasionally relishing the pleasure of pulling their goddamned survey stakes out of the ground himself—and just about a year ago he had had to face down Danny Reisman, who somehow thought he was going to turn his low-rent, bad-element lounge over on Aero Drive into some sort of casino.

Reisman had had some kind of connection to Las Vegas gambling interests who, of course, had connections with Howard Cannon, the powerful Nevada senator. Reisman claimed he had it on good authority that Page was some kind of federal no-man's-land, and that there wasn't any law on the books that prohibited him from setting up a few craps tables, a roulette wheel or two, from lining the walls of his tavern with slot machines. When Wylie had gone to see him to dissuade him from his plans, Reisman had told the project construction czar that, like it or not, the equipment was scheduled to arrive and be installed by nine the following morning. If so, Wylie had countered, his bulldozers would rumble down to Danny's Lounge at noon. And rumble down they had, but Reisman ultimately decided to spare his building by loading the gambling equipment back onto a truck in front of Wylie's watchful eye.

There had been sizable brawls in Page, some despicable but probably unavoidable domestic violence, a couple of murders, and a single suicide by now, but all things considered, this wasn't the wild camp that Wylie remembered Boulder City had been, and that was a fact he was proud of. He had hoped Page could become a model kind of community, and perhaps it still could someday; its businesses were thriving now, lawns and trees were altering the sandy specter of the mesa, nobody was sleeping in cars anymore, and the school was a virtual showplace. People seemed happy; they seemed to feel they were part of something important; and if the tin-shack look of the town still bothered him, well, maybe that would give way with time.

It had taken eleven months to pour the first million cubic yards of concrete—the job then one-fifth complete—but the second million cubic yards—reached in October 1961—had been placed in only five more months. The cold winter weather—temperatures once so low that MCS couldn't pour mud for an entire week—had slowed operations enough that in the middle of May 1962, they just now were reaching the three million mark, the interlocking blocks now standing almost 400 feet high.

As the lawns in town turned green and the spring winds began to settle, it seemed to Wylie that the job was demanding more of him than it had in half a decade. He had just helped MCS and the five unions skirt another strike, this time the several parties needing only a day and a half to agree on a master contract that called for a sixty-cent wage increase spread over the next three years. In concert with the Salt Lake and Denver offices, Wylie also was in the process of letting contracts for the supply and installation of the dynamos, turbines, transformers, switchyards, and transmission lines that would have to generate and deliver this hydro-dam's electricity. Escorted by Art Greene, the governors of Utah and Arizona and dozens of lesser officials and cronies were about to make a "Governors' Farewell to Glen Canyon," a VIP float trip from Hite to Kane Creek to kiss the canyon good-bye, and they wanted Wylie to join them. He lately had agreed to give the commencement address to the forty-one seniors about to graduate from the high school, knowing his speech would be titled "The Price of Leadership" but not having taken a minute yet to consider what he ought to say. And if all that wasn't enough in addition to the daily business of adding a few feet of mud to the block—Saturdays tacked onto the schedule now to speed things up a bit—his co-workers were about to host a dinner for him: it had been thirty years since the greenhorn engineer from Albuquerque had talked his way onto a survey crew down in the bowels of Black Canyon, thirty years since the fifty-six-year-old fellow had taken up government service, almost half a lifetime already proudly carrying the hod of the Bureau of Reclamation.

As part of the prime contract's specs, the Bureau had stipulated that the project's mixing plant—"batch plant" in the parlance of the construction trade—would have to be capable of producing 420 cubic yards of wet concrete every hour around the clock, a greater volume of concrete than any single operation ever before had produced. The plant would have to deliver one of twelve different concrete mixes within seconds of being requested, each mix weighed and measured accurately to within 2 percent for all materials. It would have to work without a hitch for nearly four years, breakdowns being absolutely out of the question.

As soon as it was awarded the prime contract, Merritt-Chapman & Scott had turned the design and manufacture of the batch plant over to the Noble Company of Oakland, California, and Noble's engineers quickly had designed and built a prototype plant to one-eighth the scale of the final structure. Based on the tests the company had conducted on its working model, it had refined its design and calculated the myriad stresses and demands on the big plant, then had milled it piece by piece and shipped it to Arizona aboard forty flatcars.

MCS drilling, blasting, and mucking gangs long since had cut a bench for the plant's foundation 400 feet high on the west wall of the canyon, just upstream from the keyway, when the disassembled plant began to arrive in April 1959, and the structure was only partially erected as the strike commenced. But beginning work again in January 1960, Noble's crews were able to complete what was the world's largest concrete plant by the end of April, to test it throughout the month of May, then to jam it into high gear in the middle of June, the plant spilling virtually perfect concrete twenty-four hours a day, five and then six days a week, for thirty-nine breakneck, miraculous months.

Dry cement was, by its nature, fairly warm stuff. Sitting in a silo on the baking rim of Glen Canyon, it kept to a fairly constant 150 degrees during the summer months. Reclamation engineers had estimated that if mixed with 120-degree pozzolan, aggregate that reflected the air temperature (averaging about 87 degrees), and the tepid water out of the river, a batch of concrete would average 94 degrees as it was placed, then would further warm as it cured—temperatures far too hot to produce concrete of the quality this dam demanded, meaning concrete that wouldn't crack into millions of pieces, concrete that wouldn't crush under its own incredible weight. The need, as determined by the fellows in Denver's concrete control division, was for concrete that could be placed at temperatures between 40 and 50 degrees Fahrenheit, maintained at less than

50 degrees while it cured, with each block in the dam lowered finally to 40 degrees before its joints were grouted. And the solution to the need, of course, was refrigeration, the mechanical cooling of the materials that went into the concrete mix, the mechanical cooling of the blocks themselves, cooling accomplished by a companion refrigeration plant built by Lewis Refrigeration of Seattle, a mammoth operation in itself composed of eighteen ammonia compressors, eight condensers, ten chillers, and twenty-two ice-making machines, with a power installation of 6,275 horsepower and a refrigerating capability equivalent to the production of 6,000 tons of ice a day—a snow cone an hour for every kid on earth.

The batching process began as a system of screw conveyors and air slides delivered cement and pozzolan from their silos to storage hoppers in the plant, and as sand and aggregate, moving along a wide conveyor, was sprayed by icy water, the water pumped from a series of cool-water wells and chilled in a tall refrigeration tower. After passing through layers of separating and recleaning screens at the top of the plant, the aggregate tumbled down chutes to insulated storage bins, where it was further chilled by fan-forced cold air.

Inside a dust-free control room, with a console that looked like something the space program might have devised, the plant operator—one of only eleven men at work in the plant at a given time could view the contents inside each of the bins on a bank of closed-circuit television monitors, and with the aid of an array of dials, buttons, gauges, and switches, could mix the materials according to any of twelve predetermined recipes, a standard four-cubic-yard batch of concrete—just enough to fill the space taken up by a king-sized bed—containing about 5½ tons of the several sizes of aggregate, 1½ tons of sand, 200 pounds of flaked ice, 500 pounds of water, 750 pounds of cement, 375 pounds of pozzolan, and 32 ounces of Protex, an additive designed to minimize air bubbles in the mix.

Once the mix was selected, clamshell gates at the bottom of the several bins would swing open, spilling their contents into rubber-lined "batchers," which would weigh the incoming aggregate and other materials and signal the gates to close when the proper weights were reached—six aggregate, two sand, one cement, and one pozzolan batcher, plus screws and hoses delivering precise amounts of ice and water, each one in turn filling and discharging into a massive collecting hopper, all in a total of about ten seconds. By means of a fast-moving, swiveling chute, the mix then would be dropped into one of six tilting mixers for a furious, deafening two-and-a-half-minute tumble before the mixer would empty the ready mud (now 47 degrees on the dot) into a holding hopper, which, when opened, would drop it into a transfer car on an electric railway for a short journey to a loading dock, where next it was dumped into a waiting cableway bucket, then whisked away into open space—the plant producing 24 cubic yards of concrete every three minutes, an average of 335 cubic yards an hour over the project's duration, a record of 10,217 cubic yards in twenty-four hours, one-five-hundredth of the dam's total volume in just that single astonishing day.

With the new wage agreement in place by the middle of 1961, the men operating the draglines out at the Wahweap aggregate pits now were making $4.67 an hour—good money for the middle of nowhere, good money anywhere in America at the moment; the bottom-dump drivers who hauled aggregate, cement, and the pumice pozzolan were making $4.02; the several batch plant operators were at $4.67 as well; the cableway operators, the men who swung the 30-ton buckets out of the sky, were worth $4.92; and the sandblaster nozzlemen, the ironworkers and carpenters, cement finishers, jackhammer men, vibrator operators, and all the rest who labored down in the hole took home biweekly wages that ranged from $3.59 to $5.40 an hour, depending on what their unions had done for them lately. And by the time the dam had climbed out of the canyon and they had hitched their trailers to their trucks again, every one of them would be earning forty cents more per hour, thanks to the lingering scare of a subsequent strike.

Wages themselves, however, didn't determine a hierarchical system among the men. Everyone knew that on this job and every other, what you got paid wasn't necessarily what you were worth, and there were certain jobs at Glen Canyon that carried with them a prestige that had nothing to do with money. As they had been since they had drilled the powder holes for the first blast way back in what seemed like historical times, the Glen Canyon high-scalers were a separate and distinguished breed, young men (usually) who were willing to walk backward off the lip of the canyon, to hang from a rope in yawning space, to free-fall 50, 100 feet at a time before carefully setting their brakes, and in the midst of that bizarre predicament, to trim the walls with pry-bars, to drill deep holes while they hung, to set charges and salvage equipment and paint survey markers and do a hundred other things only acrobats could do, all for about the wages a driver or a serviceman made. Iron-workers, like ironworkers everywhere, also sat near the top of the bragging stack; so did the electricians and crane operators, the meticulous cement masons and the nearly magicianlike mechanics who could fix any machine that had ever been made. And if you worked on the canyon rim—digging, hauling, screening, or even mixing the mud—you somehow didn't compare with the boys actually down on the blocks, the men who *knew* what a hole it was, who constantly breathed the chalky, slightly acrid, and not unpleas-ant aroma of wet cement, who stood for years in the sagging mud as well as on the dry and now damn near eternal concrete. It was the men who rode the monkey-slides who somehow knew they were something special.

Before the mud had begun to rain out of the sky in June 1960, the riverbed had been scraped down 137 feet, down to bedrock, far enough down to expose a small inner canyon cut millions of years before, bedrock which had to be grouted—injected with watery cement applied by high-pressure hoses—the grout packing into myriad small cracks and fissures in an effort to make them imper-meable and to keep the river from crawling under the dam. The abutment walls in the keyways also got grout where it was necessary, got rock-bolts where the sandstone was slightly suspect, got a trimming a barber might have admired before the first forms were set and the batch plant first stirred up its recipes. Then it was time to *go and highball!*, as the bellboys down on the blocks would holler into their radios, signaling to the cableway operators up on the rim that one load of concrete was dumped and another load was needed.

From the moment the mud spilled out of a transfer car on the west-wall trestle and into a cableway bucket, it belonged to a single man, a guy who never got his hands dirty and never soiled his shoes, one who much of the time couldn't even see where he was sending his six-ton bucket filled with 24 tons of soggy concrete. Positioned inside a glass control booth at the end of the trestle, the cableway operator could see the bucket loading for himself, could see when his load was ready to be lowered into the hole, but it wasn't until the blocks almost had reached the canyon rim that he could get a good view of the bucket's journey or the quick flushing of its contents.

The two 50-ton-capacity cableways had been erected early on and had been used to transport a variety of materials and equipment into and out of the canyon. A third, 25-ton cableway had gone into service late in 1960, not long after concrete placement had begun, the smaller highline doing the yarding and logistical work now, as well as delivering wet concrete to the power plant, freeing up the two larger highlines—their cables a single, locked-steel coil fat as a man's fist—to do nothing but move mud to the dam, or on occasion, to work in tandem to lower mammoth objects such as the 20-foot sections of the penstocks into place. Each cableway operator had needed four minutes to deliver his load and bring the bucket back to be loaded again in the early weeks and months when the concrete was being placed more than 500 feet below the level of the batch plant, but by the time the dam stood alongside the loading dock, the cycle had been reduced to two minutes, the two operators by then able to move 720 tons of the stuff an hour, enough concrete set down in Glen Canyon every sixty minutes, between the two of them, to pave thirty city blocks.

But before either cableway operator could send concrete on its way, the boys at the bottom of the hole, down on the growing blocks, had to be ready, and a signalman (called a bellboy since back in the days before the advent of two-way radio) had to call and request it. The first step down there on the receiving end of things was to erect the forms that would hold the concrete, forms the size of the specified blocks, the biggest bottom blocks as large as 60 feet wide by 210 feet long—half the square footage of a football field. In the cramped and crowded spaces down at the dam's foundation, single-use wooden forms had been employed initially to shape the blocks, but soon reusable steel forms—two miles of them end to end, enough at any one time to begin a new pour on each of the dam's twenty-six blocks—were put into operation.

In accordance with Louis Puls's specs, the two arching rows of blocks were connected by offsetting longitudinal joints to create an interlock, and the transverse joints were placed along lines radial to the arch, giving each block the faint suggestion of a pie shape when you looked down from the rim of the canyon, the size of each block in relation to its neighbors appearing rather random but actually determined by foundation and abutment characteristics, by the geometry of the arch itself, and by an engineer's armload of stress and thrust and weight-bearing calculations. Large crews of laborers—the legions of unskilled men who were indispensable to a variety of tasks—were required to move the heavy forms, to strip them away from lifts where prior concrete had set, to wash and scrape and haul them to new positions, to raise the 7½-foot vertical walls of the forms, ten men using A-frame scaffolding and manual-crank winches to lift each form member into place, interlocking each sheet to the ones beside it, anchoring each to bolts embedded in the previous pour.

With the form for a new pour in position—creating a high rectangular fence now—sandblasting and washing crews crawled down onto the hard gray floor inside it trailing fat hoses behind them, first scouring the bottom concrete—some of it as fresh as seventy-two hours old—then washing away the accumulated grit, flaked concrete, and dirt with water jets. Next came the pipefitting crews, quickly laying a web of one inch aluminum cooling pipe— 852 miles of it before the dam was finished—across the floor on 30-inch centers, pressurizing the pipe to test it, then connecting it to the giant chillers that sent a steady stream of 40-degree water coursing through the coil. Last, a thin layer of dry grout was applied to the floor with brooms.

The forms prepared and the surfaces cleaned, the coils and grout in place, a seven-man placing crew—several crews scattered across the stair-stepping blocks waited for concrete. On a signal from the bellboy, who used a radio as well as an electronic beeper, the cableway operator above would begin to pull and punch the panel controls, kicking and thumping foot pedals as he worked, moving the huge cableway towers slowly along their tracks, their maze of cables whirling through massive flywheels now, the twenty-wheeled carriage assembly rolling out across the track cable, the hoist cables lifting the concrete bucket away from the loading dock, then down, lickety-split, 700 feet per minute, almost 8 miles an hour, blazingly fast as the men on the block watched the bucket fall toward them, the bellboy talking it down all the way, talking it down to a target no bigger than the bucket itself, at last signaling for it to stop—the cableway operator working blindly now—just six feet above an otherwise certain disaster.

With a quick pull on a rope lanyard, hydraulic gates on the bottom of the bucket opened and 12 cubic yards of concrete—a piddling amount in relation to a lift that on that day might have been 60 by 120 by 7½ feet in volume—slumped onto the grouted surface, the bucket pulling away even as the last of its mud was ejected, the workhorse bucket highballing back up the wall of the canyon.

Each one armed with a weighty, unwieldy air-driven vibrator—a tool that looked like a fat fence post attached to an umbilical hose— three crew members in tall rubber boots, likely as not young Navajos, next climbed onto the mound of concrete, jamming, fighting their whining probes into the slag, spreading it thin, vibrating out the bubbles, squeezing it into the corners of the forms, sweating, struggling to stay in control of the oscillating sticks until the concrete lay at a depth of about 20 inches, stepping it back from the wet mud below it to interlock the layers, doing undoubtedly the dirtiest job on the dam site, getting their guts shaken out, a job so taxing that three men rested while three others pounded down a 12-yard load, the first three going back to work again in as little as a couple of minutes, once the bellboy called down another load and the big bucket dropped it into their laps.

Relentlessly the buckets swung down to the swelling lift, the mud coming continuously, 167 bucket-loads needed before that 60-foot-by-120-foot block was filled to the top of the form, crews needing all of one shift and at least half of another to top it out, the completed lifts on the biggest blocks taking twenty hours or more, the crew who began placing a lift one morning sometimes completing the pour as its men pounded their vibrators into the slag on the subsequent day.

There were times when a fellow had the shitty luck simply to be standing in the wrong place, other times when carelessness led to trouble. It didn't happen every day, but regularly men got hurt— broken bones and lacerations, pulled muscles, eyes and lungs irritated by chronic exposure to the dust. Once, a high-scaler nursing a hangover neglected to tie onto his line properly and plunged 600 feet to his death; during the course of the project, several men down on the river level were mortally injured by falling objects; a few men were crushed by heavy equipment, fewer were crushed by the weight of wet concrete; two men were killed when the length of pipe they were welding happened to be harboring a stick of dynamite. Like the folkloric legends from Hoover Dam, stories maintained that there actually were bodies buried in the Glen Canyon mud, entombed in the dam for all time, but the stories were the stuff of fantasy. Even in the handful of instances in which workers were trapped beneath a mound of mud, there was never too much of it to move to recover their bodies or, as was far more common, to rescue them when they were stuck.

Nonetheless, eighteen men were killed during the dam's construction, and a total of 365 injuries to employees over the course of the project were serious enough to take them away from their jobs—slightly more than seventeen disabling injuries for every million man-hours worked—few enough that Merritt and Reclamation both were proud of their safety records. Enough, though, that people in Page dreaded the sound of the accident siren that would send Drs. Kazan and Washburn reaching for their bags and hard hats and rushing down to the dam, that left kids at school waiting nervously to see if the door to their classroom would open with unwelcome news, that left both wives in the Bureau section and wives in the trailer town terrified if soon after the siren, the white safety truck stopped at their addresses.

It was that concern about injury and about the remote possibility that your husband or your dad might not come home from his shift one afternoon, one quiet and still-cool morning, that principally troubled people in Page these days. The manifold concerns of the strike months now were ancient history; everyone who wanted to work was working, and Page was a bona fide boom camp. The shops and markets and service businesses were making real money now, and new entrepreneurs were hanging out open signs seemingly every other week. Water, at last, flowed from the taps dependably, and it looked like water instead of tea. Brownouts and blackouts were rare now that the generators no longer had to be relied upon; a 15-kilovolt transmission line—something of a coals-to-Newcastle project—had been strung across the high desert to supply the town, which soon would return the favor and offer Arizona more electricity than it knew what to do with. Although the drive-in theater seemed to be doing a spotty business (perhaps too many people still remembered the *free* movies on the concrete slab), the indoor Mesa Theater had become a kind of social center; the country club was thriving, the swimming pool was a frenzy of activity eight months out of the year; and the Big Dam Rodeo, Navajo Ceremonial Days, and a spirited, union-town Fourth of July celebration now punctuated the community's calendar. "Page is a modern, attractive city," boasted Chamber of Commerce President Gerald Harding in a promotional brochure. "It isn't just a sand hill any more."

The post office and the banks and the bowling alley were still too crowded—absolute chaos about an hour after each shift change—but those were the problems of prosperity, emblems of how well things were going, and townspeople took them in stride, understanding the alternative, proud of what Page had become. Some people, it was true, were worried about the growing problem with flies—the mesa that had seemed absolutely insect-free in the beginning, so sterile that folks trying to grow a tomato plant or two had had problems with pollination, now teemed with flies and not a few mosquitoes. Others were rather more fundamentally concerned about the threat of nuclear attack, but the Coconino County civil-defense chief had assured them recently that the project's access tunnel—curling through the canyon wall from near the sanitation plant and the number-three green on the golf course down to the bottom of the dam—probably was the best fallout shelter in Arizona, maybe in the entire United States. There was plenty of room for everyone in the 2½-mile-long tunnel, and as long as people were careful to stockpile their canned goods, there seemed to be no cause for alarm, no need for backyard bunkers in Page.

The truth of the matter was that at the end of 1962 *the* biggest problem in Page was that everybody wanted to be in pictures. Hundreds of employees were ignoring their regular work stocking grocery-store shelves, waiting tables at the steakhouse, pumping gas at the Chevron station, or selling surfer shirts and long-life batteries in hopes that they could get parts in what people were saying was going to be the greatest movie ever made.

The excitement had begun back in April when Hollywood producer-director George Stevens and members of his company had flown into Page to scout film locations for *The Greatest Story Ever Told,* the renowned moviemaker already certain that his exterior scenes would be shot somewhere in the desert Southwest, not simply to save money, but because the canyonlands country somehow *felt* more like the ancient Holy Land must have felt than did the contemporary countries in what once was ancient Palestine. It hadn't taken Stevens and company long to find what they were looking for—sites for Jesus' baptism by John the Baptist, for the walled city of Jerusalem, for Bethlehem and Lazarus' tomb—all just a few miles from Page in the Wahweap and Kane creek drainages, land that soon would be submerged by a rising reservoir, splendid locations that could be used once for making a film, *this* film, then never again. Stevens had been thrilled, and the maze of machinery for a major Hollywood motion picture soon was set in motion.

By the time Charlton Heston stepped into the icy Colorado River near the mouth of Kane Creek—wearing a concealed rubber suit to help this John the Baptist withstand the water's late-October chill—nearly 3,000 residents of Page were part of the production. They worked as caterers, cooks, waitresses, cabin maids, carpenters; worked as wranglers to tend the exotic Middle Eastern animals imported for verisimilitude; worked as extras, shorn of their watches, eyeglasses, and shoes and wearing rough-fabric smocks that didn't allow for underwear; a few fortunate individuals actually got to speak a line or two, got to chat, while they endlessly waited for the cameras to roll, with John Wayne, with Van Heflin and Shelley Winters and Ed Wynn and Pat Boone and Sal Mineo and Max Von Sydow—the Swede, however, appearing so much like Jesus that few were brave enough to trouble him with small talk.

For weeks, progress on the dam seemed like yesterday's news; what kept half the community focused for sixteen hours a day—and the rest intrigued and a little envious—were the sets, the shoots, the movie stars eating hot-beef sandwiches over at the Empire House just as if they were regular people. Morning after morning, the locally based extras were required to report to wardrobe before dawn, to get their beards and sandals and sackcloth, then to ride in buses the 15 or 20 jostling, jolting miles to the sets representing Nazareth, Bethany, or Tiberias, where they would wait, then be told to be ready and wait some more, finally strolling through cobbled streets or laying out fishing nets or briefly loading donkeys before darkness descended and the buses rolled back to Page.

It was almost Christmas before the scenes by the riverside had been shot, just a month before the diversion tunnel would be closed and this River Jordan would begin to resemble the Sea of Galilee. Then January snows succeeded in delaying scenes scheduled to be filmed on the benchland above the creeks, above the quickly invading slackwater, until well after Lem Wylie had locked the west tunnel gates and the water indeed had begun to rise—the last images of that part of primeval Glen Canyon ironically captured in a film about the Bible. By the time the set crews burned Bethlehem in

April—Stevens and his cameramen getting their final outdoor footage—the gates in both diversion tunnels had been screwed down and a motionless lake nearly reached Jerusalem.

While their wives were employed in the movie business—coming back to their trailers at night with swooning stories about how *handsome* Charlton Heston was—the men in the hole, a chasm that had filled for more than two relentless years now, had placed the four-millionth cubic yard of concrete, the highest block in the dam reaching a height of 582 feet on November 19, 1962.

The *Saturday Evening Post* lately had come back to town, the magazine taking its second look at how Lem Wylie's project was "Taming the Colorado." Feature writer Jack Goodman noted that Wylie was so central to this project that his visage might well have been painted on the dam's sloping downstream face, noted that MCS's Jim Irwin seemed to feel the weight of the economic pressures on him the way the fat cables felt the weight of the concrete buckets. Goodman interviewed high-scaler Zug Bennett, bellboy Johnny Vezzoso, and concrete control inspector Vic Gezelius, each one saying a bit of the dam belonged to him as well. But if the *Post* wanted a real story, Wylie wondered why it hadn't waited a little longer, waited to come to town in January to watch the three slide-gates in the west tunnel close, to watch as the river rose against the rocky mound of the cofferdam and to see how men and machines could turn back a river that had flowed through the canyon since before time began. *That* was going to be the story Lem Wylie wanted to read, not just how they were building this mammoth dam, but how the bastard had begun to do what it was supposed to do.

Not that there was any sort of celebration on the morning of January 21, 1963; not that a busload of VIPs or reporters had to be herded down into the hole somehow to make it official; it was far more workaday than that. A crew of laborers simply went to work chipping the ice out of the tunnel gates' tracks, clearing them so that the gates could fall and seat themselves. The weather was cold, the river seemingly colder still, and the work was bitter, ice tending to form everywhere water touched metal, preventing the ironworkers from making much progress in screwing down the massive guillotinelike steel leaves. But after two days of effort, of chipping ice and nursing the reluctant steel, the gates at last were closed tight against the low flow of the river, against the current that had sought out that tunneled detour for almost four years now, the river having nowhere to go all of a sudden, stopping and rising instead, climbing up the cofferdam as if it knew there was another, higher tunnel dug into the opposite wall of the canyon, another possibility for escape.

By the time in early March when the water had risen 33 feet, enough to slip into the east-wall tunnel and through the system of outlet gates installed inside it, the captive river briefly escaping again, the plugging operation in the west tunnel was going well, a temporary concrete plug already poured into place behind the slide-gates to secure them, the first of three 50-foot-long, 41-foot-diameter permanent plug sections—each one keyed into the tunnel wall so that no amount of water pressure could ever push it out of place—now getting its concrete, its cooling tubes and grouting tubes, the first step in what ultimately would be 300 feet of contiguous plugs and backfill concrete, surely enough to dissuade the river—the reservoir now—from ever attempting to seek out the tunnel again.

But Lem Wylie wasn't anxious only to keep the Colorado's water in check; by getting the gates jammed shut and the first of the concrete in place behind them, Wylie also was warding off the conservationists. They increasingly were crying about how Glen Canyon ought to be spared and about how the water legally couldn't rise until Rainbow Bridge had been protected from it, and as Echo Park had proved, their cries occasionally turned into action. As far as Wylie was concerned, the politicians were welcome to settle the fate of Rainbow Bridge. It would be a hell of a long time before Lake Powell would rise high enough to flow under the giant arch, and if a decision was made in the meantime to build a barrier dam to keep the water out, well, the construction engineer didn't much care one way or the other. What he did want to make damn certain of, however, was that there was going to be a reservoir behind his dam. He hadn't devoted six years to this project just to see it stand in the sun like some sort of dinosaur, the river flowing around it forever just because of the bird-watchers. With Stewart Udall sitting behind the big desk at Interior, and with Floyd Dominy at the top of Reclamation, Wylie really wasn't too worried that the conservationists would get their way and keep Glen Canyon empty, but a few million pounds of prevention didn't hurt either. With the west tunnel forever sealed, there *had* to be a reservoir of at least some minimal size behind the dam. The east tunnel alone, even with its gates wide open, couldn't handle the river's high early-summer flows, to say nothing of its occasional monstrous floods. With the west tunnel already harder now than the rock that surrounded it, the Man Upstairs himself would have a hard time keeping the river—all of the river—running down-stream.

On March 13 they did schedule a bit of a ceremony, some Bureau brass from Salt Lake City and Denver as well as a pack of reporters following Wylie down into the gate chamber above the east diversion tunnel, where three sets of tandem steel gates were in position and where the whole flow of the Colorado, swelling now with the earliest runoff, was passing between their bronze and cast-steel framework. To Wylie himself went the task, the privilege, of pulling the levers that sent double-gates 1 and 3 slicing through the brown water that coursed through the tunnel somewhere below his feet, their gate leaves seating against their base plates, suddenly sending all of the surging water through the middle pair of gates. Wylie waited until his watch struck 2:00 P.M.—this was an occasion of some real moment, after all—then he manipulated the lever on the hydraulic hoist that narrowed the number 2 gates to within 50 inches of closure—to the precise level that, with the water pressure provided by the current height of the lake, released only 1,000 cubic feet of water per second to continue to flow downstream. The rest of the river—21,000 cubic feet per second and swelling daily—slowed and swirled and eddied, then grew still.

The gaping mouths of the penstock tubes—not yet extended on to the power plant—protruded from the toe of the dam. Above them, a latticework of wooden scaffolding still clung to the steeply sloping concrete, the precarious catwalks and shelves and ladders hanging at intervals upward for more than 500 feet, the smooth, streaked concrete of Block 2 over near the eastern wall of the canyon rising precisely to elevation 3,715 feet, as high as the dam would reach, a lighted and tinseled Christmas tree rising six feet farther. It was the first week of April, the winter holiday long since passed, but when the vibrating and finishing crews had topped out Block 2 on March 30, the first block in the dam to be completed, the Christmas tree had been the emblem the workmen chose to signal this first success—one down, or rather up, and twenty-five still to go.

By April of 1963, the crews were hauling and mixing and placing concrete faster than ever before, the dam far thinner now near the top and rising a lift a day, a streamlined work force of about 1,600 men highballing twenty-four hours a day, six days a week, the smell of completion, the taste on the tongue of it, as real as the acrid concrete. Out at the aggregate pits, the draglines now dropped their loads into two 70-yard mobile hoppers instead of directly into dump trucks, the hopper feeding the trucks now, eliminating idle time, significantly speeding up that first phase of the operation. Between the aggregate plant and the stockpiles at the rim of the canyon, tandem bottom-dump trailers now were hauling 75 tons of rock in the same time it used to take to move just 25 tons, the 100-foot-long double rigs pulled by new Mack tractors that seemingly could have moved Navajo Mountain. Down on the blocks, hydraulic cranes were raising forms in half the time it used to take crews of laborers cranking hand winches, and new concrete buckets, 6,000 pounds lighter than their predecessors and with dump gates operated by low-maintenance air pressure instead of hydraulics, were delivering the last of a total of 425,000 loads of mud.

Yet even when this big gray plug was completed, its ten million tons of concrete cured and cool and hard as almighty hell, there was going to be plenty of work still to do. The dam itself needed elevators and gantry cranes, parapets and trash racks, drainage galleries and gutters. The penstocks needed wheel gates; the spillways needed radial gates; the outlet works needed bulkhead gates at their intakes and hollow-jet valves where they emerged beside the power plant. MCS had built the structure that would house the power plant, but the turbines and dynamos, transformers and switches, and literally thousands of items that ultimately would allow them to make electricity had yet to be installed under a separate contract. The 230-kilovolt transmission line that would connect Glen Canyon to Reclamation's western power grid was taking shape but wouldn't be completed for almost a year, and the 345-kilovolt line that would deliver electricity 240 miles to the Pinnacle Peak substation outside Phoenix (the city whose army of air conditioners would consume the biggest share of the project's power) had only lately gotten going. The building housing the visitors' center and offices for the perma-nent Bureau staff, which would be perched on the western rim above the dam in the shadow of the shaved beehive, wouldn't even be started until well after the batch and refrigeration plants, the cableways and conveyors and tracks were dismantled and hauled away for salvage. This operation couldn't begin to produce power for almost a year and a half; all eight of the generators wouldn't begin to spin for three years, maybe more; Lake Powell wouldn't fill for ten years, maybe twenty.

But it all would happen, it all would be completed someday, and in April of 1963 more than ninety inspectors under Norman Keefer's charge were trying to make sure that the Bureau of Reclamation ultimately would own a dam that was worth the $250 million or so it was spending to see it finished. Out at Wahweap, inspectors were checking the sand and the sea of aggregate; at the silos, inspectors peered into every truckload of pozzolan and cement; there were inspectors in the batch plant measuring the wet concrete, inspectors in labs testing its strength once it had cured. Down in the hole, insistent inspectors examined the placing and cooling and grouting of concrete, the installation of 1,540 strain meters, stress meters, joint meters, and resistance thermometers; they oversaw the construction of the dam's electrical and plumbing systems, the fabrication and installation of penstocks, pipes, gates, hoists, scaffolds, and stairs; inspectors examined it all.

But despite their best efforts, Glen Canyon Dam cracked, of course. Despite repeated inspections at dozens of venues throughout every day and night, hordes of horizontal and vertical cracks had begun to appear in the blocks way back in 1960. There were hundreds more cracks in 1961, but by 1962 the cracking had begun to decrease. By 1963, by the time the blocks were comparatively wafer-thin, they were curing almost without incident, without the stress fractures that for three years had been painstakingly packed with lead wool, then injected with high-pressure grout, sealed in much the same way that the control cracks—the contraction joints between the blocks—had been grouted, stuffed with a cement gruel that would harden within the minuscule crannies that formed the dam's lines of latitude and longitude, making the twenty-six blocks, each built out of almost a hundred separate lifts, resemble a single block, a monolithic thing that forever would hold back the Colorado, but never *all* of the Colorado. Water always would find circuitous routes through the web of grouted joints, through the grouted cracks and the concrete-sandstone contacts in the keyways, through the joint-free sandstone in the canyon walls themselves. Minerals in the water slowly would seal some of the fissures and watery passageways, but the dam always would leak, always give way to a little water, and a drainage system had been designed and built into the dam to carry the leaking water away. The dam would literally rise an inch or so once enough of a reservoir was pressing against it, and when the sun bore down on its downstream face, its arch would expand upstream by about an inch and a half. It would weigh ten million tons, but it would always move around a bit, "plastic," the engineers called it, as it grew comfortable in its canyon.

The crews started making bets at about the time Lem Wylie shut the diversion gates down to 1,000 feet per second. The reservoir, 40 feet above the riverbed now, would start to rise very quickly, the runoff spreading laterally against the canyon walls, climbing the cofferdam, spilling over it soon and at long last pressing against the vertical concrete of the dam itself. The earthen cofferdam stood 165 feet above the riverbed, 300 feet above the foundation of the adjacent dam, and you could get good odds that the water would breach it before July. A lot of money rode on each of the days in the month of May, and although it was a risky wager, some guys were convinced that the cofferdam would go under before the end of April. And sure enough. At 7:00 A.M. on April 18, the water in Lake Powell, already backing 80 miles upstream, beyond Forbidding Canyon, beyond Hidden Passage and Music Temple, past Hole-in-the-Rock and the mouths of the San Juan and Escalante rivers, found a low spot in the crest of the cofferdam and clawed its way through it, carried away clay and gravel and earth, digging a formidable trench, carving a sudden canyon out of the cofferdam, water pouring through it and into the trough separating the two dams, dark water rushing down to the concrete dam's foundation, surging against the concrete and then stopping, rising, climbing up the great gray arch in seemingly no time at all, the cofferdam disappearing.

For some reason, the color footage of the first contact of Lake Powell with the upstream face of the dam didn't make the final version of Reclamation's second film, this one entitled *Operation Glen Canyon*. Perhaps it was because there was just too much else to show in the short production—the early work blasting the canyon walls and building the bridge, the cutting of the keyways and tunnels, the construction of a town, the ceaseless digging, conveying, trucking, mixing, and placing of mud, the key personnel, the milestones. There was footage of Fred Seaton dropping the contents of the first bucket of concrete back in 1960, a few seconds of film marking the completion of each million cubic yards thereafter, footage finally of Lem Wylie standing atop Block 25, depressing a lever on the air hose that released the final 12 yards, vibrator crews knocking down the mound, pressing it into a corner of the forms, finishing once and for all.

The weather was rainy on the morning of September 13, 1963, the mood a mix of euphoria and relief, a strange nostalgia for what never had been anything but hard work, as well as the simple anticipation of moving on. The Merritt-Chapman & Scott corporate flag—a galloping black horse on a snow-white field—flapped in the wind, a big sign painted on plywood announced the occasion, but only a few local and regional officials looked on as Wylie was handed the bucket's gate controls by bellboy Johnny Vezzoso. With little fanfare and no speech making whatsoever, the concrete slumped out of the bucket and onto the block, and three vibrator operators moved in to settle it. As they did so, Wylie removed the old and tattered fedora he had worn since he came to Glen Canyon in the summer of 1956. He tossed it onto the mound of mud, a whining vibrator snared it, and it disappeared into his dam.

9

THE FATE OF
RAINBOW BRIDGE

It was the sort of junket a junior congressman got to take. The older pols—subcommittee chairmen and fellows with a measure of stature and influence—made regular trips to watch rocket launches down in seaside Florida or flew to Hawaii to congratulate the good citizens of the newest state, but the young and inexperienced members of the House were lucky if they got invited to an occasional swine producers' confab out in Scratch Ankle, Iowa, or, like Stewart Udall, got to endure the heat and sand and mosquitoes on a trip into southern Utah, hiking a dozen blistering miles just to look at some strange rock. If this was what Udall really wanted to do on his summer vacation, Wayne Aspinall, chairman of the House Interior and Insular Affairs Committee, certainly wasn't going to get in his way. It never hurt to send a committee member or two out to make an on-site inspection, and in the case of Rainbow Bridge, you sure didn't want to give the crazy conservationists a leg up by letting them claim they were the only ones who had ever bothered to go see the thing.

The only problem with Udall, the only one that concerned Aspinall anyway, was that he fancied himself something of a conservationist as well. Yes, he was a son of Arizona and as such had always supported southwestern water development; his battle against Representative Craig Hosmer and the rest of the Californians on the issue of whether Glen Canyon could hold water had come at a critical time, and he had worked hard to help win House approval of the Colorado River Storage Project bill back in 1956. Now, in the summer of 1960, in addition to campaigning in hopes of retaining his own seat, Udall was working hard in support of Massachusetts Senator John Kennedy's bid to become president, and Aspinall had no quarrel with Kennedy either; in fact, he was hoping to become Kennedy's secretary of the Interior in six months' time. It was only the Arizona congressman's budding preservationist sentiment that worried Aspinall, his fuzzy comments about the need to protect the land for future generations, a hands-off kind of attitude that was becoming fashionable in a few circles, but one which could cripple the development of the West, Aspinall understood. Yet in the case of this Rainbow Bridge squabble, Udall probably was the perfect member of the House Interior committee to take an active interest—he could speak the language of the people on either side, after all—and Aspinall encouraged the expedition, telling Udall to take his kids and enjoy himself, thanking his lucky stars that *he* didn't have to go sleep on the rocks and sand.

Stewart Lee Udall was born in 1920 into a prominent pioneer Mormon family in tiny St. Johns, Arizona, gateway to the middle of nowhere, hard by the banks of the Little Colorado River, a desert stream that intermittently flowed northwesterly up through Hunt and Holbrook, Winslow and Cameron to empty into the Colorado far below Cape Royal in the Grand Canyon. In 1873, his grandfather, David Udall, had led a group of immigrant Mormons across the Colorado River ferry operated by his maternal great-grandfather, John D. Lee, and down into Arizona Territory, where the settlers promptly and in clear-thinking Mormon fashion developed an irrigation system, turning the water of the Little Colorado onto the parched lands surrounding St. Johns. David King Udall ultimately was elected to Arizona's territorial council; his son Levi became a superior court judge and state supreme court justice, sons Don and Jesse became judges as well, son John served as mayor of Phoenix, and in 1954, Levi's son Stewart was elected to the U.S. House of Representatives, two years after Barry Goldwater—nine years Udall's senior and a member of an equally influential Arizona family—had gone from the Phoenix City Council to the United States Senate.

Although they labored in separate legislative bodies, although they belonged to different parties and were poles apart on most political issues, both men worked diligently in Washington to promote Arizona's interests—interests that always had been centered on a single issue, water development, and on a single, canyon-captured river, the Colorado. Goldwater's 1940 expedition with Norm Nevills through Glen and Grand canyons hadn't prevented him from enthusiastically supporting the Colorado River Storage Project or the long-dreamed-of notion of a series of dams between Glen Canyon and Hoover dams. And as he prepared to venture into Glen Canyon for the first time twenty years later, Udall saw no reason to regret his own vote in support of the project that would flood the canyon and send lake water to Rainbow Bridge.

The 1956 Colorado River Storage Project Act had included language assuring conservationists that "the secretary of the Interior shall take adequate protective measures to preclude impairment of the Rainbow Bridge National Monument." It was language meant to mollify opposition to the project's centerpiece dam, a vague promise on the part of Congress that Reclamation wouldn't be allowed to intrude into the minuscule 160-acre monument, just as it had been kept out of Echo Park. But the legislation said nothing about *how* Rainbow Bridge would be protected, nothing about when it would be protected, not a word about who would pay for the obviously expensive protection.

According to the Bureau's current schedule, reservoir impoundment behind Glen Canyon Dam would begin early in 1963; water would slip into the mouth of Forbidding Canyon almost immediately, but it would be May 1964 or thereabouts before water would rise to the mouth of its Bridge Canyon tributary. Without some sort of barrier dam in place to "preclude impairment," Reclamation estimated that slackwater would cross the monument boundary in June 1966, and when and if Lake Powell ever filled completely, water would stand 57 feet deep in the rocky channel beneath the giant arch. In May, however, the House Committee on Appropriations had deleted from the Bureau of Reclamation's budget the first $3.5 million earmarked for Rainbow Bridge protection in the current public-works appropriation bill, the committee stating flatly—and clearly counter to the CRSP's enabling legislation—that it saw "no purpose in undertaking an additional expenditure in the vicinity of $20 million in order to complete the complicated structures." Because the funds needed to build a barrier dam or dams were unavailable, at least for now, Udall, Aspinall, and the rest of the members of the Interior committee found themselves in a ticklish spot. Should they risk the big-spender label and lobby hard for the money to keep the water away from the bridge, or should they save the citizens some money and endure the wrath and the charges of double cross that would come from the conservationists? For his own part, the young Arizona legislator thought the wording in the CRSP bill was mighty clear—you could build the dam and reservoir in Glen Canyon *only if* you kept the water out of the national monument. Yet, on the other hand, it seemed a shame to build roads and powerlines and God knew what else into that incredibly remote country supposedly to protect it, ostensibly to keep it wild. When it came right down to it, Udall wasn't sure what seemed the right thing to do at Rainbow Bridge, so he thought he'd go look firsthand for himself.

In early August, in the company of his sons Tommy and Scott, as well as Pennsylvania Representative John Saylor—a congressman who staunchly had been in the conservationists' camp since the beginning of the CRSP struggles—Udall climbed onboard a boat at Hite and ventured into Glen Canyon, becoming, like so many others had before him, mesmerized in a matter of miles by the sensuous sculpture of the canyon walls, by the peaceful roll of the river, by the beckoning, secretive side canyons, by the whole wild wonder of the place. On the morning of August 9, the visitors hiked into the mouth of Forbidding Canyon, up through the Narrows and into tiny, twisting Bridge Canyon, and at last to the unbelievable arch itself, a glorious rainbow of rock, arching upward 290 feet above its foundations, spanning 275 feet, a clear, cool creek so small you could step across it trickling beneath the stone, the water at work now for fifty million years cutting the canyon's deep meanders, carving the gaping hole that shaped the rainbow, its salmon-colored rock streaked black by oxidation. The congressmen and the boys soaked their feet in the soothing creek; they walked under the arch; they climbed up and onto its top, where it was 40 feet thick and almost as wide; and they hiked a mile beyond the bridge, still farther into the wilderness on that western shoulder of Navajo Mountain, before they had to begin the return trip to the river in order to get to a meeting scheduled in Page with Lem Wylie, the man who had been charged with devising a plan to protect the bridge.

According to the Glen Canyon project construction engineer, a barrier dam at one of several potential sites downstream from the monument would have to be built to a height of at least 180 feet; a diversion dam upstream from the bridge that would send the Bridge Canyon runoff through a tunnel and into Aztec Canyon would have to be 50 feet high. Even with an upstream diversion dam in place, Wylie assumed that some sort of pumping plant would be required in the area just upstream of the barrier dam to remove seepage and rainwater that would otherwise pool there. Yes, Wylie said, material for the earth-fill dams was available on the high mesas in the vicinity of Rainbow Bridge, and neither dam would pose undue construction challenges. What would be a bit of a trick, however, would be to build roads to access the construction sites—at least 15 miles of them from the nearest existing dirt tracks, across some of the roughest topography on the planet. As Wylie saw it, those roads likely would cost more to construct than would the dams themselves. But it all could be accomplished, he insisted. If Congress gave him the money, he'd be glad to keep Lake Powell from lapping beneath Rainbow Bridge.

By the time he sent the report of his inspection trip to Interior Committee Chairman Aspinall on August 27, Udall was convinced that any dam-building scheme in the Rainbow Bridge area would be a disaster. Representative Saylor favored building only the upstream diversion dam and tunnel, as a means of preventing silt and rubble from washing downstream and accumulating within the boundaries of the national monument, where, he worried, they would be trapped by the slackwater of the lake. But his colleague now believed that even that much intrusion into the region would necessarily destroy the Rainbow Bridge wilderness in a misguided effort to save it.

The boundary of the national monument—a simple 160-acre, quarter-section square—had been designated entirely arbitrarily when the monument was created back in 1910, Udall pointed out in his report to Aspinall, and the boundary bore no particular relationship to the arch, the canyon it spanned, or the surrounding redrock maze. While dams that would keep Lake Powell out of the monument certainly could be constructed, Udall argued, their construction would destroy enough of the surrounding wilderness to render the monument essentially meaningless. "Although the lake water offends a basic principle of park conservation," Udall wrote, "it is my conviction that the construction of any man-made works within five miles of the present monument boundaries would do far greater violence to the first commandment of conservation—that the great works of nature should remain in their virginal state wherever possible. The natural setting of Rainbow embraces a much larger area than the box-like artificial monument, and it is a gross mistake to detach the arch itself from its environment."

Calling the great bridge "unquestionably the most awe-inspiring work of natural sculpture anywhere in the United States," the representative offered Aspinall two basic recommendations: The Congress should clearly and directly resolve the issue of whether the lake would intrude into the monument, not settle it by default simply by refusing to appropriate protective funds. And second, the boundaries of the monument should be extended, perhaps many miles in every direction. The enlarged monument would include no roads or other developments, but a thin fjord of the reservoir would reach into it and serve as its principal access. Udall acknowledged that all the land surrounding the monument belonged to the Navajo Nation, but he was hopeful that the tribe might be willing to discuss a land exchange similar to the one that had given the federal government control of the land surrounding the dam site and the town of Page.

It was a bold notion, something even the major conservation organizations had yet to propose. It would be complicated, and it certainly wouldn't satisfy everyone, but it would do a far better job of protecting the extraordinary wilderness that bounded Rainbow Bridge than would those dams. To make the plan work, the conservationists would have to give a little and allow a bit of lake water to intrude into this bigger monument, and although they hadn't compromised the last time a similar intrusion was at issue, surely, thought Udall, these circumstances were different.

Navajos had known about Rainbow Bridge since a few of them first had fled into the fringes of the Glen Canyon country in the early

1860s in an effort to conceal themselves from Kit Carson's marauding troops. *Nonnezoshi,* the rock rainbow, seldom was visited in the succeeding decades, but it always was greatly revered. Tribal myths long had contended that the source of rain was a spring high on *Naatsis'aan,* Head of Earth Woman, known otherwise as Navajo Mountain, so it made obvious sense that the rainbows that accompanied and blessed the rain rose into the sky from nearby Nonnezoshi, actually two rainbows, male and female, arching in sexual union across the little canyon, their offspring floating away from that singular spot to every corner of the Navajo country.

Although it seems certain now that prospectors had encountered the arch as early as 1882 in the midst of their searches for mineral wealth in the region, the "discovery" of Rainbow Bridge, its first recorded sighting, belonged to two separate and competing 1909 expeditions, one group headed by University of Utah archaeologist Byron Cummings and escorted by a Paiute scout named Nasja Begay, the other organized by U.S. Land Office surveyor William Douglass, with a Paiute named Mike's Boy leading the way. The two groups chanced on each other during the course of their wanderings, agreed to join forces after a bit of acrimonious debate, and finally encountered the now-legendary bridge after many days of drudgery and several conclusions that they were utterly lost. It was August 14, 1909, and at last the bridge belonged to history.

Following the appearance of an article by Cummings in the February 1910 issue of *National Geographic,* President William Howard Taft created by executive order a 160-acre national monument with Rainbow Bridge at its center, and thenceforth a few intrepid travelers visited the site each year. Nasja Begay returned in 1913 with western writer Zane Grey in tow, the Ohioan making much of the fact that his group was only the second to successfully reach the giant arch, describing it as "the one great natural phenomenon, the one grand spectacle which I had ever seen that did not at first give vague disappointment, a confounding of reality, a disenchantment of contrast with what the mind had conceived."

By the time Norm Nevills, Barry Goldwater, and company reached Rainbow Bridge by hiking up from the river in 1940, several hundred people briefly had sojourned in Bridge Canyon, most of them traveling to the monument on foot or horseback over a difficult 13-mile trail from Rainbow Lodge, an idyllic, 500-guest retreat on the southwestern slope of Navajo Mountain, the lodge owned and operated by Goldwater's friends Bill and Katherine Wilson. The steady, six-mile, canyon-bottom climb up from the river to Rainbow Bridge was far easier, but next to no one was floating the river in those days, and a few days' stay at Rainbow Lodge, climaxed by an overnight trip to the arch, including a camp in its shadow, became a popular vacation for the southwestern cognoscenti, a far more impressive thing to do than to ride a burro to the bottom of the Grand Canyon.

Barry Goldwater and his wife, Peggy, later became partners with the Wilsons in the lodge, but it disastrously burned to the ground in 1951, and without accommodations at the trailhead, the trail itself saw far less traffic in the following years, the same years that commercial boating entered a minor boom and Boy Scouts and sedentary businessmen, teachers and nurses and grizzled river rats began to make the 12-mile round-trip trek to Rainbow Bridge the one virtually obligatory side trip of every Glen Canyon expedition. Ken Sleight, Frank Wright, Art Greene, and others escorted hundreds of people to the storied bridge; and boatwoman Georgie White, with what you might consider a bit of ill-considered enthusiasm, went so far as to hack a succession of footholds into the sandstone at one abutment of the bridge, hammered a piton into the rock and strung a rope from it, in order to make it easier for still more visitors to climb onto the top of the arch. By the time Ed Abbey and Ralph Newcomb floated Glen Canyon by the water-soaked seats of their pants in 1959, they were able to tell which side canyon would lead to Rainbow Bridge by the fire pits, tin cans, abandoned socks, and assorted detritus they encountered at the mouth of Forbidding Canyon, and Abbey noted that his name was number 14,468 in the park service's visitor register he found cached beneath the wondrous slickrock arch.

By the summer of 1960, Rainbow Bridge just wasn't that remote anymore, Abbey understood, and it would become downright accessible as soon as water filled the canyon behind the dam and curled up Forbidding Canyon and into the narrow crack out by Bridge Creek. Motorboats by the armada would be able to dock within spitting distance of the once secreted and sacred place; the silence, the utter silence, would be replaced by the growls of outboards and the squeals of darting teenagers. Nonnezoshi would be defiled just as surely whether water backed all the way underneath it or just to within a few hundred yards; if slackwater undermined the bridge's abutments and it collapsed, it would be a needless and stupid tragedy, but wouldn't it somehow equally be destroyed by hordes of daily visitors who would reach the bridge by jetboat, by people packing transistor radios who would take a quick look, then say it wasn't much?

The Sierra Club *Annual*, published in November 1958, contained a long, elegiac article on Glen Canyon by Charles Eggert—a New York State resident who earlier had assisted the club with its film *Wilderness River Trail*, a production that had been of real value in the battle for Echo Park. Eggert's article was followed by a fourteen-photograph essay by Philip Hyde entitled "The Last Days of Glen Canyon." But it wasn't until 1960—the funds for protection of the monument still not forthcoming—that the organization's voice grew strident, evidencing a sense of betrayal, of outrage, an anger focused on the threat to Rainbow Bridge but fueled with emotive force by the imminent loss of *all* of Glen Canyon. "Something that dollars cannot replace—the national park system—is being endangered again to avoid making good on a promise accepted in good faith," David Brower wrote in the *Sierra Club Bulletin*. And Brower made it plain that he believed the ultimate goal of the opposition still was to secure a dam for Echo Park: "Secret hopes rise anew," he wrote, "to build Echo Park dam in spite of solemn promises to preserve its wonderful canyons and protect the national park system from needless destruction. 'Break the agreement and invade the national park system at Rainbow,' the thinking seems to run, 'and we can do the same to Echo Park.' The arguments for the invasion are specious, *but the invasion threat is real.*"

In the spring of 1961, Congress again refused to appropriate money for the construction of barrier dams in Bridge Canyon, and Brower, feeling increasingly as though he and his allies had been had, offered an alternative course of action: if the secretary of the Interior had no money to begin protective works at once, he simply should "direct that further construction at Glen Canyon Dam *cease* until funds are provided."

John F. Kennedy won the presidency by a fraction of a percentage point in November 1960, then somehow overlooked Wayne Aspinall as he subsequently worked to assemble his cabinet. His choice for secretary of the Interior, a choice alternately hailed and harshly criticized, was a surprising one. Kennedy wanted Congressman Stewart Udall, the clever, up-and-coming young politician from Arizona, to become the chief steward of the nation's lands and natural resources. Washington insiders—a miffed Wayne Aspinall among them—openly wondered whether Udall had amassed enough experience to handle the complexities and competing constituencies of the job; grazing and mining and timbering interests worried that the laissez-faire days of Doug McKay and Fred Seaton were gone forever; but conservationists were enthusiastic, thinking Udall just might turn into the best Interior secretary since Harold Ickes. He appeared to be an unrepentant water man, it was true, yet somehow you had to believe him when he talked about the need for material progress and the maintenance of the life-giving land to proceed hand in hand. Even Dave Brower averred early in 1961 that Kennedy's choice for the Interior chief had been a good one. He and Udall had tangled a few times during the CRSP hearings half a decade before, but their confrontations never had been venomous, and the two had talked privately often enough that Brower had begun to like something about this fellow with the slow, southwestern drawl and eyes that never hesitated to look squarely into your own. Brower knew Udall was someone he could work with, guessed he would be someone who could understand the enormous responsibilities inherent in his job, prayed he would recognize right out of the box that he had no alternative but to suspend the construction of Glen Canyon Dam.

But Secretary Udall disappointed the Berkeley conservationist, disappointed everyone who was hoping against hope that the Rainbow Bridge issue somehow would rise up to spare all of Glen Canyon, or even a part of it if the reservoir's level permanently were held low enough to keep water out of the monument. Udall wanted to honor the language and spirit of the CRSP compromise; he wanted to try to find an amicable and reasoned solution to the problems Lake Powell would pose at Rainbow Bridge, but as a new secretary still trying to get his feet under him, he never considered for a second telling Floyd Dominy to suspend operations on a project the two of them still very much believed in. What Udall did do instead was to host an expedition, a kind of pilgrimage of conservationists and dam builders, congressmen and reporters to Rainbow Bridge, one which he hoped would make it as obvious as the vaulting arch itself that it made no sense to build barrier dams in order to meet the terms of the CRSP legislation if, in fact, the dams would do more damage than would the lake. Dominy, he knew, certainly could live without building dams as a favor for the cockeyed conservationists; Congress, he knew, would be happy to spend its money otherwise; if only Brower and company would recognize that *doing nothing* was the best of several less-than-ideal options—especially if Udall were able to succeed in expanding the boundaries of the monument, perhaps turning it into a wilderness park, then surely some sort of balance could be struck and Rainbow Bridge might become a kind of common ground.

On a Friday evening late in April 1961, Udall, Dominy (kept on at Reclamation despite the change in administrations, the commissioner the kind of New Deal Democrat whom Kennedy was comfortable with), and National Park Service Director Conrad Wirth were ensconced in the Merritt-Chapman & Scott guest house in Page, the three Interior men hosting fifty-seven others—congressmen and congressional staffers, members of assorted advisory boards, leaders of the nation's major conservation organizations, Navajo tribal officials, reporters for *Life, Time, National Geographic, Sports Illustrated,* and a score of newspapers—temporarily in residence at the Empire House and Glen Canyon motels, all of them about to descend on Rainbow Bridge, to drop into the canyons by helicopter, making what literally would be a whirlwind tour in an effort to get a feel for the bridge and its surroundings, the barrier dam sites, the possibilities for this new national park Secretary Udall lately had been dangling in front of the conservationists like a carrot.

In a hangar out at the airport early on Saturday morning, Udall thanked Dominy and the Bureau of Reclamation for picking up the tab for the fleet of choppers (Dominy doffing his rancher's Stetson in reply), thanked the assembled participants in the tour, used a map to point out three possible boundary configurations for the new preserve that Udall already was suggesting might be called Navajo Rainbow National Park, used the same map to point out the several dam sites, tried to impress upon all the gathered parties the fact that the helicopters were a most unusual and deceptively easy means of transport into the rugged Rainbow region, then tried to pique their excitement about the day-long trip by saying that "certainly the crown jewel of the whole Colorado is Rainbow Bridge itself. I think it's the most magnificent piece of sculpture anywhere in the world."

With that, the big eight-passenger choppers from Stead Air Force

Base in Nevada began to lift into the air, flying high above Navajo, Labyrinth, Face, West, and Wetherill canyons to the edge of Cummings Mesa, where small two-passenger whirlybirds waited to begin the series of round-trips that ultimately would drop everyone into Bridge Canyon just downstream of the giant arch—an arch so large it would span the Capitol dome, Udall pointed out for the sake of the assembled Washingtonians—this awesome quirk of nature that was the subject of so much study, concern, and consternation.

Standing in the warm April air, wearing jeans and hiking boots and appearing undeniably proud of his native terrain, Udall explained that Reclamation and USGS studies both had concluded that *if* lake water ever filled the narrow streambed beneath the bridge, it would not undermine the bridge's abutments and lead to its collapse, but Dave Brower, quiet till now, politely countered that independent geologists were not equally convinced. When Udall tried to assert that reservoir water inside Rainbow Bridge National Monument—or inside a Navajo Rainbow National Park, for that matter—would not be allowed to set a precedent for the invasion of other preserves, saying "we'll make it very clear that this is a unique, unfortunate case made necessary by special circumstances," Brower responded that if you allowed an invasion once, you inevitably made it easier to allow one a second time. When Floyd Dominy spoke up, claiming that a dam at Site C, the dam site strongly favored by conservationists, almost five miles downstream from the bridge, would require five million cubic yards of earth-fill—as much as Glen Canyon Dam itself—fill that simply couldn't be found in the canyon-bottom area adjacent to the damsite, Brower simply shook his head in disgust. And another problem, Secretary Udall added, pointing his comment toward the conservationist, was that "we may already have run out of time to build at Site C," with the reservoir scheduled to rise past that point in just a year and a half.

Neither Brower nor his allies Sigurd Olson, Frank Masland, Anthony Smith, or Weldon Heald—all prominent conservationists themselves and each one a director of the National Parks Association—pressed Udall further on the subject, and he was able to turn to the issue of the expanded national park, a diversionary tactic, to be sure, but the issue that truly excited him that morning. "The people that are really your hosts today," said the secretary, "are Chairman Paul Jones and the other members of the Navajo Tribe who've joined us. Other than here at the bridge itself, these

160 acres that comprise the current monument, all of the surrounding country, north to the San Juan River and west down to the Colorado, including Navajo Mountain above us there in the east, belongs to them. And, of course, I want to quickly add that if we are going to do anything at all in the way of expanding the monument, we are going to have to do some bargaining with some very tough bargainers, the Navajos, because it is a matter of a land swap. In other words, we'll take these lands, whatever we decide upon as a park, and we will give them other lands that are suitable for their own purposes."

Paul Jones, a World War II veteran and former college professor who had returned to the reservation and won races for the tribal chairmanship in 1955 and 1959, a man who favored broad silk ties even on outings such as this one, paid inscrutable attention as Udall outlined the three proposed boundaries of the new park—the largest embracing almost all of the country from Page north and east along the Colorado and the San Juan to Paiute Canyon, then south to include all of Navajo Mountain, an area of 750 square miles, the smallest proposal containing just over 200 square miles, excluding Navajo Mountain. The new park would be bounded by the waters of Lake Powell, *invaded* by the waters of Lake Powell, if you wanted to look at it that way, in each of the three proposals.

Jones knew that Nonnezoshi itself would have to rise into the sky before the tribe would trade away Navajo Mountain, Head of Earth Woman, Naatsis'aan, but he saw no reason to say so at that moment, letting Udall and the rest of these park people speak their minds, knowing that "fair and equitable" land trades were easy to talk about but rather more difficult to bring into being. Before he and the members of the tribal council would have much in the way of comment, they would need to know precisely what federal lands the secretary planned to offer them in exchange.

But David Brower had more to say. Despite being galled by Interior's obvious stonewalling on the barrier-dam issue, Brower thought the plans for an expanded park had much to recommend them. "This is extraordinary country," said the conservationist, who now was beginning to believe that Wallace Stegner might have been right; this Glen Canyon country *might* have been better than Dinosaur. "There is only one Colorado province and one kind of terrain like this in the world. We are moving now to make a reservation for all the world, we hope for all time. And I hope that the Navajos themselves will be willing to help this be set aside as one of the great national parks."

Paul Jones offered Brower a small and unyielding smile but nothing more in reply before Udall suggested that the intrepid members of the group climb up onto the top of Rainbow Bridge to better survey the surrounding country, a few game congressmen and conservationists joining him on the trek, the Navajos and most of the scriveners staying behind, Brower quickly overtaking the secretary for the lead, the mountain climber turned advocate easily scrambling up the arching stone.

En route to Denver, and then back to Washington, the Bureau of Reclamation airplane carrying Commissioner Dominy, National Park Service Director Wirth, and their boss, Interior Secretary Udall, flew from the Page airport northeasterly above the twisting canyons of the Colorado, beyond Hite to the place on the edge of the Needles country where Dominy had designs on another dam, this one at the so-called Junction site, right below the confluence of the Green and the red Colorado. Dominy had shown Udall this site before, similarly shown it to him from the air a few months earlier, and just like the last time, Udall was impressed by the area not so much for its reservoir potential as by the certainty, my God, that this too should be a national park. Pointing out the thousand hoodoo spires of the Needles themselves, the high plateau known as Island in the Sky, a kind of lofty wedge between the two rivers, and the distant cache of canyons called the Maze off to the west of their confluence, Udall tried to solicit Wirth's opinion of the country that lay below them, Dominy rolling his eyes in exasperation with the ridiculous ideas of this new Interior chief, Wirth keeping his enthusiasm in check for the moment, cognizant of Dominy's almost certain opposition to the idea of still more canyon-country parks.

But back in Washington, Wirth soon put Udall in touch with Bates Wilson, long the superintendent of Arches National Monument, immediately north of the town of Moab, and Wilson, who probably knew the wilderness topography of southeastern Utah better than anyone else, assured the secretary that the country within a 30-mile radius of the confluence of the Green and Colorado was glorious in every direction—diverse and wild and wondrous. Wilson also reminded Udall that, together with Glen Canyon and its tributaries, the confluence country had been considered for inclusion in the national park system once before, back when FDR's Interior head Harold Ickes had tried to create an Escalante National Monument that would have reached all the way from Lee's Ferry in the south to the towns of Moab and Green River in the north, the largest preserve ever proposed in this country. All of that amazing terrain deserved park protection back before the war, Wilson said, and it deserved it still. In the case of Glen Canyon, it was, of course, too late to preserve it in its natural state, what with a kind of canyonlands Lake Mead about to fill it. But above what would become the Glen Canyon reservoir, the land remained pristine, trammeled only a little by uranium exploration and nearly a century of cattle grazing, and yes, the confluence country would make a splendid national park.

On the long Fourth of July weekend, 1961, Stewart Udall again was ensconced in the desert Southwest, again heading an entourage of administrators, legislators, and reporters into the back of beyond, again looking at the possibilities for a new national park, this time immediately upriver from Glen Canyon instead of at its flank. Agriculture Secretary Orville Freeman and Utah Governor George Clyde joined Udall this time, but Commissioner Dominy was conspicuous by his absence. For six days, the two secretaries, their families, and two dozen taggers-on floated the ruddy Colorado and the pea-soup Green, camped along their willowed banks, explored Grandview Point and Upheaval Dome on the Island in the Sky, hiked from Chesler Park in the Needles to dramatic Druid Arch, a formation so remote in terrain so wonderfully rugged that Bates Wilson himself had discovered it only four years before. The Republican governor and Republican congressmen argued all the while that despite its grandeur, this country had too much oil and gas and minerals potential simply to "lock it up"; the Kennedy Democrats countered at every turn that the economic impact of park development would be far more stable and long-term than that of industrial development. The usually cordial arguments sometimes lasted late into the campfire-lighted nights, the notion of the need for another dam and reservoir in this region getting surprisingly little lip service.

By the end of his first summer in office, Stewart Udall was as cognizant as David Brower had hoped he would be of the weighty responsibilities he held as the head of the Department of the Interior, but he was also feeling rather buoyant—confident of what could be achieved in the coming Kennedy decade, excited about working to secure an equilibrium between development and preservation, one that inevitably would leave extremists on both sides discontented, but which would best serve the nation as a whole. And Udall envisioned the Colorado Plateau—the great river slicing through its heart—as a region where that kind of balance might foremost and symbolically be struck, a place where vital water and hydropower development could work hand in hand with the protection of some of the most inspiring wildlands in the world.

On the map in Udall's office you could see it, starting there at the spot where Hoover Dam stood and following the twisting river upstream—the once-calamitous river corralled in Black Canyon, the reservoir back of the dam providing electrical power and recreation and reaching all the way to the monument and the park that protected the Grand Canyon country. Upriver from Grand Canyon, and just beyond the ferry crossing that bore his great-grandfather's name, his own middle name, Glen Canyon Dam was rising, a dam that would drown a canyon he had been enchanted by, it was true, but one which would provide myriad agricultural and urban benefits with the money its giant power plant produced. Lake Powell—beautiful in its own way, he hoped, and managed by the park service as a National Recreation Area, much like Lake Mead—would provide a negotiable waterway for millions of people into awesome canyons at its edges that would remain unmarred, some of which, like the land surrounding Rainbow Bridge, he hoped would become a vital new park in their own right. And above Lake Powell, where the rivers still would cut their primeval canyons, another park—perhaps the wildest and most wonderful of all the preserves in the Southwest—would serve as a kind of guardian of the natural order of things, a reminder of what this region had been before people tried to tame it. The map, with both Navajo Rainbow and the confluence park (which Bates Wilson had suggested they simply call *Canyonlands*) penciled in, was a map bent on *balance*, the secretary believed—cascading rivers and spreading lakes, water and power and wilderness.

Yet if that kind of symmetry were going to be achieved during the course of Udall's tenure at Interior, his attention would have to be weighted toward conservation, at least for the short term. The Colorado River Storage Project was well becoming a reality, and Reclamation was at work on plans for further development downstream—the major Arizona water project that had been dreamed of since the days before Hoover Dam; cattlemen had had the run of the federal lands for grazing so long they thought they owned them, and the extractive industries long had drilled and pumped and blasted wherever their whims had led them. In order to balance those fundamentally important kinds of development, Udall believed the Colorado Plateau now needed a strong dose of conservation, sweeping protection for the best of its strange and spectacular landscapes. The Navajo Rainbow and Canyonlands preserves would be essential in that regard, and Udall was convinced that they should be congressionally mandated parks rather than monuments created by executive order; the boundaries of the recreation area that would surround Lake Powell should reach as far away from the water as possible to protect side canyons and mesas unaffected by the reservoir; perhaps nearby Arches and Capitol Reef national monuments should be enlarged and made parks themselves; perhaps the park boundary of the Grand Canyon itself should be expanded. Imagine a succession of parks, monuments, and recreation areas reaching all the way from the blue water of Lake Mead to the slickrock heights of Dead Horse Point near Moab, 700 contiguous miles of federally protected canyon country, a winding, *balancing* ribbon of land and rivers and lakes!

Nationally, Udall knew—you could feel it in the same way you could feel the collective promise of these nascent Kennedy years—Americans were growing cognizant of the importance of conservation. They were beginning to know that they had to nurture their lands as well as put them to use, and the young Arizonan felt fortunate to be in a position to play a role in that regard. Bills first introduced back in 1957 that would create "wilderness areas" out of pristine pieces of federal lands now were making headway in Congress; a few National Park Service officials, as well as members of the Outdoor Recreation Resources Review Commission, currently were suggesting, albeit cautiously, that the nation should set aside a system of undeveloped and undisturbed river corridors; there was both idle talk and serious discussion ongoing about the creation of national wildlife refuges and the need for nationwide clean-air legislation, amazing chatter about the possibility of a new agency whose sole role would be to work to limit pollution. Speaking at the Sierra Club's Seventh Biennial Wilderness Conference in San Francisco that fall, Udall tried to capture the spirit of the country's broadening conservation sentiment by borrowing words from Wallace Stegner, telling his audience that he had had no choice but to abandon his own prepared remarks once he'd come across a copy of the letter Stegner had written almost a year before asserting that "something will have gone out of us as a people if we ever let the remaining wilderness be destroyed." Reading at length from the letter, the secretary concluded, still quoting, " 'We simply need that wild country available to us, even if we never do more than drive to its edge and look in. For it can be a means of reassuring ourselves of our sanity as creatures, a part of the geography of hope.' " Yes, *the geography of hope,* that was what Udall knew had been entrusted to him.

It wasn't long before Udall had wheedled and cajoled Stegner—now a Sierra Club director and deeply involved with a variety of conservation issues as well as his professorship at Stanford—into serving as his special assistant, as a consultant on park expansion and wilderness issues, particularly those that were current out in the Colorado Plateau country. Together, the two men toured the areas of the proposed Navajo Rainbow and Canyonlands parks, and Stegner heartily endorsed the secretary's plans. Later, this time in the company of Joe Carrithers, Wirth's assistant at the National Park Service, Stegner visited Capitol Reef National Monument, then traveled south to the canyons of the Escalante, where the writer was at once crushed by the realization of how much of that Glen Canyon tributary would be buried by the reservoir and elated by what, in fact, would be spared. Reporting back to Udall, Stegner recommended park status for Capitol Reef, recommended some sort of hands-off protection for the spectacular Escalante watershed, and he endorsed as well Udall's new notion of a book that would elucidate the idea of saving land for public purposes. With the help of Stegner and a brace of young conservation writers in the autumn of 1961, Udall got to work on the book that two years hence would become *The Quiet Crisis.* The move toward balance, it seemed, now was beginning to snowball.

But if the secretary was energetic and possessed of a vital new vision, he was also sadly naive, a kind of administrative innocent once the very conservationists with whom he felt such kinship began to attack him so stridently and so personally, once the Navajos began to accuse him of double cross, once people in southern Utah began to claim that this Arizonan was trying to steal their lands and stifle their livelihoods. Despite his new park proposals, despite his words in support of wilderness, Udall still appeared to many conservationists to be immorally in bed with the Bureau of Reclamation, and he still was doing nothing to block the coming, cursed reservoir from violating Rainbow Bridge. Writing in the October issue of the *Sierra Club Bulletin*, David Brower charged, "We now know that the life expectancy of one of America's greatest scenic resources, including the pristine approach to Rainbow Bridge, is reduced to fourteen months. The exact time is not important here. What needs to be chronicled is a flagrant betrayal, unequaled in the conservation history that sixty-eight years of *Sierra Club Bulletin*s have recorded." As Brower surmised the situation, Udall willingly was being hoodwinked by Floyd Dominy and his henchmen at Reclamation, the secretary himself actually *wanting* to build the barrier dam, but acquiescing to pressures from the Bureau when he spoke about how the intruding reservoir would do the lesser damage.

Five months later, in March 1962, in an open letter to Udall, published in the *Bulletin* and in newspapers across the country, Brower's anger had grown caustic:

> *Preclude* impairment, the law says. It doesn't say to plead
> excessive cost. Or to hustle through some kind of "geological
> whitewash." Or to arrange a series of show-me trips to lead
> editors and congressmen into believing that protection is just too
> much load on taxpayers and would tear up the countryside with
> roads and scars. . . . And when the law says *preclude impairment*,
> it spells it out in unmistakable words: "no dam or reservoir . . .
> shall be within any national park or monument." Not maybe.
> Not yes but. Just *NO*. . . . If Rainbow is not protected, it is not
> your subordinates who will be responsible. It is you. You,
> Secretary Stewart L. Udall, the man who dared to have a dream
> that others hadn't the courage or boldness to dream. And
> President John Kennedy, who you let think your dream was
> worth dreaming. Don't let him down. Don't let yourself down.
> Nor us.

Although a little wounded by Brower's charges, Udall responded quietly, saying simply that Interior recognized its responsibility under the provisions of the CRSP act, noting that protection funds had been part of the department's budget requests for the last three years, but that they had been denied by Congress each time, affirming that he would try again to get the money to do the job. What Brower didn't understand, Udall explained to associates privately, was that while, yes, he agreed with him that the CRSP language should not be ignored, what he personally wanted Congress to do was to amend the act, to admit formally it was doing so, to *legally* let the water slip under the bridge and forget about that catastrophic conservationists' dam once and for all. What Stewart Udall wanted to do instead was to concentrate on wrapping a national park around Rainbow Bridge, a national park that for some reason Dave Brower and his cohorts now were choosing to ignore.

Lately, even the Navajos had begun to rise up against him. Chairman Paul Jones was a man with whom Udall had had a warm relationship since his days as a congressman, and following the helicopter trip to Rainbow Bridge, Jones had told him privately, as well as the assembled reporters for attribution, that he would do everything he could "to come to an agreement to trade other lands for acreage in the area." Now, however, less than a year later, Jones somehow had changed his mind. "Development of recreation and tourist attractions are certainly desirable," the chairman had written to the secretary, "but . . . they will not afford job opportunities. They will not materially assist in the development of reservation resources which must be accomplished in order to give the people an alternative to their less than subsistence grazing operations."

Jones's letter had been in response to an official Interior department communication with the Navajo tribe proposing the exchange of 274,000 acres surrounding Rainbow Bridge (basically the largest of the three park areas suggested earlier, minus Navajo Mountain) for equal acreage the Navajos long had been known to covet. But now that the trade officially was on the table, Jones had responded that he did not feel he "would be justified in recommending cession of such a substantial portion of the Navajo Reservation for park purposes in consideration of any, either, or all of the 'trading stock' you mention. Actually Stew," Jones wrote, personalizing the matter much the way David Brower had done, "I am profoundly shocked that you would propose as an item of 'trading stock' the Church Rock–Two Wells area," a region long the subject of a separate and, as far as Jones was concerned, entirely unrelated land exchange. To make matters worse, Jones was outraged that his old friend had publicized the details of the exchange proposal before Jones had had a chance to respond to them. "I do not know what possible purpose you had in mind in releasing a copy of your letter to the press before I had an opportunity to reply. I cannot think of any useful purpose served thereby, and I am deeply incensed as a result of your having followed this course."

If Jones had thought it through, it seemed to the beleaguered Udall, he might have realized that the proposal had been publicized, in part, in an effort to mollify the screaming conservationists, a vain attempt to convince them that they might indeed end up with something better than a barrier dam at Rainbow Bridge. But, based on his letter at least, Paul Jones wasn't interested anymore in issues pertaining to the reservoir, the bridge, or the possibilities for the park. All Paul Jones and the Navajos were willing to say was that they were out of the trading mood.

If it wasn't enough that Dave Brower now seemed to consider him a traitor to the conservation cause, that Paul Jones had decided he was just another conniving white man, the people in southern Utah, his kindred Saints, had begun to shout that Udall was trying to lock up any chance they had for economic prosperity, that this Canyonlands National Park scheme was just another federal land grab that would be battled with bullets if it had to be, regardless of whether a Mormon was playing the front man. Utah's Republican Senator Wallace Bennett was aghast at the idea of yet another park, rightly convinced as he was that it would "forever remove the land from the possibility of further development." Republican Governor George Clyde told the secretary he was not opposed to tourism, only to the creation of what he called a Frankenstein that would smother future development. And the commissioners of Grand, San Juan, Wayne, and Garfield counties—Republicans and Democrats in roughly equal numbers—were wasting no time on words, instead instructing their road bosses to crank up the Cats and to go blade dozens of miles of new "county roads" they theretofore hadn't seemed to need, roads slashing like knife cuts across Udall's boondoggle park. He and Bates Wilson were damn near alone, it was beginning to seem to the secretary, two voices crying in a suddenly threatened wilderness, and his harmonious notion of land use seemed to be teetering on the brink.

Down on a sloping shelf of land at the flank of Wahweap Creek, Art Greene was ready for the lake to rise. Like the secretary of the Interior, Greene fervently believed that tourism was the future of the whole of this slickrock country. Like Udall, Greene wanted to see the creation of the Canyonlands park up north in the Needles country as well as a new national park surrounding Rainbow Bridge. Like Udall, Greene wanted the Glen Canyon diversion gates to close and the reservoir to rise toward the legendary arch that long had been his livelihood.

Art Greene had hated the idea of Glen Canyon Dam when he first heard about it back in the years soon after the war, but by the time Bureau geologists, hydrologists, and engineers were scouring the canyon in the 1950s, Greene successfully had negotiated the long-term lease of 3,840 acres of land that one day would lie at the reservoir's edge—beating the goddamn dam, he would tell you, by moving upstream of it—then had bulldozed in a rough road and an airstrip, built a cafe and eight stone cabins, and started planning the boat ramp that in a decade would provide access to open water. But during the subsequent years that Greene had spent tangling with the Bureau of Reclamation and the National Park Service over whether he and his family could stay in the Wahweap area, his tour boats still somehow had to get to Rainbow Bridge, so Greene, his son Bill, and three sons-in-law had cranked up the dozers again and carved a 24-mile road connecting their little Wahweap Lodge with a good launching point on the river at the mouth of Kane Creek—48 miles of dust and bounce and rattle an unfortunate part of every excursion to show tourists the arch.

Now, with the summer of 1962 approaching, with a 200-foot boat ramp ready to get wet on Wahweap Point, and with the park service and the boys at Reclamation resigned to the fact that the Greene clan might be around as long as their dam would be—the Greenes' Canyon Tours, Inc., finally having become the officially sanctioned Wahweap-area concessionaire—the future was looking bright. The reservoir would start to rise in eight months or there-abouts; water quickly would slip into Wahweap Canyon and climb up the rock toward the lodge and the trailer park; and trips to Rainbow Bridge *and back* would be feasible in just a single day. No one in the Greene family, least of all the wizened, toothless patriarch himself, was eager to see the canyon go under, to lose the favorite haunts of more than twenty years on the river, but the reservoir—a canyon filled with water instead of one cut by water—was the coming thing, and the Greenes long since had demonstrated their determination to make the best of it.

On May 26, in fact, Art had made good money escorting a big group of politicos through the length of the canyon, a "Governors' Farewell to Glen Canyon on the Colorado River" (a governors' good riddance, if you wanted to know the truth), hosted by Utah Governor George Clyde and Arizona Governor Paul Fannin. "Though we are losing a river and a canyon," Clyde told the press corps that had accompanied the tour, "we gain a beautiful lake 186 miles long with 1,800 miles of shoreline in the most spectacular of settings. Lake Powell is destined to become the nation's most scenic recreational playground."

During three days on the river, Art Greene, his son, and sons-in-law had shown the governors and their guests California Bar and its gold-mining relics up near Hite, the Stanton dredge, listing badly in the chocolate river channel, Anasazi ruins in Lake Canyon, the Mormon steps at Hole-in-the-Rock, Music Temple, and Hidden Passage, and helicopters had been pressed into service to save the group the six-mile trek from the river to Rainbow Bridge. Once in the shadow of the arch on the morning of the final day of the trip, Jim Eden—already on board as superintendent of the fledgling Glen Canyon National Recreation Area—had outlined the park service's plans for the place: bids soon would be let for the construction of a floating marina capable of withstanding the regular fluctuations of the reservoir in the little canyon, a facility complete with permanent houseboats for rangers and service personnel, a public information center, snack bar, gas pumps, and a tie-down dock, plus a diesel-powered generating plant and water and sewage systems. Scheduled to be in operation by June 1963, the marina would be assembled and tested near Wahweap, then towed to its anchorage at the mouth of Bridge Canyon, a mile below and out of sight of the bridge. Together with marinas that initially would be constructed at Hite, Hall's Crossing, Bullfrog Creek, and Wahweap, the park service would spend $36 million to develop recreation at Rainbow Bridge and throughout the reservoir area, Eden told the governors' group. Rainbow Bridge would become truly accessible for the first time, according to the park service representative, easily approachable by hundreds of thousands of people a year on a smooth highway of water, and with a hand-rolled cigarette stuck to his lips, the lips spreading into the hint of a grin, Art Greene had listened to Jim Eden's enthusiastic words, knowing that a hell of a lot of those people he was referring to were going to be riding Greene-family boats to the bridge.

At the same time that the governors and their guests were saying good-bye to the canyon that most of them had never seen before, the Sierra Club similarly was encouraging its members to go take a last look at Glen Canyon. Unless something, or some*one*, soon saved it, David Brower wrote in the *Sierra Club Bulletin*, the gates of the diversion tunnels would close in January, and Glen Canyon would become just "one more fluctuating reservoir instead of one of the most exquisitely beautiful wild canyon experiences in all the world." Three separate club-sponsored trips were scheduled for the coming summer season, and the *Bulletin* explained that numerous commercial outfitters could be contacted directly to arrange trips, and people who had their own boats and river gear simply could head for Hite. "It is not comfortable for anyone to contemplate what is about to happen to one of the most superb exhibits in all the plateau province of John Wesley Powell," Brower continued. "But before the lake that is to bear his name drowns the great places he discovered, you owe it to yourself and to the future to know and to remember these things lost."

Through June and July, the river swelled with the winter's runoff and surged unimpeded through the canyon, carrying more people on its broad back than it ever had before in a single summer. By August and September, the river's flow had subsided, but still more people were making week-long pilgrimages to the canyon—a dozen boats pulled ashore at the mouth of Forbidding Canyon on the day that the National Parks Association, on behalf of several allied conservation groups, filed suit in U.S. District Court against Interior Secretary Stewart Udall in an effort to force him to forestall the closing of the diversion tunnels until and unless Rainbow Bridge had been protected, and other boats tied fast at that spot on the day soon thereafter when the court ruled that the conservationists did not have legal standing to sue the government. Only ice floated the frigid river on January 18, 1963, the day Interior Department solicitor

Frank Barry delivered his conclusion to Secretary Udall that "the provisions originally included in the Colorado River Storage Project Act calling for protective measures at Rainbow Bridge National Monument have been suspended by the Congress and are no longer operative. Under the present state of the law applicable to Glen Canyon, it is the intention of the Congress that construction and filling of the reservoir should proceed on schedule without awaiting the construction of barrier dams at Rainbow Bridge."

With that legal opinion in hand, Udall informed Floyd Dominy of what Dominy had known in his bones since the beginning of this fuss—that the gates would be closed on schedule—Dominy in turn alerting Ole Larson in Salt Lake City and Lem Wylie in Page that nothing had higher priority than getting the west tunnel gates shut. David Brower in turn flew to Washington, hoping to see the secretary—waiting outside Udall's office on the morning that Lem Wylie's crews began to struggle to lower the slide-gates—Brower believing that an impassioned personal plea somehow might dissuade the secretary with whom he not so long ago had felt he had much in common, so many kindred sentiments. But Udall wouldn't see him, couldn't see him because he and Dominy were about to meet the press, announcing Reclamation's plans for two new dams downstream from Glen Canyon, dams which likewise would invade a national monument and even a national park. With Stewart Udall's acquiescence, Brower observed in horror, and with the precedent they had succeeded in establishing at Rainbow Bridge, Dominy and his beaver-boys now wanted to dam the Grand Canyon.

Lake Powell had risen 200 feet above the riverbed, and it lapped at the dam just 15 feet below the intakes of the river outlet tubes on the June day in 1963 when the secretary of the Interior received the book from David Brower. Advance copies of *The Place No One Knew* just had come back from the bindery, and Stewart Udall was one of the first people in the country to see the finished book, accompanied by a long letter from the Sierra Club administrator. Two thousand miles away from Udall's office in Washington that day, the Colorado had surrendered its current in Glen Canyon, slackwater already reaching more than 100 miles upstream, far beyond the confluence of the San Juan and the Escalante, burying Stanton's dredge, submerging sandbars and gravel beds, surrounding the rough-barked trunks of cottonwood trees with water, forcing rabbits and rodents, coyotes, foxes, and deer to move to the talus slopes, sending this strange new waterway far into Forbidding Canyon, into the Narrows near the mouth of Bridge Canyon, the muddy, debris-laden lake now just a mile from the abutments of Rainbow Bridge. Udall read in the first pages of Brower's undeniably beautiful book his charge that on January 21, 1963, "the last day on which the execution of one of the planet's greatest scenic antiquities could yet have been stayed, the man who theoretically had the power to save this place did not find a way to pick up the telephone and give the necessary order."

"Because the public and the Congress did not know then what they know now," Brower wrote in his companion letter to Udall, "Glen Canyon Dam was authorized, and the destruction of a major recreational resource is now under way. . . . The mistake has been made, however, and can no longer be corrected. Our Glen Canyon book documents a tragic and unnecessary loss. . . . I hope you will take the time to read every word of the Glen Canyon book. It will take you about three hours to do so, but I believe you must, simply because the nation's chief conservation officer ought to have read all that is said. Wallace Stegner says the book is not a voice in the wilderness, but a chorus of voices for the wilderness. It presents some of the most important things Americans have yet said about the crisis in conservation."

David Brower had given up. For the first time since he had come to rue his role in the decision to dam Glen Canyon, for the first time since he had been shocked by the realization of what glories were secreted there, he now had acknowledged that the tragic mistake had been made and no longer could be corrected. Unlike the Echo Park struggle of a decade before, the battle for the fate of Rainbow Bridge had been lost, and with it, Glen Canyon now was gone. Brower, the 20,000 members of the Sierra Club, and their allies in conservation organizations across the country had been beaten this time, beaten by an Interior secretary who they had been hoodwinked into believing was a brother-in-arms, by a Reclamation commissioner who was maniacal in his determination to dam the West's great canyons, by a spitfire project construction engineer out in Arizona who had made sure that his career-capping job had been completed on time, the incredible Colorado subdued on schedule.

Instead of being one of Brower's trademark calls to action, one of his impassioned pleas that something had to be done *now* to save a piece of the vanishing wild, *The Place No One Knew* plainly and rather reverently was a eulogy, a lovely, heartbreaking lament for the loss of something extraordinary. If in seeing this book, the secretary of the Interior came to regret what he had been a central party to—much the way Brower himself had come to hate his own ignorance and inaction at the end of the 1950s—then perhaps the book and Glen Canyon itself would serve a cautionary role in preventing a similar tragedy from happening ever again. Glen Canyon was gone, Brower now admitted, but because of its awful loss, perhaps other wild places might survive the Kennedy years and beyond, might survive forever.

From his perspective, it appeared to Stewart Udall that day in June that Brower was in danger of losing sight of the whole forest of conservation causes by focusing so relentlessly on the tree that was Rainbow Bridge, on the glen that was Glen Canyon. Although he truly was disappointed that Congress never had gone on record saying it wouldn't sanction the building of a barrier dam, Udall remained convinced that the right thing had happened in Bridge Canyon. Although his hopes for a Navajo Rainbow National Park had all but disappeared—the Navajos resolutely refusing to discuss the land trade further—at least the original national monument would remain, surrounded by Navajo-owned land so rough and remote it was in virtually no danger of development, surrounded too by the federally controlled lands of Glen Canyon National Recreation Area. The park service's marina would float outside the monument boundary; there would be no gas pumps or snack bars inside it (as there were at most *other* monuments, by the way). Apart from the old trail descending to the bridge from the ruins of Rainbow Lodge, the only evidence there of humankind would be the thin finger of the reservoir, an intrusion on wondrous Nonnezoshi, yes, but how much more of an invasion than the roads and developments that were parts of every other park preserve?

As for Glen Canyon itself, Udall's 1960 float trip down the length of the canyon, plus his several subsequent excursions into the region, had given him a clear and lamentable image of what would be lost beneath the lake. But the reservoir that would fill Glen Canyon *was* needed—needed to support important irrigation projects in the upper basin, needed to provide dependable flows to the lower basin, needed to supply a million kilowatts of peaking electrical power. And regardless of whether you considered motorboating, waterskiing, and bass fishing lower forms of recreation than hiking or river running, Lake Powell would serve hundreds of thousands of people, providing water-borne enjoyment and entertainment instead of solitude, but providing a very popular kind of leisure activity nonetheless.

Two years into his tenure at Interior now, Udall still believed in balance, in weighing the development needs of the nation against the equally important need to keep parts of the planet primeval, and he and his people were making progress toward that end. If a park around Rainbow Bridge now looked unlikely, at least the Canyonlands park appeared to be on the brink of becoming a reality; congressmen such as John Saylor and Udall's brother Morris (who had succeeded him as an Arizona representative) were working hard to overcome the lobbying efforts of the oil and gas and minerals industries, as well as the more parochial opposition of Utah's own congressional delegation. The plan to make a park out of that rocky and spectacular place already had more than enough support for passage in the Senate; a bill probably would pass the House with only a little more legwork; and equally promising, it now seemed certain to Udall that House and Senate versions of the wilderness legislation originally drafted by the Wilderness Society's Howard Zahniser and shepherded by Minnesota Senator Hubert Humphrey would reach a conference committee before the current legislative session was completed at the end of 1964, and President Kennedy surely would sign both bills into law as soon as they reached his desk.

Brower had written to him about "the crisis in conservation," and Udall very willingly agreed that a crisis existed, but he did not believe that the Sierra Club or its beautiful coffee-table books somehow had cornered the market on drawing the public's attention to the dimensions of the problem. Biologist Rachel Carson's *Silent Spring*, published a year before, had shocked Americans everywhere into a realization of the dire threat posed by chemical and industrial pollution; Udall's old friend from Tucson, Joseph Wood Krutch, and his informal adviser, Wallace Stegner, together with T. H. Watkins, Alvin Josephy, and Supreme Court Justice William O. Douglas were writing splendidly, it seemed to him, about the need to preserve some of the frontier intact, and Udall's own book, *The Quiet Crisis*, out just now, was intended to paint a broad portrait of the condition of the American land and to call for its careful husbanding. There had been losses of wilderness and failures of resolve on the parts of conservation-minded people during the course of the Kennedy administration, Udall willingly would admit, and a few more were likely to follow. After all, water development, mineral and agricultural development, the use of the land for housing and manufacturing, and for traveling from place to place had to continue if the nation were to survive and prosper. But as long as he played a part in it, Udall wanted people like Dave Brower to know, there would be companion efforts, efforts absolutely in tandem, to see that healthy lands and wildlands survived as well.

The vigorous promise of the Kennedy administration came to an end in the early afternoon of November 22, 1963, but at Interior, in Congress, and throughout much of the country, the slowly crescendoing awareness of the importance of conservation—conservation as an imperative rather than just some sort of clubby, fly-fishermen's frivolity—continued. In 1964, during the first full year of the Johnson administration and the final session of the eighty-eighth Congress, Canyonlands National Park came into being, a preserve that by park service edict would remain essentially primitive, contiguous with the northeastern boundary of the Glen Canyon National Recreation Area, the park encompassing 338,000 acres, more than 500 square miles, in the region surrounding the deep, canyon-bound, seemingly subterranean confluence of the Green and Colorado rivers. In September of that year, the Wilderness Act became law, immediately designating as wilderness—forever off-limits to the hand of humankind—9.1 million acres of national forest, national park, and national wildlife refuge lands, and establishing a system for the review and possible inclusion of additional lands in the succeeding years. And during the same span of months, Congress began grappling with a bill that would preserve portions of a select group of rivers around the nation in their primitive states, ribbons of water that would be managed not as parks per se but as wild corridors, floatable but otherwise inviolate. By early 1965, President Johnson was calling for approval of the wild rivers legislation in his State of the Union address, and Interior Secretary Udall's book *The Quiet Crisis* was being hailed as a kind of blueprint for the new *environmentalism*, a word of recent coinage that seemed to encompass collective concerns about America's land and air and water.

Elsewhere on the publishing front, the Sierra Club's *In Wildness Is the Preservation of the World* had become the best-selling trade paperback book in the nation; Eliot Porter's second exhibit-format book, *The Place No One Knew*, still in hardcover, already was in reprint, and a new book, *Time and the River Flowing*, focusing on the Grand Canyon and the threat that the Bureau of Reclamation now posed there, just had been released. Since *This Is Dinosaur* had been published back in 1955, books had played a surprisingly important role in alerting people to the dangers inherent in the Bureau's big dam plans, in drawing attention to the swelling urgency of a variety of land and water issues, and the Sierra Club's beautiful books in particular now were bringing a kind of emotion to bear that the conservation movement never had been assisted by before. If the Sierra Club's own internal strictures had attempted to muzzle David Brower as a direct governmental lobbyist, he had become eloquent as a publisher, editor, writer, and occasional photographer instead, his success in swaying opinion never more evident than in the fact that Floyd Dominy now was countering with a book of his own, an official Bureau of Reclamation publication, printed by the Government Printing Office at public expense, a flashy, full-color coffee-table tome in its own right entitled *Lake Powell: Jewel of the Colorado.*

Photographed in color by California-based landscape photographer Josef Muench and by the ebullient Dominy himself, the text of the book belonged solely to the commissioner, the burly, big-drinking, swaggering head of the outfit that *built things* waxing poetic as he posited the notion that

TO HAVE A DEEP BLUE LAKE
WHERE NO LAKE WAS BEFORE
SEEMS TO BRING MAN
A LITTLE CLOSER TO GOD.

Much as the conservationists always had, Dominy too tried to tinge his words with a touch of theology, continuing to note that "there is a natural order in our universe. God created both man and nature. And man serves God. But nature serves man." In the case of Glen Canyon, Dominy explained, "man has flung down a giant barrier in the path of the turbulent Colorado in Arizona. It has tamed the wild river—made it a servant to man's will," creating in the process a wonderland of blue water, a place where once-remote splendors such as Rainbow Bridge at last had become accessible. "Now all of you can see it—easily. Your boat will moor to floating docks at the entrance to Rainbow Bridge Canyon. Then you take a walk on a trail along the canyon's side. You'll find the bridge undamaged by Lake Powell's waters . . . and you can marvel at its arched and graceful beauty. . . . If I sound partisan toward Lake Powell, you are correct," Dominy confessed. "I am proud of this aquatic wonder and want to share it with you."

Just seventy-five cents would buy a paperback copy of *Lake Powell*, and thousands were distributed free of charge during the months that led up to the August 1965 congressional hearings on Reclamation's proposals for two new dams in the still unimpounded canyon that separated Lake Powell from Lake Mead, dams proposed for the great Grand itself, copies of the book going to every representative and senator on Capitol Hill, to hundreds of newspapers and magazines, to chambers of commerce and water boards and the heads of western corporations, a calculated attempt on the part of the commissioner to tell the *truth* about Lake Powell, to counter the conservationists' alligator tears about the death of Glen Canyon with the affirmation of the birth of something better, a shining blue jewel in the desert, "the man-made rock of the dam become as one with the living rock of the canyon."

But instead of being brought up short by Dominy's picture book—their distortions and false accusations at last exposed to the light of governmental reason—Brower and the conservationists responded with ridicule, lampooning *Lake Powell* in the *Sierra Club Bulletin* as the work of an embattled agency using "poetry and pictures to sell its religion. But a bible isn't written overnight. The ring is hollow. It is man worshiping mammon, and, in particular, one man worshiping himself." The commissioner was bowing before technology, it was clear to the conservationists, when properly he should have been praising the pristine earth. Strangely, however, it wasn't Floyd Dominy who seemed to them to be the true betrayer, his hands now bloodied by printer's ink. Dominy was Dominy, acting out another of his water-impounder's fantasies. Instead, the man who genuinely incensed the opposition, who seemed so duplicitous he made the conservationists want to scream, was

Stewart Udall, the supposedly environment-minded secretary of the Interior who had written in a foreword to Dominy's book that "once in a blue moon we come upon almost unbelievable beauty. Such was my reaction at my first sight of Lake Powell . . . an exciting new concept of conservation: Creation of new beauty to amplify the beauty which is our heritage."

Was Udall duped into letting his name be attached to such drivel? Or worse, did he *believe* his ridiculous words? You simply couldn't praise Lake Powell at the same time that you willingly were surrendering Rainbow Bridge and working to secure dams in the Grand Canyon and still claim you sought some sort of equilibrium. If that was balance, it seemed to the conservationists who sadly watched the water rise—the branches of the old cottonwoods now looking like the desperate hands of the drowning—the secretary of the Interior's scales were badly skewed.

10

FLOODING THE
SISTINE CHAPEL

After thirty-two years of government
service, all of them but his stint in the
South Pacific during World War II spent
with the dam-building Bureau of Reclamation, Lem Wylie was ready to retire. He had joined the Bureau in 1932
as a chainman on a survey crew plotting
the progress of Hoover Dam's gargantuan diversion tunnels, and his career
had seemed to parallel Reclamation's
resolute rise to power and eminence
throughout the arid West. It was with
Hoover Dam that Reclamation had begun its emergence from a bungling,
bankrupted branch of Interior intended
to pour a little water onto parched and
marginal desert farms into the world's
crack civil-engineering outfit, a billion-
dollar-a-year agency that built the biggest and most impressive dams and
hydropower facilities ever conceived,
and did so with a workaday calm and
assurance that was nothing short of
stunning. In little more than three
decades, Reclamation had constructed
Hoover, Bonneville, Shasta, Grand
Coulee, Fort Peck, and now Glen Canyon, each dam entering the record

books at or near the top of the list of the world's highest, longest, and most massive structures. Glen Canyon Dam was the one-hundred-ninety-fourth of the Bureau's water-impoundment projects so far this century, Lem Wylie playing a part in six of them, playing the czarist central role, the part of the man who shouldered the son of a bitch, in his last one.

During the past seven and half years, Wylie had overseen the transformation of a wasteland into a place now marked on the maps—a place where a monumental plug blocked the course of the Colorado, where a scrappy little city rose above the blowsand, broad highways and high bridges reaching this region now, huge powerlines stretching away, ready to deliver the product that soon would flash from the dynamos. And this was the way to hang it up, Wylie knew, not waiting until they kicked you so far upstairs you had nothing to do but sharpen your pencils, but building a magnificent dam, pouring more mud than anyone could have imagined, doing it right, building a structure that would last a millennium, maybe two, then getting out while you still had some grit and some self-respect.

Yet although he was leaving the Bureau of Reclamation, Wylie was not going to quit working. "I have no intention of ever retiring," he told the many people gathered at the Empire House for his farewell dinner on January 24, 1964. On the wall behind the dais was a large drawing of the Department of Interior seal, accurately executed except that the bull bison who normally stood stalwartly in profile now was lying in the grass, asleep. It was an insider's kind of humor, and Wylie thought it was funny all right, but it compelled him to say something else: "I'm a guy who doesn't know how to do anything but get up early and go out to the job. I'm a guy who's got to have a job or he doesn't know who he is."

In a month's time, Wylie and his wife, Maxine, would have moved out of their pleasant house on Navajo Drive, put much of what they owned in storage, and Wylie would be at work again, an employee of a Colombian construction company this time instead of the government of the United States, working as project construction engineer—working again as the guy at the top of the heap—on a huge dam and hydropower project just under way on Colombia's Río Nare, the desert abandoned for the tropics, the paltry bureaucratic pay of the Bureau of Reclamation abandoned for some real money for the first time in his life, the Spanish he learned as a boy in New Mexico serving him in good stead now at age fifty-eight.

But Wylie wasn't thinking much yet about Colombia. On the evening he said so long to his co-workers and friends in Page, he assured them that this job had been the highlight of his career; no other had been so full of challenges, none had succeeded so well. "It's all a question of people," he told them, "and here at Glen Canyon I've been privileged to work with some of the best." Wylie reminisced about the first time he and Louis Puls had seen this windy, sandy country, had peered down into the deep hole for the first time, his excitement immediate; about the early days in the old school in Kanab and the long, kidney-busting ride to the canyon rim; about the first monkey-slide with a single brakeman up on the rim who had to watch for a mark on the cable to know when to keep the cage and its occupants from augering into the riverbank below; about the strike and how the townspeople had toughed it out, showing him a kind of resolve he'd always remember; about how the men who worked in the hole had left their signatures on a piece of history, and about how he knew that all the people in front of him that night were thankful that that filthy hat of his had found a home in the dam.

Page and the Glen Canyon country had a bright future, Wylie wanted them to know. The first of the eight generators would go on line in about six months, and, in a way, that would be a bigger milestone than topping out the dam had been. Under his successor Vaud Larson's able command, the rest of the turbines would begin to spin on schedule, and although the town's transient population would continue to dwindle as each phase of the project was completed, Page, as a place where people wanted to live and where tourists would flock on their way to Lake Powell, seemed certain to prosper. More than 10,000 people had visited the recreation area in 1963, and Reclamation and the park service had estimated that half a million people a year would vacation here by the end of the century. Tourism would provide a stable economy over the long term, and for the short haul, the Bureau still would be very much on the scene, completing Glen Canyon over the course of the next three years, as well as getting to work on Marble Canyon Dam, 40 miles downstream from Lee's Ferry, much of the construction boom it would create surely spilling upstream to Page. Floyd Dominy and the fellows at Reclamation still had a lot of work to do to fully harness the Colorado, and Wylie assured his listeners that he would be back from time to time, making sure they did it right.

It was the Supreme Court's ruling on the eleven-year-old lawsuit, *Arizona* v. *California*, that had led to the current plans to construct Colorado River dams in Marble Canyon and in lower Granite Gorge near the mouth of Bridge Canyon, both sites well within the geologic, if not the political, boundaries of the Grand Canyon. Back in 1952, with the upper basin slated to get its share of the water divided by the Colorado River Compact via the newly proposed CRSP, and with California cavalierly diverting several million acre-feet of the Colorado by now, Arizona, given the back of the hand of the water wizards for three decades already, finally went to court. What the state of Arizona needed was a dam and diversion project that would send a major share of the Colorado, coursing along the state's northern plateau and western boundary, to the populous and quickly growing Phoenix-Tucson corridor to the south, where subterranean aquifers were being depleted at an alarming rate. But since Congress had blocked a bill authorizing such a project in 1950, the state had had to settle for a lawsuit instead of something concrete, a suit that Arizonans hoped at least would prove their ownership of the water until they somehow got a dam built that would allow them to divert it.

California, employing a battery of lawyers instructed to stall for time as much as to try to persuade the courts of the rightness of that state's own claims—water flowing westward all the while—had succeeded in tying the issue into a complex legal knot until early in 1963, when the Supreme Court, to the astonishment of water experts everywhere, upheld Arizona on almost every one of the suit's myriad counts, most importantly on the contention that the state did have an entitlement to 2.8 million acre-feet of the Colorado, regardless of the fact that California had been using that water for a generation now. It was a ruling that shocked and frightened the Californians (how the hell were they supposed to replace the water?), that sent the Arizonans into absolute spasms of delight, and that played right into the hands of Commissioner Floyd Dominy and the Bureau of Reclamation.

On January 21, 1963, Dominy and Interior Secretary Stewart Udall had announced a monumental new multibillion-dollar water plan that would do dozens of things—dam and divert water from northern California's Trinity River into a system of tunnels and canals that would deliver it south to the San Joaquin Valley and the Los Angeles metropolitan region, where it would replace water currently supplied by the Colorado River. That Colorado water in turn would be pumped east instead of west now, part of the Central Arizona Project that had been dreamed about since octogenarian Arizona Senator Carl Hayden had been a mere pup of a representative back in 1912. The water, 1.2 million acre-feet of it, would be pumped out of an existing Colorado River reservoir, Lake Havasu, downstream from Hoover Dam, but the electricity needed to power the pumping would come from two new "cash register" dams in the Grand Canyon at sites that had been mapped and tested and pronounced suitable since the days before Hoover Dam was built, the surplus power of these new dams providing enough cost-offsetting income to make the entire project look like something other than economic sleight-of-hand.

To Dominy's mind, this was the kind of grand and complex scheme that made it worthwhile to head up Reclamation, to endure the million headaches, the piss-ant politicians, and the crybaby conservationists. This project would go a long way toward solving the increasing Southwest water shortage, and it possessed a terrific mix of large-scale dams, long and complex delivery systems, and massive hydropower facilities, all bound together by Bureau ingenuity. Although Stewart Udall was less enthusiastic about the proposal, he didn't evidence that fact on the day he and Dominy addressed a brace of reporters to announce it, the secretary—still a loyal Arizonan despite the eight years he had spent in Washington by now—one who realized that at long last here was a means of getting his state its share of the Colorado, a means of preventing his native state from literally going dry.

At the moment, Udall was in the midst of a protracted struggle with the conservationists over whether water from Lake Powell should be allowed to reach Rainbow Bridge, and he didn't relish the idea of doing battle with Dave Brower again—he could see him sitting scowling there at the back of the room—over the issue of another perceived invasion of the parks, but politics sometimes was a nasty business, and if Udall could shepherd the Pacific Southwest Water Plan into being—imagine those beautiful wide canals brimful with water as they skirted the edge of Phoenix—while at the same time continuing to work for those aspects of the conservation agenda that seemed to him to make good sense, it would be worth all the rancor and name calling and outright cat fighting he knew he would have to endure.

Brower and company would hit the ceiling over this proposal, to be sure, but Udall was ready to counter their expected barrage of arguments. Conrad Wirth, Udall's director of the park service, already had forwarded to the secretary a statement that Bridge Canyon Dam would have little or no deleterious effect on Grand Canyon National Park; Carl Hayden, way back in ancient times, had made sure that the park's enabling legislation expressly allowed "the development and maintenance of a government reclamation project" within its boundaries; and even the Sierra Club once had approved a plan to build Bridge Canyon Dam, David Brower himself voting in favor of a 1949 resolution to that effect. Although both Bridge Canyon and Marble Canyon dams would be big dams, they would be relatively low in relation to the depth of the canyon—their reservoirs rising no more than a quarter of the way up the 5,000-foot walls of the canyon at their deepest points. Instead of filling a portion of the Grand Canyon virtually to its rim, as Lake Powell unarguably would do in parts of Glen Canyon, these new reservoirs wouldn't do much more than flatten out the riverbed and calm some thunderous rapids. There would be a fight all right, but Udall was confident that Congress ultimately would respond to reason.

That day in January 1963, in fact, the secretary had had a more ticklish issue to contend with than the coming furor over the Grand Canyon dams: he had to find a way to fill the fledgling reservoir behind the dam that was now nearing completion in Glen Canyon. When it was full, *if* it was ever full, Lake Powell would contain 27 million acre-feet of water, roughly two years of the river's total flow. If none of the Colorado's water currently had been in use, it would have been a straightforward enough matter to fill the reservoir—you simply would have shut the river off for two years' time, finally letting water escape downstream through the penstocks once your pool was full. But there were two complications that made that procedure impossible. First, well over half of the river's total flow already was being consumed downstream, used to fill the reservoirs behind Hoover, Davis, Parker, Imperial, and Laguna dams, used to fill canals that angled west to California, used in its ancient riverbed to deliver a small measure of its water to Mexico, used to produce power that millions of people depended upon. The secretary obviously could not fill Lake Powell at the disastrous expense of the Colorado's downstream users. And second, he knew, there was the matter of the need to start generating power, minimal power at least, at Glen Canyon as soon as possible to begin to meet power-supply contracts and to begin to *pay* for the billion-dollar Colorado River Storage Project. But, if you generated power, in the process you also necessarily took water out of the reservoir whose size you were trying mightily to increase. It was a complicated problem, and plainly the Colorado just wasn't big enough to meet its many commitments.

Udall's decision to order the closing of Glen Canyon's west-tunnel gates on January 21, 1963, was part of an attempted solution, Interior's effort to assuage the river's several constituencies: under the terms of a plan dubbed the Lake Powell Filling Criteria, Lake Mead never would be allowed to drop below 1,123 feet above sea level, the height at which Hoover Dam's eighteen generators still could operate at maximum capacity, and at all times at least 1,000 cubic feet of water per second would be released from Glen Canyon Dam, much more than that amount periodically as Lake Mead levels dropped close to the 1,123-foot mark. Lake Powell in turn, or so the Bureau's hydrologists predicted, would need a full year to rise 350 feet up the face of Glen Canyon Dam, up to elevation 3,490, the minimum height above sea level at which its own power generation would become possible. If everything went precisely as the filling criteria orchestrated, downstream water users would never have to suffer more than a 10 percent cut in water supplies, Hoover Dam's power output wouldn't be depleted, Lake Mead would drop lower than it ever had before but wouldn't have to be drained, Glen Canyon's power plant would begin generating electricity on schedule in the autumn of 1964, and the secretary of the Interior would avoid making legions of new antagonists.

But the water wouldn't cooperate. The Rocky Mountain snow-pack was light in the winter of 1963, and the spring runoff came on correspondingly low, the river carrying less peak-level water than it had in many years. Lake Powell wasn't filling as fast as it should have, Lake Mead was dropping far quicker than anyone had expected, and Stewart Udall—as was becoming commonplace—found himself in the midst of another fight. Throughout the dry summer and autumn, Lake Powell only crept upward while Lake Mead continued to drain like a bathtub. By the end of March 1964, with another subnormal runoff now seeming certain, Lake Powell had impounded only half the water needed to begin generating power, and Lake Mead was already down to the 1,123-foot level. With the filling criteria giving him no other option, Udall ordered the Glen Canyon gates open, and for more than a month, 10,000 to 18,000 cubic feet of water per second poured through the Grand Canyon en route to its recapture in Lake Mead, upper-basin politicians screaming all the while that their water was being taken from them, Utah Senator Wallace Bennett demanding that Udall "stop the Interior Department's water pirating policy," Colorado Senator Ed Johnson describing the secretary's action as "arbitrary, dictatorial, positively unnecessary, and perfectly stupid."

Lake Powell had dropped 20 feet, Lake Mead had risen some, and the firestorm of criticism from the upper basin was getting nasty by the time, on May 11, Udall changed his mind and announced that not only would Glen Canyon Dam releases be cut back to 1,000 cubic feet per second again, but the filling criteria also were being amended. Henceforth, Lake Mead would be allowed to drop as low as elevation 1,083 feet, the level required for *minimum* power generation at Hoover Dam. Predictably, the upper basin hailed the secretary's good sense while it was the lower basin's turn to tar and feather him. Udall himself acknowledged the gamble he had decided to take—namely that Lake Powell, capturing almost all the Colorado now, would rise high enough to begin producing power before Lake Mead dropped too low to do so.

High above the river, but far below the rim of the Grand Canyon, a small plane droned overhead on the afternoon of May 12, then dropped a bag onto the riverbank not far downstream from Lava Falls, a message inside alerting the river runners to the fact that the release from Glen Canyon Dam had been cut from 12,500 to 1,000 cubic feet of water per second the day before, the river's flow at Bright Angel Creek, 103 miles downstream from the dam and 100 miles above the boaters, already reduced to 9,000 cubic feet per second. The message from Jim Packard, chief ranger at Grand Canyon National Park, suggested that the group of nine abandon their boats and hike out of the canyon via a trail at Whitmore Wash because of the decision to shut the river down, but it was precisely that kind of caprice on the part of the Interior secretary and the

Bureau of Reclamation—their willingness to turn the wild Colorado into some sort of plumbing system—that initially had spawned this river trip, and the boaters defiantly decided to try to wash on out of the canyon on what remained of the higher water, knowing they risked getting utterly stranded in the process, but preferring that risk to giving in to the very people who planned to destroy *this place* too, now that Glen Canyon was gone.

Sixteen days earlier, in the midst of a spring during which virtually no one else was running the Grand Canyon because of the erratic and stingy water supply, the nine had set out from Lee's Ferry on a river trip that was truly a mission, a journey that had taken them past the ominous metal scaffolding that climbed the Redwall limestone at the Marble Canyon dam site, then on into the wondrous womb of the Grand, a sojourn that would be recounted in words and photographs in a new Sierra Club exhibit-format book, a kind of companion to *The Place No One Knew* that was being prepared for publication as quickly as possible. This Grand Canyon book, however, was intended to help save a reach of wilderness river rather than to decry its death.

Rowing three wooden boats they called dories—flat-hulled, broad-beamed boats swept upward at bow and stern that were a backlash of sorts against the growing popularity of synthetic rubber boats—were veteran boatmen Martin Litton, a Sierra Club director and editor at *Sunset* magazine, Pat Reilly, an early river runner and protégé of Norman Nevills now employed by Lockheed Aircraft, and painter–park ranger Bill Jones. Along for the ride—the ride of their lives, they anticipated—were Reilly's wife, Susie, herself a Grand Canyon veteran, photographer Philip Hyde and his wife, Ardis, biologist Joseph Hall, and François and Patience Leydet. Plans called for Litton, Pat Reilly, Hall, and Philip Hyde to contribute photographs to the book, which would be accompanied by epigraphic quotations from a variety of writers on the Grand Canyon itself and on the idea of wilderness (these blurbs now a trademark of these Sierra Club tomes), as well as a narrative text by François Leydet, lately the author of the Sierra Club's *The Last Redwoods*. Although he had been unable to join them on the trip, David Brower, as always, would be very much a part of the finished book, his design and editorial instincts having played the principal role in making the seven previous books in the series something of a publishing phenomenon.

What the many people involved with the current book hoped to convey in the end was a sense of the river and the canyon as a symbiotic entity, a vital environment that was much more than majestic rock and churning water, a place that was *alive* in a way that would be destroyed by silting, stagnating reservoirs. And they hoped too to show many Americans that a trip riding the river through the Grand Canyon was an elemental kind of experience, a journey into the earth and back in time, a slow sojourn that no one could experience without being enlightened, renewed, and enriched, the twisting course of the river and the risky unpredictability of its rapids contributing to the sense that here, at least, the world was still untrammeled and uncontained.

The looming battle against the Bureau's proposed Grand Canyon dams was not going to depend solely on a single book, however. It was going to be fought with every conceivable weapon and with every plausible argument by hundreds of committed conservationists and, they hoped, by millions of Americans who would shun that label but who knew, nonetheless, that the Grand Canyon, perhaps foremost of all of the nation's natural treasures, should be left alone. A decade earlier, America's small but burgeoning conservation movement had discovered during the fight for Echo Park that the case *could* be made that the land did have a kind of legal standing, that arguments claiming certain special places should remain forever primeval didn't always have to be losing arguments. Conservationists had begun to discover that they could muster clout. In the succeeding years, they had grown aware in Glen Canyon that compromise can be a dangerous and complex business, and had experienced the wrenching emotional toll of losing something they might have saved, something they considered precious. With those collective lessons now in hand, there was no question that the battle against the Bridge Canyon and Marble Canyon dams would be an impassioned one, and it was a battle the conservationists intended to win.

Yes, the Grand Canyon had great value as a refuge, as an invigorating, sustaining wilderness; it had great geological and biological value—it was a kind of living laboratory; but the conservationists were determined to fight the dams on another front as well: neither of them was needed. Their primary purpose was to supply electricity to power the pumps of the Central Arizona Project, but cheaper electricity could be obtained from modern coal-fired generating plants, the opposition intended to argue, and the promise of nuclear power plants was very bright. At well over a billion dollars for the set, the two dams would have a hard time ever paying for themselves, let alone the rest of the massive Pacific Southwest Water Plan; the current problems associated with filling Lake Powell proved that the Colorado simply didn't carry the capacity to fill and maintain *two more* evaporating, seeping, water-wasting reservoirs; and the reservoirs' purported recreational benefits were redundant at best, sandwiched as they would be between Lake Mead and Lake Powell, the two largest man-made lakes in the nation. Plainly, these were two dams that shouldn't be built, even if they had been slated to plug something other than the Grand Canyon. Floyd Dominy and the boys at the Bureau of Reclamation finally had taken their empire building too far; Stewart Udall and a slew of western politicians finally had encountered some dams they couldn't sell. You could feel it in the air, the conservationists encouraged themselves; they had lost Glen Canyon, but this time it was their turn to win.

Up in Glen Canyon in the early autumn of 1964, Stewart Udall's May decision to screw down the diversion gates had resulted in Lake Powell rising dramatically despite the meager runoff, slackwater reaching all the way to Hite now, reaching into the mouth of the little canyon that sheltered Rainbow Bridge, the reservoir gaining 94 vertical feet in a little more than three months, its water storage up from 2.5 million acre-feet to more than 6 million, the eight penstock intakes on the upstream face of the dam 20 feet below the surface of the water now, ready for their wheel gates to open and the captive water to plunge down to the power plant.

After sixteen months of installation, eight more months of careful calibrating, balancing, and synchronizing of its giant dynamo, then two final weeks of testing, Glen Canyon Generation Unit 1 was pressed into service just before midnight on September 4, 1964, a few days short of eight years since the project had begun, Colorado River water at long last hurtling through the dark bowels of the dam itself, falling 330 feet through a fat steel penstock, falling with enormous force to the powerhouse, encountering deep within it a turbine as big as an army tank, spinning the turbine into 150 revolutions a minute and creating 155,000 horsepower (the dynamo mounted directly above it and connected to it via a whirling shaft suddenly producing 112,000 kilowatts, enough power for 100,000 people), the water finally falling through a tailrace and back to the river from whence it came.

The contract to complete the power plant, transformer deck, and switchyard had been awarded when Lem Wylie was still in command back in June 1962, the $7.9 million job going to Ets-Hokin and Glavan, Inc., of San Francisco, that company also winning a $13 million contract to construct the 345-kilovolt transmission line that would carry Glen Canyon power south to Phoenix. Despite a seemingly endless series of squabbles with its subcontractors, resulting in lawsuits heaped upon lawsuits and more than a year's delay in the construction of the powerline, Ets-Hokin had been able to ready the power plant on schedule for the installation of the first of the eight generators General Electric was supplying to the Bureau at a cost of a little more than $1 million apiece. Unit 2 was now scheduled to go into service in three weeks' time, Unit 3 sometime before Christmas, the last of the eight early in 1966, the entire project pronounced complete in a day or two less than ten years if glitches could be kept to a minimum in the meantime.

Although a horde of hard hats had clustered on the visitors' walkway above Unit 1 on the night they threw the power switch—a similar group gathered inside the secreted control room—and although the powerhouse was chaotic around the clock these days with machinery, men, and noise, the rest of the hole by now had grown strangely quiet. During the summer, crews had been at work dismantling the wooden catwalks that had hung from the downstream face of the dam during the two and a half years it had climbed out of the canyon. The two largest cableways had come down, their fist-thick cables unstrung, their towers disassembled, battered old bucket number 6 set aside to become a museum piece. The cement silos were gone; the aggregate stockpiles had been bladed down to the bedrock; the bench on the canyon wall where the batch plant had stood now was under water. Even the wire-mesh footbridge that had spanned the canyon rims since the very beginning—always a kind of scrappy symbol of the project itself to the people who had crossed it a thousand times, white-knuckled but somehow willing as they bounced and swayed 700 feet above the canyon bottom—now was only a memory, the highway bridge and the maintenance driveway along the narrow crest of the dam now the only means of traveling across the chasm. Merritt-Chapman & Scott chief Jim Irwin's work force, which had comprised 1,600 men when the final concrete had been poured a year before, had dwindled throughout the finish and clean-up phase of the operation and now was reduced to office staff, the trailer court emptying in a steady, month-by-month exodus—a few spaces periodically refilled by men who were part of the Ets-Hokin crew, but the presence of the multi-national New York City company now nearly unnoticeable.

Although it was within days of completing its obligations out in God-awful Arizona, things were not going well in 1964 for the corporation that had had the wherewithal back in 1957 to take on the $108 million prime contract solely on its own. Involved in shipbuilding, marine salvage, metallurgy, and chemicals production as well as big-scale construction, Merritt-Chapman & Scott had earned a net profit of $24 million the year before it won the Glen Canyon contract, but that was the year things had begun to go bad.

On the grounds that the company had milked it out of millions, the federal government in 1956 had refused to renew its contract with MCS subsidiary Capital Transit, which for many years had operated the Washington, D.C., bus and trolley system. Soon thereafter, MCS chairman Louis Wolfson, a horse-breeding entre-preneur as well as the corporation's flamboyant senior executive, had failed in his bid to take over Montgomery Ward, MCS stock falling sharply once the widely publicized attempt fell through. Then in 1958, the company had pleaded *nolo contendere* to charges that in the early 1950s it had bribed a Grant County, Washington, official in order to help secure the construction contract on the county-financed Priest Rapids Dam project; and that same year, the Secu-rities and Exchange Commission had charged Wolfson with personally attempting to drive down the market in American Motors after he had sold short his sizable holdings of the automaker's stock.

Although MCS had won a $48 million navy submarine-building contract in 1963, its net earnings that year had fallen to less than $4 million, its presumably ongoing losses at Glen Canyon, as well as problems in other corporate divisions, prompting the SEC to begin an investigation of whether the company was juggling its books in order to show a paper profit and keep its stockholders from jumping ship. By the time Jim Irwin, weary from three and a half years of breakneck work, and worried too about what was going to become of his outfit, closed down the Page office and began to pack up his house a few days before Christmas 1964—eager for a few months of fishing before he put on a hard hat again—the Bureau of Reclamation had paid MCS installments totaling $134,699,717, the nearly $27 million more than the contract's bid price representing the cost of twenty-five orders for construction changes, undisputed increases in wage expenses, as well as a paltry $289,000 the Bureau finally had agreed to pay in the aftermath of its wrangle with the company over whether the post-strike fifty-cent wage increase was, in fact, a kind of premium pay. But even with almost $135 million in Merritt's pockets now, nearly every Bureau accountant familiar with Glen Canyon believed the company had taken a bath there on the Colorado, losing perhaps as much as $20 million. At corporate headquarters in New York City, MCS officials didn't care to discuss the economic particulars of the project, and out in Page, Arizona, no one seemed concerned. With Lem Wylie, Vaud Larson, and their subordinates watching like hawks all the while, Merritt-Chapman & Scott by now had built one of the world's great dams, built it well, and had finished the job just six months after the originally specified completion date, and that was all that seemed to matter on a cold and blustery New Year's Eve, 1964, as Larson signed the forms acknowledging that all work specified under the terms of the prime contract now was complete, relieving the company of further obligations, jolly Jim Irwin leaving town long before midnight.

Barry Goldwater, son of the rocky Arizona sod, had run for president in the fall of 1964, arguing during the campaign that the United States should bomb Hanoi and should get the hell out of the United Nations but should not build Marble Canyon Dam. It was the first time an Arizona politician ever had broken ranks on the issue of Colorado River water development, and it was reflective of a span of months during which public sentiment seemed split indeed on whether dams and their reservoirs were bounteous blessings or the evil works of engineers.

Goldwater, who still owned land on the flank of Navajo Mountain where Rainbow Lodge had stood, and who fancied himself very familiar with the country surrounding it, had gone on record as being still fully supportive of the decision to dam Glen Canyon, supportive as well of the Bureau's plan to build a dam at the long-coveted Bridge Canyon site at the lower end of Grand Canyon, but he was convinced that Marble Canyon, "exactly what a canyon should be," according to the senator, ought to be left unmolested. "I recall the objections to the construction of Glen Canyon Dam," Goldwater had told a congressional subcommittee, "and had I looked on this in a selfish way and remembered the six times I traveled through that beautiful canyon, it would have been quite easy for me to have voted against it. But I think since the time that it was completed and the lake is now filling, the hundreds of thousands of people who are visiting there, seeing sights they could never see except by an expensive journey down the river, justifies the construction of it."

Of the Bridge Canyon proposal, Goldwater likewise was certain that it would "not violate the grandeur of the lower gorge of the park or monument, but . . . open an area of unmatched scenic value to the visitor of limited financial means." But Marble Canyon, on the other hand, and for reasons the senator had a hard time making abundantly clear, was different. *Its* scenic beauty, its grandeur, its towering cliffs of Redwall limestone surely ought to be preserved in their pristine state, and historic Lee's Ferry, at the spot where Glen Canyon ended and Marble Canyon began, surely should not be flooded. Reflecting on Marble Canyon, the senator said, made him long for the future day when new technologies would make it unnecessary to dam any more free-flowing rivers.

And if Barry Goldwater was not of one mind about Colorado River dams, neither, it seemed, was the nation. At the same time that Glen Canyon Dam was winning an engineering award and the travel sections of newspapers across the country were touting the splendid new sea that was rising behind it, magazines like *Life* and *Holiday* were questioning the justification for Lake Powell as well as the proposed reservoirs downstream, and television documentaries were raising the specter of public monies being unwisely spent out west.

It wasn't surprising that the American Society of Civil Engineers took a favorable opinion of Glen Canyon Dam, but it indeed was a feather in Floyd Dominy's hat—as well as a tribute to Louis Puls, Lem Wylie, beleaguered Merritt-Chapman & Scott Corporation, and the more than 6,000 men and women who had devoted a year or two or ten to getting it built—when the association voted it the outstanding engineering achievement of 1964, the unadorned plug in a distant desert river being selected for the honor over New York City's Park Avenue–spanning Pan Am Building, over the twin, circular Marine City towers in Chicago, over dramatic new Dulles Airport outside Washington, D.C., the jury of engineers noting that of the eight nominees for the citation, the dam "demonstrates the greatest engineering skills and the greatest contribution to mankind."

Now that the dam was complete and its power plant was nearing completion, that recognition was likely to be one of the last times anyone paid much specific attention to the seemingly workaday dam itself, but its reservoir quickly was emerging as the hot new topic on the travel and recreation front. From *Sunset* to the *Page Signal*, from the *Los Angeles Times* to *Time*, writers were heralding the nearly 2,000 miles of shoreline Lake Powell someday would possess, enthusiastically describing the cool blue waters supplied by its feeder rivers— their silt settling out as their currents slowed, then ceased—the water crawling into coves and long fjords, spreading into broad bays and shallow lagoons, creating a wet and certainly surreal world of shimmering blue surrounded everywhere by bald and rounded rock. Making little note of what Glen Canyon had contained prior to its inundation, travel writer Maxine Brown Roberts observed in the *Denver Post* that "as you float past noble promontories, majestic amphitheaters, frowning escarpments, into shadowing narrow channels and vast, echoing caves, for mile after breathtaking mile, you can't get too excited about what lies drowned under the shimmering surface." Art Greene, the longest-tenured non-Navajo in the Glen Canyon country, a man who knew well what lay beneath the water's surface, also was discovering the reservoir's beauty: "Thought maybe you and your boss, Mr. Dominy, might like to hear from an old river rat who gave you a pretty good but losing fight over the building of Glen Canyon Dam," Greene wrote to Vaud Larson in Page, "from a fellow who with a tear in his eye and a hurt in his heart saw Sentinel Rock, Outlaw Cave, Incendiary Urn, Music Temple, and others slowly covered by Lake Powell, from a fellow who now sees where one monument was covered, ten more were brought into view, where once 150 visitors were considered a good year, now that many folks in one week can share this beauty with us. I feel sure Lake Powell is destined to be one of the outstanding recreation areas in the world."

Even Wallace Stegner, writing in *Holiday* magazine, was able to concede that "though they have diminished it, they haven't utterly ruined it. Though these walls are lower and tamer than they used to be, and though the whole sensation is a little like looking at a picture of Miss America that doesn't show her legs, Lake Powell *is* beautiful.

It isn't Glen Canyon, but it is something in itself. The contact of deep blue water and uncompromising stone is bizarre and somehow exciting." But Stegner had come away from his first visit to the new and vastly more watery version of the canyon, where smooth water stunningly was juxtaposed with standing rock, believing that "in gaining the lovely and the usable, we have given up the incomparable."

Dozens of mail sacks a day were arriving at the Washington offices of the Bureau of Reclamation now, as well as at the offices of the men who sat on the House and Senate Interior committees, thousands of letters, then *hundreds* of thousands of letters, all with one theme: the goddamn government should keep its hands off the Grand Canyon! None of the correspondents seemed to know or to care that Marble Canyon Dam would be built 12.5 miles upstream from Grand Canyon National Park and wouldn't impact upon the park. The reservoir behind Bridge Canyon Dam, it was true, would bisect Grand Canyon National Monument, but only 13 miles of the upper end of the reservoir would even *touch* the national park at a place where the river itself formed the park boundary. Dominy himself had flown the canyon at the high-water elevation of each of the two reservoirs, taking photographs every mile, the widely distributed photographs (retouched with simulated lake water) proving that neither reservoir would come close to filling up its canyon. One hundred and four miles of river would continue to flow unimpeded between the two projects—Reclamation would even offer to construct an elaborate elevator system to lower rafts and gear to the toe of Marble Canyon Dam to facilitate the launchings of river trips. Motorboating tourists in turn could ply the reservoirs, visiting parts of the canyons otherwise inaccessible to them, and, most important, power plants below the two dams would produce plenty of combined kilowatts to pump Colorado River water up and over the Buckskin Mountains and on to Phoenix, where it was much in need and demand. Radical conservationists like David Brower might want Arizona to shrivel up, its thirsty residents simply to pack up and leave, but surely, assumed Floyd Dominy, the sensible majority of Americans would realize that these projects had to be built, that they would add more in recreational benefits than they would subtract by inundation, and in no wise would they destroy the Grand

Canyon, where, Congress had said back in 1919, reclamation projects were perfectly acceptable anyway!

Yet the opposition to the dams didn't abate, and the tide didn't turn. Public opinion on the proposed Pacific Southwest Water Plan was negative enough, in fact, that by the time the House Subcommittee on Irrigation and Reclamation opened its hearings in August 1965, the plan had a new name and had been relieved of much of its weight in hopes that it might float. The new "Lower Colorado River Basin Project" legislation, as co-sponsored by virtually everyone in the Arizona and California House and Senate delegations (Barry Goldwater, defeated in a landslide of unprecedented proportions in his bid for the presidency, now was out of office), made no mention of the proposed system of dams and diversions in northern California—those would be dealt with elsewhere. Under the terms of the revised legislation, Arizona would limit its diversions via the Central Arizona Project to 1.2 million acre-feet a year; California would reduce its current Colorado River consumption from 5.1 million to 4.4 million acre-feet a year; Reclamation would begin extensive studies of a means of importing additional supplies into the Colorado basin from the water-wealthy Pacific Northwest; and this drastically scaled-down plan would be paid for by a single hydro-electric dam in Marble Canyon. Even before Congress had begun to study and debate the proposal, Bridge Canyon Dam had been scuttled, "deferred pending further study" in the words of Secretary Udall, his euphemism meaning that now or in the future Bridge Canyon Dam wasn't bloody likely.

As Stewart Udall and Floyd Dominy sat down to face the members of the subcommittee chaired by Colorado Representative Wayne Aspinall to argue in favor of the streamlined new proposal, Udall made it clear "at the outset that there is nothing pending in my office as important as this is," Dominy echoing him, telling the members of Congress that "I consider the measures before you the most important legislation on which I have had the privilege of presenting testimony during my tenure as commissioner of Reclamation." Each man testified at length about the Lower Colorado River Basin Project's benefits, about the dire needs for its water and power, about its favorable economics and widespread support, but other than Dominy's reference to the fact that Marble Canyon Dam would be well upstream from the boundary of the national park, neither man mentioned the Grand Canyon.

When it was his turn to take the witness's microphone, however, David Brower wanted to talk about nothing else. He requested that copies of *The Place No One Knew* and the book that was the product of the François Leydet expedition, *Time and the River Flowing: Grand Canyon*, be entered into the subcommittee's file on the proposed legislation. He passed out individual copies of both books to every subcommittee member, announced that a new Sierra Club half-hour sound and color film entitled *Glen Canyon* was available to them for viewing, then told them that the proposed Marble Canyon Dam would turn the Colorado in Grand Canyon into "a tame trickle, metered through valves at the Bureau of Reclamation's pleasure. . . . We do not believe that the American public would tolerate such an invasion of Grand Canyon and the national park system if the public really knew what is being proposed and what the alternatives are." As far as Brower and his allied conservationists were concerned, the federal government had the blessings of every major conservation organization in the nation to build as many coal-fired or nuclear plants as it needed to pump currently stored water wherever it was lacking, but those organizations would *never*, Brower wanted the subcommittee members to know, compromise on the issue of the destruction of the Grand Canyon.

When it was his turn to question Brower, subcommittee member Morris Udall—the secretary's brother, a man known for his strong conservation record in Congress, but known equally well for his fervent support for the Central Arizona Project and the Grand Canyon dams—wanted to zero in on the Sierra Club's impressive, but to his mind deceptive, new piece of propaganda. Udall indeed already had read *Time and the River Flowing* and had studied its seventy-nine photographs at length, preparing for the subcommittee his own written analysis, which concluded that the book played fast and loose with its photos. "We finally get down to analyzing these 79 pictures," Congressman Udall said, summing up his analysis, "and find only 12 of the scenes that would be inundated by the new lakes. Six of these twelve are behind Marble, which is completely outside the Grand Canyon, and only six are affected by Bridge Canyon Dam."

Arguing that the book didn't allege to portray only those parts of the canyon that would be affected by dams and reservoirs, Brower responded that the book's overall objective simply had been to "stress the importance of a living river, and also, in part of the argumentative portions of the text, what would happen to that living river if the dams were constructed."

Wasn't his so-called living river already destroyed by Glen Canyon Dam? Mo Udall wanted to know.

"No, not by a long way," Brower contended. Even at its lowest releases, Glen Canyon Dam would supply enough flow to keep the downstream river alive.

Hadn't the Sierra Club itself once supported an even higher Bridge Canyon Dam?

Indeed it had, said Brower, but the fact that the club now opposed any Grand Canyon dams "represents an evolution in our own thinking. There was a time . . . when the Sierra Club was for the Bridge Canyon Dam. Ten years ago I was testifying in favor of a higher Glen Canyon Dam and I wish I had been struck dead at the time. We found out how wrong we had been. I would just stress that over these years our own thinking has evolved, and I still hope that Mr. Udall's will."

Witty, congenial Morris Udall, a Tucson lawyer and banker, two years his brother's junior, his replacement in the House of Representatives in 1961, distrusted David Brower. Although he often agreed with the conservationist's stands on natural-resource and wilderness issues, Udall thought Brower and his collegial cohorts had made a bad habit of the kind of exaggeration and misrepresentation that he found evident in *Time and the River Flowing*. In the case of the Lower Colorado River Basin Project, in particular, Mo Udall was angered and frustrated by the success Brower had had in confusing Grand Canyon National Park with those sections of the canyon complex that lay outside the park boundaries and that, therefore, had no mandate for preservation. When Udall used the words *Grand Canyon*, he referred only to the lands within the park system, but when Brower uttered them or wrote them down, they were meant to envelope every inch of the river corridor from Lee's Ferry to the slackwater of Lake Mead, and that, the congressman believed, was fundamentally deceptive.

In a perfect world, Mo Udall wouldn't have chosen to build dams in those canyons himself, but the world wasn't a perfect place and compromise clearly was all anyone could hope to accomplish. As Udall saw things, his jobs as an Arizona congressman and as a member of the House subcommittee on reclamation were to keep working toward some sort of common ground, some kind of solution to the current mess, and to keep Dave Brower honest, even if it meant fighting him on his own exaggerated terms, even if it meant debating him on the issue of the dams standing before press and public on the south rim of the Grand Canyon itself, the gaping, many-colored chasm spreading below them in all its splendor—an assignment, as Udall saw it, that was a lot like "debating the merits of chastity in Hugh Hefner's hot tub."

It was the *Reader's Digest*, that bastion of American conservatism, that had organized a workshop and press conference on the issue of the Grand Canyon dams, the magazine flying scores of conservationists and media people to the canyon's south rim at the end of March 1966, for the openly stated purpose of calling attention to the insanity of flooding even a fraction of this, perhaps the most important of all the nation's scenic resources. Crying foul, crying bias—which the magazine openly admitted it possessed—Mo Udall and a dozen more congressmen, Arizona state politicians, and water officials had invited themselves to join the gathering to present the counterpoint facts, truths that they felt certain would deflate the opposition's public-relations extravaganza.

The *Digest*'s publisher, DeWitt Wallace, who hadn't much liked the Echo Park proposal ten years earlier either, recently had read an article in *Audubon* magazine by Colorado College physics professor Richard Bradley—one of the several sons of longtime conservation stalwart and canyon defender Dr. Harold Bradley—decrying the Bureau's designs on the canyon. Persuaded and much moved by the article, Wallace had decided to condense and reprint it in his own magazine, to editorialize against the proposed dams, and to try to drum up additional editorial opposition in the nation's periodicals by footing the bill for the canyon-rim conference.

From the outset, the pro-dam contingent found itself in decidedly antagonistic surroundings, the Sierra Club having widely publicized the gathering, encouraging as many club members as possible to attend. After enduring an afternoon of sometimes impassioned, often eloquent speech making on the subject of the many ways in which the dams would destroy the canyon, Morris Udall gamely had a go at the kind of "balance" his brother the Interior secretary often spoke of, telling the crowd that there was no canyon-rim viewpoint from which motorists even could glimpse either of the two reservoirs. But he summarily was shouted down, the gathered conservationists hollering "Toroweap!" until the congressman admitted that, yes, you probably would be able to see the Bridge Canyon reservoir from Toroweap Point. And when Northcutt Ely, special counsel of the Colorado River Board of California, the lawyer who so successfully had kept the state of Arizona so long tied up in court, tried to gain a kind of rhetorical upper hand by asking, "How many of you, if you had the power, would do away with Glen Canyon Dam?"—somehow supposing that *that* impoundment was looked on benevolently by these butterfly chasers—the shouts in the affirmative fairly echoed out of the canyon. It was the worst drubbing the dam proponents had endured to date, and coming as it did in front of a corps of national reporters, with the Grand Canyon itself serving as the dramatic backdrop, Udall found himself wondering if some kind of consensus *could* be struck in this case, if in fact any compromise was possible in the context of this cherished canyon.

Holding out hope that Congress itself hadn't been swayed by all the emotion, Morris Udall two months later had to carry the reclamation ball for his brother and for Floyd Dominy when the second set of House subcommittee hearings on the issue convened, the proposed legislation again amended to make it more palatable to some of the solons, to hide a pig in a fancy poke. As Udall explained it, five comparatively small reclamation projects in Wayne Aspinall's western Colorado district were being added to the bill, and the word *lower* therefore was being removed from its title, making it the "Colorado River Basin Project" now, an attempt to enlist both basins in its support, a blatant attempt to keep Aspinall—who didn't much like the idea of Arizona getting all that Colorado River water, Supreme Court decision or no—safely in the fold. The bill still included the colossal Central Arizona Project; it still charged Reclamation with finding a new source of water somewhere outside the basin, still mandated the Marble Canyon Dam; and it recommended the study and eventual authorization of a second mainstem impoundment called Hualapai Dam, named after the Hualapai Indians, whose reservation adjoined Grand Canyon National Monument. Hualapai Dam would greatly benefit the tribe, explained Mo Udall, in addition to supplying needed hydroelectric power for the surrounding region, the congressman neglecting to mention that the dam would be located near the mouth of Bridge Canyon and would, in virtually every respect, be identical to the dam that had been shelved nine months before.

With the help of sympathetic Congressman John Saylor, it didn't take the conservationists long to win admissions both from Udall and from Assistant Interior Secretary Kenneth Holum that, well yes, Hualapai Dam and Bridge Canyon were one and the same, except, of course, for the many ways in which the dam now would assist the Indians. And from that inauspicious opening to the proceedings, matters only seemed to get worse for the legislation's advocates: a barrage of legislators from the states of the Pacific Northwest rose to say they doubted very much whether they could ever support a bill that encouraged Reclamation to steal their surplus water; they doubted very much whether the people of their region would stand for such a travesty. Next came a trio of young MIT graduates—a mathematician, an economist, and a nuclear engineer—each one gunning for the Grand Canyon dam or dams, each loaded with ammunition. Mathematician Jeffrey Ingram presented a paper demonstrating how the payback plan the Bureau had devised for the new hydropower facilities had been built around a numerical house of cards, one that included the transfer of power revenues from Hoover Dam (once it had paid for itself thirty years hence), monies that by law were supposed to revert to the Treasury. Rand Corporation economist Alan Carlin used several yardsticks to show that coal and nuclear alternatives to the dams could deliver water to Phoenix and Tucson less expensively. And Laurence Moss, an Atomics International employee, touted the coming marvels of clean, cheap, safe nuclear power.

Before the six long days of hearings were concluded, Reclamation Commissioner Dominy had gone to Capitol Hill and grudgingly agreed that the Central Arizona Project was "theoretically feasible" without either Hualapai (Bridge Canyon) or Marble Canyon dams; David Brower had taken center stage to charge that the Interior Department was muzzling people within its own ranks who opposed the dams, producing as proof park service and USGS memos telling employees to keep their mouths shut on the issue and to destroy the memos once they were read, Brower finally concluding his testimony by assuring the subcommittee that conservationists in the hearing room and around the country "do not want people to go without water. We are concerned about other things, however. . . . We think it is not necessary to put hydroelectric dams in the Grand Canyon in order to get water to Arizona. We are doing the best we can to show that it is not a choice between water for Arizona and Grand Canyon, but to have both—to have a more imaginative plan."

The looks of anguish, of downright exasperation on the faces of Morris Udall and Wayne Aspinall, even on the broad visage of the ever-confident Dominy himself, convinced David Brower that it was time to pull out the stops. The dam proponents were very much on the defensive now, they were worried, and Capitol Hill insiders claimed that they had decided to try to move the legislation forward as quickly as possible to forestall further erosion of their support. The conservationists needed to do something dramatic, something national in scope again; they needed a means of rallying the millions of ordinary Americans, who, they were convinced, clearly were on their side. What they needed was a show of strength spectacular enough that even Floyd Dominy would have to surrender and leave the canyon alone.

Working with San Francisco–based advertising consultants Jerry Mander and Howard Gossage, Brower quickly embarked on an ad campaign designed to precipitate a hail of opposition to the dams falling from every congressman's and senator's mailbag. Brower envisioned the first of a planned series of newspaper ads as another of his open letters to Stewart Udall, this one a firm but measured analysis of the reasons that the dams shouldn't be built, and a plea to the secretary to stop them. Mander, on the other hand, told Brower that there wasn't any point in spending good money on an ad unless it was going to stir up conversation, unless the ad itself could become a kind of event. What intrigued Mander was an advertisement that would read in big, bold type, "NOW ONLY YOU CAN SAVE GRAND CANYON FROM BEING FLOODED . . . FOR PROFIT." The body of the ad would put a finer point on the issue, sure, but it would conclude, "Remember, with all the complexities of Washington politics and Arizona politics, and the ins and outs of committees and procedures, there is only one simple, incredible issue here: This time it's the Grand Canyon they want to flood. *The Grand Canyon.*"

Brower was wary but willing to be persuaded, finally talking the advertising departments of the *New York Times, Washington Post, San Francisco Chronicle,* and *Los Angeles Times* into splitting their press runs, his open letter running in half of each paper's June 9, 1966, editions, Mander's rather more incendiary ad running in the other half, their costs to be borne by the Sierra Club, although Brower hadn't yet consulted the club's board of directors about the sizable expense.

The copy of the *Washington Post* that Mo Udall read with his morning coffee contained the Mander flooding-for-profit ad, and his quick outrage was in no way abated once he got to Capitol Hill and was shown its companion piece, an ad that he felt unfairly singled out his brother for disdain in addition to wildly distorting the truth. Once recognized by Speaker Sam Rayburn on the floor of the House, Udall stood before his colleagues on the morning of June 9, explaining to them that he was rising "to express shock and indignation at the dishonest and inflammatory attacks made in Washington and New York newspapers this morning against the Colorado River Basin Project Act. While I have a high regard for many of the people who comprise the Sierra Club, the sponsor of these advertisements, I have seldom, if ever, seen a more distorted and flagrant hatchet job than this." Udall went on to list the ads' misstatements of fact, their exaggerations and innuendos, calling them phony, irresponsible, utterly and completely false, and at the end of the time Rayburn had allotted him, the Arizona representative railed, "Let there be no misunderstanding. Without these revenue-producing dams, there will be no aqueducts for Arizona or California. There will be no reclamation. We may have more water, but it will be out there in the river, not on our farms or in our city water systems!" And the congressman was still angry as hell, spitting bullets and still incredulous, wondering "How in the hell can the Sierra Club get away with this?" at the end of day as he incredulously showed the ads to Sheldon Cohen, commissioner of the Internal Revenue Service, over a highball at the bar in the Congressional Hotel, the Arizonan ever more disgusted by the tactics to which David Brower's organization willingly had stooped.

At four in the afternoon on June 10, a messenger from the San Francisco district office of the IRS delivered a letter to the offices of the Sierra Club, warning that the club could no longer be sanguine about its status as a tax-exempt organization. It was the opinion of Internal Revenue, the letter continued, that the club was engaged in substantial efforts to influence legislation—something IRS regulations didn't allow—and that henceforth the organization's donors should be aware that their contributions might no longer be allowed. The following day, with a copy of the IRS letter wired to him in Washington, Brower counterpunched brilliantly, hitting back in the only way he knew how, telling the reporters for the nation's major news media all about the recent communiqué.

If the country's citizens hadn't liked the idea of building dams in the Grand Canyon, they *really* didn't take kindly to the notion of the arrogant IRS flexing its muscles in the face of a little 39,000-member conservation organization. Front-page articles in newspapers around the country about what the IRS claimed was simply a "routine investigation," plus editorials in the *Wall Street Journal* and the *New York Times* denouncing it as "an extraordinary departure from [Internal Revenue's] snail's-pace tradition" and "an assault on the right of private citizens to protest effectively against wrong-headed public policies," led to a brand-new barrage of letters and telegrams opposing the dams, these coming not from people who were irrational posy pickers standing in the way of progress, but from hundreds of thousands who had their own bones to pick with the IRS, who viewed the Grand Canyon proposal not as the dastardly doing of the Bureau of Reclamation (an agency most had never heard of), but as something the hated IRS supported, and therefore as something to be despised. What seemed particularly damning to the IRS's claim that it had no interest in the Grand Canyon issue per se was the fact that it had announced that if it did indeed rule that contributions to the club could no longer be deducted from individual tax returns, the nondeductibility would be in effect not from the date of the ruling, but rather from June 10, 1966, the day after the ads had appeared—a statement that certainly appeared to be designed to discourage a sudden flood of sympathetic contributions to the Sierra Club.

For its part, although its directors were less than thrilled that Brower had initiated the ads—spending a reported $15,000 out of his discretionary fund to pay for them—the club's board publicly stood behind the advertisements, as well as its executive secretary, and vowed that now the club would battle the Grand Canyon dams more vigorously than ever before. In late July, a Mander-inspired and Sierra Club–sponsored advertisement again appeared in the nation's major newspapers, this one inquiring, "HOW CAN YOU GUARANTEE THESE, MR. UDALL, IF GRAND CANYON IS DAMMED FOR PROFIT?," the text that followed listing Reclamation dam and reservoir proposals already on the board that would encroach on Arches National Monument, and Big Bend, Glacier, Grand Teton, Kings Canyon, Mammoth Caves, Yellowstone, and Yosemite national parks. Clearly, the ad maintained, if the Interior secretary would allow the Grand Canyon to be destroyed, then nothing was safe on his watch. In August came another ad (Mander was right, they were becoming *events* now), this one designed to counter the government's claim that reservoirs behind the Grand Canyon would make it easier for millions of visitors to view and enjoy the canyon's splendor, one which rhetorically wondered, "SHOULD WE ALSO FLOOD THE SISTINE CHAPEL SO TOURISTS CAN GET NEARER THE CEILING?" It was caustic, it was cynical, and it was by far the club's shrewdest advocacy to date, employing a measure of ridicule for the first time, the club and the whole of the conservation movement evidencing a cocky kind of confidence now, the cutting rhetorical question asked repeatedly on editorial pages across the country, the *New York Times*, the *San Francisco Chronicle*, *Life*, and even the children's newspaper *My Weekly Reader* taking vehement stands now against any impoundment of the Colorado anywhere between the backwash below Glen Canyon Dam and the silt flats that marked the beginning of Lake Mead. What seemed to matter most at that moment was not how water would be pumped to central Arizona, how high the proposed dams would be, whether they would rise inside the park itself, or whether they would offer splendid recreation. What seemed fundamentally important to people from New Hampshire to Arkansas and on to Oregon—many of whom might never actually see Grand Canyon—was the assurance that it remained unspoiled, that the river was still at work, cutting the canyon ever deeper.

By late summer, Brower and his allies were buoyant, sensing a growing possibility that this Reclamation monster actually could be corralled. Public opinion was adamant in opposition; the dam builders were on the defensive, and the politicians on Capitol Hill were getting worried. But it was no time to ease up, and it would be foolish to assume that victory would flow their way as easily as the river slid past Lee's Ferry. "Raise a storm of protest in any way you know how," exhorted *Sierra Club Bulletin* editor Hugh Nash in the August issue. "Above all, act soon—the issue may be settled for all time within a few weeks or months. Remember, it's the Grand Canyon they propose to dam. *The Grand Canyon!*"

Flying westward, high above the green heartland of the nation, where broad rivers flow across the land instead of cutting deep inside it, the secretary of the Interior considered his several options, none of which seemed particularly promising. He could forge ahead, shoulder to shoulder with Floyd Dominy (who somehow still didn't see that the battle was being lost), and pray that sage old Wayne Aspinall and debt-calling Carl Hayden could shepherd this "Colorado River Basin Storage Project" through to passage in their respective houses of Congress, with Marble Canyon Dam winning authorization despite the conservationists' impressive struggle in opposition. Alternatively, he could endorse the attempt at compromise lately being pushed by his brother and several allied congressmen—a rather drastic change of course that would involve abandoning Marble Canyon Dam and extending the boundaries of Grand Canyon National Park all the way upriver to Lee's Ferry in exchange for the approval of Hualapai (Bridge Canyon) Dam and the abolition of Grand Canyon National Monument. Hualapai would be the better power-producer of the two dams, said the men at the Bureau; it would store more water, evaporate less, and, if its reservoir no longer backed 80 miles through a national monument, it would in no way "invade" the national park system, merely touching a few miles of the western boundary of Grand Canyon National Park. The conservationists would get an expanded park and would claim that they had "saved" Marble Canyon; Carl Hayden and the Arizonans would win the dam at Bridge Canyon they had longed for for fifty years, and the Central Arizona Project finally would get under way. But the whole deal would depend on the conservationists' willingness to compromise, something that might not interest them with an outright victory now in view. A third way to proceed—one Stewart Udall hadn't even dared to mention to Dominy yet—might be to abandon both dams, simply forget them now and forevermore, and to power the CAP with electricity from some non-Reclamation source. The canyon would be spared, water would wend its way to Phoenix and Tucson nonetheless, and a blow would be struck for balance. Brower and company would accept that option, of course; Mo and the rest of the Arizona politicians would complain but finally say okay, and Dominy's inevitable shit fit, the secretary knew, would be unpleasant but would be survived.

Crossing into the dry quadrant of the country, the Boeing 727 with the presidential seal painted on its fuselage flew over the lofty spine of the Rockies, already dusted with the first of the autumn snow, then high above the confluence of the Green and Colorado rivers, that buff-and-red-colored country far below a national park now, and away to the south, the secretary could see the bright reflective surface of the water that wound throughout Glen Canyon, but he couldn't make out the place in the far distance where they would stand in two days' time, Stewart Udall and his wife, Lee, accompanying Lady Bird Johnson on a four-day swing through the West, the first lady scheduled tomorrow to dedicate California's new Point Reyes National Seashore, as well as the recently designated Big Sur Scenic Highway before, on Thursday, September 22, she officially dedicated Glen Canyon Dam.

Ten years after the CRSP had passed Congress at the end of a long legislative battle similar to the one now being waged over new Colorado River dams, Glen Canyon Dam was done, the dedication coming just twenty-three days shy of a decade since President Eisenhower had pushed the symbolic button that sprayed fractured rock away from the canyon wall, initiating a project that would utterly transform the heart of that region of rock and sand and sky. Lake Powell—8.7 million acre-feet of water in September 1966, water trapped in transit from the moist slopes of the Rocky Mountains to the sea—now spread throughout the Glen, slipping within a mile of Rainbow Bridge, the reservoir standing 383 feet deep against the cool wall of the dam, 40 feet above the penstocks' intakes, still 177 feet below its concrete crest. In the powerhouse at the toe of the gargantuan dam, one by one the generators had gone on line, three of them before 1964 was finished, three more in 1965, all eight finally spinning in a strangely muffled but nonetheless formidable unison, the massive concrete-and-steel structure surrounding them set into subtle, perpetual vibration—you could *feel* the surging power—on February 28, the Ets-Hokin crews that had completed the power plant, transformer and transmission structures, plus a canyon-length list of detailing and clean-up jobs, calling it quits seven months later, only a few days before the secretary of the Interior and the president's wife arrived.

Every schoolchild from Flagstaff to Fredonia, from Kayenta to Kanab had been invited to come to Page to greet the first lady, and the legion of yellow school buses had begun rolling down Seventh

Avenue early on Thursday morning, 3,000 youngsters cordoned behind ropes out at the airport by the time the chartered Frontier Airlines turboprop (the runway far too short for the presidential plane) wheeled to a stop outside the little terminal. In the two years since her husband had become president, Mrs. Johnson had emerged as a strong advocate of community and highway beautification—pushing successfully for funds to tear down billboards and to erect tall fences around the nation's sprawling junkyards, hoping to make America look and *feel* like a country its citizens cared about—and she remarked in the midst of the short motorcade through Page that new towns like this one were fortunate in that they didn't have to contend with blight and rampant ugliness left over from earlier generations, the first lady making appreciative note of the hundreds of trees rising above roofs on that mesa somewhere out west where it was obvious that trees never had grown before.

The town on Manson Mesa that Lady Bird Johnson saw that Thursday morning did look more like a *town*, like a place where surely babies were born and where people could happily live out their lives, than it ever had before. The streets were paved, the gutters and sidewalks swept, the hospital and school and the big new Babbitt's market attesting to the fact that this was 1966 and Page was keeping current. The lawn in the park was rolling and wonderfully green; the churches, standing side by side on the south edge of town, seemed to form a kind of city wall keeping out ungodliness; stores and shops were housed in permanent buildings, complete with festively painted signs and plate-glass windows showing their wares, making it seem likely that if you couldn't buy it in Page these days, you probably didn't need it; and the houses on the west side of Seventh seemed to sit comfortably on the mesa now, seemed anchored and a part of the place, grass and gardens and shade trees sheltering them from the desert.

But except for the airport, the forlorn Windy Mesa Lounge, the Bureau of Reclamation's big warehouse, and a tin-sided welding shop, the part of Page that not long ago had reached a dozen blocks east of Seventh Avenue somehow had disappeared. A thousand trailers once had filled that side of town, and now a thousand trailers had been hitched up and pulled away, leaving little behind but the concrete pads and power hookups, streets that no longer went anywhere, and a battered toolshed or two. The community that had seemed to be bursting its seams three years ago—more than 6,000 people at home for a time on the little mesa top—now was home to only 1,600. There never were waits anymore at the banks or the barbershop; the stores seemed strangely quiet; whole blocks of classrooms at the school had been abandoned; and dozens of friends and acquaintances—most of the people who had *made* this place and had seen to it that that dam now plugged the canyon—had had to move away.

But Page at the end of the project wasn't like the town that had nearly come unstuck during the long months of the strike. Everyone—the Bureau personnel, the contractors and their crews, the merchants and teachers and service people, Anglos and native Navajos—had known all along that this kind of slump would come. But it wasn't a bust, and Page, Arizona—still a government camp but well on its way to emancipation—was going to make good: the Bureau had estimated that half a million people a year would visit Lake Powell by the turn of the century, but that many already had been enticed by the lake in 1965! Lake Powell *was* the jewel of the Colorado, and Page was going to become its service center, a tourist town, a good town, small but clean and cordial, the kind of community the first lady of the land would visit.

The combined members of six high-school bands—short of seventy-six trombones but not by many—lined the service road from the east rim of the canyon down to the crest of the dam, Mrs. Johnson and the Udalls and the additional dignitaries receiving a rousing musical salute as their cars rolled down the asphalted slope and onto the man-made mountain of concrete. The first lady took a moment to peer over the retaining walls down to the captive lake on the upstream side of the dam, then far below on its downstream face to the big and boxy power plant. Accompanied by Secretary Udall and Buzz Bennett—the acting commissioner of Reclamation for the duration of one of Floyd Dominy's regular junkets overseas, off helping other nations dam their raging rivers—Mrs. Johnson marveled at the depth of the canyon, the lovely ocher and orange striping of its walls, and she wondered about the flat field that appeared to be nothing but dirt separating the dam from the powerhouse. It *was* nothing but dirt, Bennett explained, earth removed from the downstream cofferdam then backfilled there on top of the massive penstock pipes. The first lady nodded, then wondered again. Wouldn't that patch of earth—more than an acre of it, as it turned out—look lovely planted with wildflowers? Texas bluebonnets would be perfect, she said, not letting her beautification efforts lag.

With the dignitaries and special guests, Bureau employees and politicians, the townspeople of Page, and hundreds of proud citizens of the great states of Utah and Arizona gathered on Glen Canyon Dam now, Buzz Bennett went to the podium and introduced Stewart Udall, the Interior secretary serving as master of ceremonies for the dedication. Udall in turn introduced governors, senators, and congressmen from the two adjoining states; introduced Raymond Nakai, the man who had become the Navajo tribal chairman in 1963 when he had defeated long-term incumbent Paul Jones; the secretary called attention to Richard Mynatt, a New York executive representing Merritt-Chapman & Scott—the corporation in ruin now and in the process of liquidation, its chairman, Louis Wolfson, indicted on new stock-fraud charges just three days before. Finally, the secretary acknowledged the presence of Lem Wylie, lately back from a two-year tour of duty in Colombia and actually considering retiring now; Wylie, who more than any other man had seen this thing through, stood as his name was mentioned, offered a wave and a satisfied smile, but said nothing, assuming the concrete he was standing on spoke eloquently in his stead.

For his part, Udall wanted to note that after a decade of the toil of men like Wylie and many thousands more, this key unit of the Colorado River Storage Project now was storing water and producing maximum power; he wanted to acknowledge the roles absent Wayne Aspinall and ailing and hospitalized Carl Hayden had played in getting it constructed; and he wanted everyone to know that Lake Powell surely *was* the most beautiful man-made lake in the world. Then he introduced Mrs. Johnson.

"This is a unique kind of country, and I don't have to tell you, it's *my* kind of country," intoned the first lady, her Texas-entangled voice ringing off the concrete and the encircling sandstone. "As our plane glided down into the immense canyon landscape, I could see from afar this plug in the river, the place where Glen Canyon Dam stands. And now that I am close, I can get a feeling for the vastness of what has been done here.

"Those of you who have worked for months and years to bring this about must have a justifiable pride in your accomplishments. My hat is off to the men who figured the stresses and strains, and who diverted the river during construction, and to all the rest of the dreamers and doers who brought this project, live-born, into its rocky cradle. . . .

"I'm sure you all have seen, as I have, those disfigurements of rock or tree where someone with a huge ego and a tiny mind has splashed with paint or gouged with a knife to let the world know that Kilroy or John Doe was here. As I look around at this incredibly beautiful and creative work, it occurs to me that this is a new kind of writing on the walls, a kind that says proudly and beautifully, 'Man was here.' . . . I am proud that man is here."

With that it was all accomplished—this dam dreamed up, designed, constructed, and set to work, dedicated now to a promising American future. But before the bands struck up again, before the first lady was hustled back to the airport en route to Santa Fe, Secretary Udall took the microphone again, surveying the crowd and seeing beyond the sea of sunlit faces to the canyon that cut away to the south, his great-grandfather's ferry crossing around only nine bends of the river, the spot where Marble Canyon Dam would rise just a dozen more downstream. "May I just say in closing," said the secretary, pausing to capture the crowd's attention, his words filled with surprising emotion, "and I know this so much better now than I knew it five or six years ago when I became secretary of the Interior, there is something very precious and very special that all of us have here. This heritage that we have is so much richer than we know, and we must conserve it and use it wisely."

The many hundreds of people standing before him in the warm September sun applauded him politely, but it was hard for them— and even somehow for the secretary himself—to know whether he was referring foremost to the great dam on which he stood, to the canyon—now captive and filled with water—that stretched upstream behind him, to the canyon on the opposite side of the dam that angled away unfettered, wild still, untouched for 300 miles, or whether, perhaps, his words were meant to encompass it all.

11

A CENTURY AFTER
THE *EMMA DEAN*

Glen Canyon Dam had been dedicated as the middle years of the 1960s boiled into dissension and conflict, as the meaning and proper role of environmentalism was becoming an increasingly acrimonious topic. It wasn't until Glen Canyon Dam was done—its concrete arch at long last spanning the walls of the canyon, its power plant finally churning a million kilowatts—that the dam fully emerged as a potent symbol of that angry environmental debate. And it wasn't until a long reservoir rose behind the dam—a kind of enormous flood that wouldn't recede—that the fight over whether to build similar dams in the Grand Canyon was settled once and for all.

By early 1967, as far as Floyd Dominy was concerned, it was time to fish, this bait-cutting business having gone on long enough. The second set of House hearings six months before had been a basic disaster—the wolf-crying conservationists seemingly convincing several congressmen that the Grand Canyon dams weren't needed, didn't belong, and shouldn't be built. The IRS sanction against the Sierra Club had backfired disastrously, and now, Dominy was convinced, something decisive had to be done to get this project back on track. All was well in the Senate, at least; the several versions of the proposed project had had votes aplenty in that chamber right from the beginning, but if dam-building legislation was ever going to pass in the House, either it had to be scaled down enough to mollify the fence-sitters or the opposition somehow had to be crushed. Taking what obviously was the easier course, Dominy decided to do some trimming.

As the Ninetieth Congress convened in January 1967, the members of the House Reclamation subcommittee were greeted with new Bureau of Reclamation recommendations for the further development of the Colorado River basin. The grandiose Pacific Southwest Water Plan, proposed with great flourish four years before, now was just a shadow of its former self. Nowhere in the Bureau's current documentation were water projects within the state of California even mentioned; nowhere did the agency even hint at diverting water to the Southwest from the Northwest's Columbia River basin; Wayne Aspinall's five western Colorado water projects were still on board (because Aspinall was the critical player in all of this); but rather stunningly, Marble Canyon Dam had disappeared.

Eight months earlier, Reclamation had countered the growing opposition to the Grand Canyon impoundments by suggesting that Bridge Canyon Dam be put on indefinite hold and that critical Marble Canyon Dam be constructed with all possible haste. Now the commissioner was proposing precisely the opposite: Marble Canyon Dam would be axed, the idea abandoned forevermore, and Reclamation would offer no objection if Congress saw fit to extend the boundaries of Grand Canyon National Park all the way upriver to Lee's Ferry. At the opposite end of the canyon, however, Grand Canyon National Monument would be abolished, and heretofore-on-hold Bridge Canyon (called Hualapai) Dam would become the power producer that would supply the Central Arizona Project. It was a flip-flop, yes, but one that surely would succeed. Sympathetic congressmen wasted no time in backing the Bureau's new proposal, convinced that they finally were going to beat back the conservationists—or so it appeared for the better part of a fortnight, so it seemed until the secretary of the Interior pulled the rug right out from under them.

Floyd Dominy had just left town late in January on his annual overseas tour of water projects being funded by the Agency for International Development and engineered by the technical experts at the Bureau of Reclamation. No sooner had the commissioner's plane passed outside American airspace than was Secretary of the Interior Stewart Udall announcing to his own high-level subordinates at Interior, as well as to Acting Commissioner Buzz Bennett and a few trusted insiders at Reclamation, that, with the commissioner at a safe distance for a few weeks, it was time to cut *both* Grand Canyon dams loose from the Central Arizona Project and the rest of the pending Colorado River basin legislation. Water would never flow toward Phoenix if it depended on Grand Canyon dams to do so, Udall had become convinced during the Christmas holiday, and now he was ready to act. After examining some thirty-eight different Interior department schemes that alternatively might have supplied power for the CAP, the secretary had seized upon one that wasn't perfect, by any means, but which would satisfy several constituencies and, at long last, get some construction going out in Arizona.

At a February 1 news conference, with obvious pleasure and no small amount of relief spread across his farmer's kind of face, Udall announced that the Johnson administration was withdrawing support for Reclamation's most recently amended version of the Colorado River Basin Project. In its stead, the secretary said, the administration now recommended only a kind of lean and mean Central Arizona Project, one which wouldn't make Colorado River delivery guarantees to the state of California, which wouldn't consider water importation from the Northwest, a plan ignoring the five western Colorado projects altogether, the secretary mentioning the Grand Canyon dams only long enough to say that *neither* should be built.

In order to provide power to pump CAP water to Phoenix and Tucson, Udall explained, the administration now favored the formation of a Bureau of Reclamation partnership with a consortium of private electrical utilities already planning a huge coal-fired power plant somewhere in the northern part of the Navajo reservation. The Navajos recently had signed long-term leases with the Peabody Coal Company to strip-mine the tribe's vast Black Mesa coal deposits; mining already was under way on the high mesa that lay southeast of the Kaibito Plateau; a pipeline that would deliver slurried coal from the mesa to the new Mojave power plant on the California-

Nevada border was under construction; and the Navajos actively were seeking new markets for their coal. As Udall envisioned the plan, its cost trimmed down now to a little more than $700 million, the Bureau of Reclamation would offer $80 million or thereabouts as an advance purchase of 400,000 kilowatts of electrical power to the group of southwestern utilities known as WEST, the Western Energy Supply and Transmission Associates. The federal money would be used to finance start-up construction on a 2.5-million-kilowatt generating station, a monstrous power plant that would suck thousands of acre-feet of water annually from nearby Lake Powell, which by now could spare a little water, and that would be situated on the outskirts of Page, Arizona, a town which by now could use a little business.

Reaction to the secretary of the Interior's new proposal was sudden, and it was far from passive. Wayne Aspinall, whose congressional district would get nothing from this administration version of the Colorado River legislation, vowed before the day was out that Udall's plan never would live to see the light of day. California Representative Craig Hosmer—who had fought hard against Glen Canyon Dam a decade before, but who now was an arch proponent of the Grand Canyon dams as a means of settling his state's long water war with Arizona—scoffed that "possibly Secretary Udall will next propose two large gambling casinos at Las Vegas as a substitute for Hualapai and Marble Canyon dam revenues." State of Colorado water officials claimed that the new administration proposal would be the kiss of death for any and all Colorado River legislation, and Arizona state legislators promptly charged that Udall had "double-crossed his own state" and had "caved in to the ultra-conservationist Sierra Club." Even the secretary's brother Morris was unsettled by the administration's change of course, telling reporters that he was standing by the renewed Hualapai Dam proposal, and reminding his older sibling that it was Congress, not the executive branch, that wrote the nation's laws and spent the people's money.

If Secretary Udall had neglected to inform his brother beforehand about the new tack he was taking, he had taken the time to brief senators Henry Jackson of Washington, Clinton Anderson of New Mexico, and Arizona's wizened Senator Carl Hayden—the key members of the Senate Interior Committee—before he announced his plan publicly, and the three, realizing that the alternative simply was to continue the losing fight for the Grand Canyon dams, had offered the secretary their immediate support.

Floyd Dominy returned from his foreign travels to discover that Reclamation, likely as not, was going to be entering the coal-burning business, an idea that disgusted him, an end run that infuriated him, but one that he now knew he could do little about—the commissioner gamely testifying in support of Udall's plan at the third and final set of House hearings in March 1967, yet explaining to people privately that he had abandoned the Grand Canyon dams only because, as he put it, "my secretary turned chickenshit on me." Dominy was a man who was used to getting his way, a fellow who loved to win these congressional rounds of combat, commissioner, he liked to say, of an outfit that got things accomplished. But now his Pacific Southwest Water Plan was a grand scheme come very nearly to naught, and even if the CAP survived, it would be powered by coal-fed steam generators instead of by falling water. Sure, electricity from a new Navajo reservation power plant could supply the Central Arizona Project, Dominy nonchalantly told the members of the House Reclamation subcommittee, hating his words as he spoke them, openly admitting only that the Udall plan was politically expedient, one that would get things moving at long last, the burly dam builder from Campbell County, Wyoming, resolutely refusing to consider the possibility that this time the conservationists plainly and simply had beaten him. Finally, on July 31, 1968, more than four years after this process had begun, Floyd Dominy paid only passing attention as House and Senate conferees agreed on a $1.3 billion Colorado River Basin Project bill, one which plainly sanctioned the government's participation in the construction of a coal-fired power plant at Page, Arizona, and specifically prohibited the future construction of dams on the Colorado anywhere between the arcing concrete walls of Glen Canyon and Hoover dams. In September, President Johnson signed the bill into law.

Poor Page, a town that not long ago had boasted that it was the "drinkingest, gamblingest, churchiest town in Arizona," looked in 1968 as forlorn and empty as did dozens of other dots on the western landscape, victims of booms that had turned to bust. Just five years before, Page had been the kind of camp that never slept, carefree and hearty and irrepressible as an eight-year-old, which is what it was. Nowadays, however, you could take a nap in the middle of Seventh Avenue, if you wanted to, and know you'd wake up alive.

The Bureau of Reclamation was estimating that 1,200 people were still at home on the top of Manson Mesa, but its number seemed wildly inflated on the days when Grant Jones waited hours on end for someone to walk into the shoe shop, on the evenings when eight out of ten lanes at Ted Selna's bowling alley stood idle. The drive-in theater long since had shut down, as had the Navajo Cafe, Danny Reisman's tavern, and all but one of the beauty salons. Dr. Kazan lately had begun flying his small plane to outposts in southern Utah a couple of days a week in search of patients; dentist LaVon Gifford similarly had hauled a trailer down to the reservation town of Tuba City, where it would serve as an occasional clinic. Sand by now so completely had powdered the trailer court that you could barely make out the former gridwork of streets, and although Ted Selna, ever optimistic, just had signed a purchase agreement with the Bureau, hoping to turn the land into a trailer court again—this time for summer tourists—right now it belonged again to the desert. Lawns in front of empty houses on the west side of town had withered and been overtaken by tall weeds; plywood covered picture windows in an effort to keep kids from further reducing the value of houses federal appraisers already said were worth next to nothing. Membership levels at the country club had dropped so low it was going to be hard to keep the golf course open, and even the Navajos, for whom Page had become an important supply center, were curtailing their trips to town, or so it seemed in comparison with the days back when Page was the proud focal point of all the Colorado Plateau and its residents hardly owned a complaint.

But none of that meant that Page was going to become a ghost town; the camp was alive, if rather more somber than it once had been. There were signs these days that it might even be on its way to a rebound, and people who might have chosen to give up on the place had decided to stick around. Many were employed by the Bureau's operations and maintenance division—making this dam and power plant work now instead of endeavoring to get it built—the effervescent Arkansan Harry Gilleland staying on as special-services officer, proudly showing this engineering marvel and the miracle that backed up behind it to a never-ending succession of reporters and VIPs; Elmer Urban staying on to manage the town until the Bureau decided it was ready for emancipation; former job foremen and steelworkers, high-scalers and mud-men making the switch to motel management, to piloting tourist boats and opening auto-repair shops; people staying on the mesa top because now that the sand was settling and the tin shacks were steadily being replaced by permanent structures, you plainly could see that there were a lot of worse places to live than this one. These were the valley days; this was as quiet as Page, Arizona, was going to get, and the future surely was full of promise.

The fact that the number of annual visitors to Lake Powell had reached more than half a million people thirty-five years before the Bureau of Reclamation had estimated this would happen surprised all sorts of people, but it only proved to the commissioner of Reclamation that he'd been right all along. Floyd Dominy loved this lake; he had rhapsodized about it in his picture book; he spoke effusively about it at every conceivable venue and in hundreds of casual conversations. Lake Powell *belonged* to him in an inarguable kind of way; when people told him it was beautiful, he would offer them his thanks; when they thanked him for creating it, he'd tell them they were welcome.

Glen Canyon Dam and the reservoir that now slowly climbed the walls of the canyon had been the cornerstones, the gemstones, the jewels of Dominy's career at Reclamation. Although the project had been authorized as part of the CRSP back before he had settled into the commissioner's chair, that was a trivial detail as far as he was concerned. Dominy had kept this dam from dying an untimely death on Capitol Hill, as he remembered it; he had seen that good people were put to work on it and had monitored every step of its rise from the river. He had marked the day the diversion-tunnel gates were screwed down as a milestone, a bona fide feast day, and had seen this lake evolve from a murky, driftwood-encrusted pond into a glistening, shining sea. There was nothing else like it in the world— an impossible realm of spreading blue water and towering slickrock—and Dominy still was certain, as he had been back before the reservoir had begun to rise, that in not too many more years three million people annually would visit this place he had named for Powell, reveling in its beauty and splashing in its captive water. *National Geographic*, after all, had devoted thirty-one pages to Lake

Powell in its July 1967 issue—proof both that this was a marvelous place and that millions more people now would want to see it.

But not everyone was as enchanted as Floyd Dominy was by the Glen's watery transformation. In the summer of 1967, eight years after his single passage through its pre-dam depths, novelist Edward Abbey returned to Glen Canyon as a National Park Service ranger, a seasonal lackey complete with the chalk-gray, epaulet-shouldered shirt and the Smokey the Bear hat, working at Wahweap, at the marina 90 miles uplake at the mouth of Bullfrog Creek—what now was Bullfrog Bay—and at Lee's Ferry, all three locales part of the Glen Canyon National Recreation Area. Abbey already was well versed in the employment requirements of the federal government, having spent three summers as a ranger at Arches National Monument near Moab and several others as a fire lookout posted above the rim of the Grand Canyon. The Lake Powell assignment, however, was his first acquaintance with the dictates of a recreation area, as well as his initial reencounter with the canyon he had floated in 1959, flooded now up to its nipples. Abbey, the self-proclaimed "wild preservative," the rebel raconteur and increasingly angry defender of deserts, already took a decidedly dim view of the ongoing digging, blasting, scraping, gouging, burning, and paving of the spare Southwest landscape, and he quickly discovered along the shorelines and the shimmering surface of Lake Powell that neither was he much impressed by the consequences of canyon reservoirs.

Less than a year after his stint beside Lake Powell, Ed Abbey's *Desert Solitaire: A Season in the Wilderness*, appeared. A memoir of his three summers spent at Arches and of his two decades of explorations into the sere slickrock country of southern Utah and northern Arizona, it was a personal, impassioned account of his ramblings through this unique and, as he observed it, desperately endangered part of the planet. Slowly at first, then ever more insistently, the book became word-of-mouth required reading, a kind of *Catcher in the Rye* for the coming-of-age of the environmental movement, the author's testimony of concern for a harsh and sometimes hostile landscape resonating fiercely for many thousands of people who were learning—or wanted to learn—to love the desert.

Abbey at Arches, Abbey in the Needles and the Maze and the

Island in the Sky, on the main street of Moab and in the slender heaven of Havasu Canyon—*Desert Solitaire* introduced its readers to splendidly desolate places seemingly safe from development for dozens of years to come, to glorious landforms already in dire trouble, and, in the case of Glen Canyon, to a place that no longer existed. His single sojourn there, he wrote, recounting his float trip with his friend Ralph Newcomb, had been a journey into "Eden, a portion of the earth's original paradise," and the canyon's damming and filling up with water would have been an unparalleled tragedy were it not for the certainty that the lake was only a temporary condition, just a flooded moment or two in time before the day when "some unknown hero with a rucksack full of dynamite strapped to his back will descend into the bowels of the dam; there he will hide his high explosives where they'll do the most good, attach blasting caps to the lot and with angelic ingenuity . . . ignite the loveliest explosion ever seen by man, reducing the great dam to a heap of rubble in the path of the river."

It was an astonishing, if wistful, notion—the idea that somehow you could *get rid* of ten million tons of dam, and that after a few dozen years of scouring by floods and wind-blown sand, Glen Canyon could become a wondrous *canyon* again, cut by a free-flowing river. And although Abbey wrote nothing more about the possibilities for the dam's destruction, suggesting that likely as not they were only futile daydreams, his readers, a legion of them before long, quick converts to what he called the "cult of the wild," began to imagine what was heretofore unimaginable, their copies of *Desert Solitaire* stowed in daypacks or thrown onto dormitory bunks, the most wonderful, wild, irreverent, incendiary passages underlined, dog-eared pages pointing them out, the bearded, pipe-smoking philosopher of stone and baking sun exhorting his readers to hold out hope, somehow convincing them that *Glen Canyon Dam had to go!*

Abbey had written in *Desert Solitaire* that "the impounded waters [in Glen Canyon] form an artificial lake named Powell, supposedly to honor but actually to dishonor the memory, spirit, and vision of Major John Wesley Powell." It wasn't enough that Floyd Dominy and his fellows had flooded Glen Canyon; they had added insult to the awful injury by naming their impoundment after the storied explorer—and Abbey was far from alone in his objection to the choice of appellation. From the beginning of the struggle for the future of the Colorado River, the reclamationists, water developers, and dam builders on one side and conservationists, river outfitters, and peripatetic desert rats on the other all had held up Powell as a kind of hero, a founding father of their separate and distinct visions of the desert Southwest, its limited water, and the ways in which people properly should live in the empty land. And each side long had resented the other's claim that *its* ideas, efforts, and actions were those the major would have endorsed.

Development-minded, reclamation-minded men and women had always endorsed the choice of the reservoir's name; Powell, after all, had been fascinated by the possibilities of bringing water from the rivers to the dry lands, arguing in 1878 that "conquered rivers are better servants than wild clouds." And, as if he had foreseen the day when giant dams would rise out of the Colorado's canyons, Powell had written in 1890, "If the waters [of the Colorado] are to be used, great works must be constructed costing millions of dollars, and then ultimately a region of country can be irrigated larger than was ever cultivated along the Nile, and all the products of Egypt will flourish therein." Clearly, Powell was a water man, claimed the latter-day reclamationists, and indeed his memory would be honored by the reservoir called Lake Powell.

But, much in the same way that his ideas had been ignored or only partially accepted during his lifetime, they had been too little understood in the decades since his death, according to the opponents of Floyd Dominy's grand plan for the stair-step damming of the Colorado's canyons. Although it was true that Powell had favored irrigation in the arid West, he had known that, at best, the region's rivers were capable of watering only 2 or 3 percent of its potentially irrigable lands. He had proposed that no western agricultural land be transferred to private ownership unless it possessed an indissoluble water right; he had suggested that state and county boundaries be laid out along natural drainage divides in an effort to avoid incessant water conflicts and political wranglings—and had been ignored on every count, of course. Powell actually had been a *conservationist*, the conservationists of the latter half of the twentieth century argued, a man who understood that while some lands and waterways should be altered and made to serve humankind, others necessarily should be left alone. "Lake Foul," some people began to call Glen Canyon's reservoir, both to express which side of the issue they were on as well as to leave the memory of the conservation-minded major unmolested.

A century after the *Emma Dean*, Powell's chair strapped to it like a kind of crow's nest, had floated the river's ripples and its many miles of placid water in Mound and then Monument canyons—the major later changing his mind and naming the two the Glen—both the people who hailed Glen Canyon's dam and reservoir and those who hated them as if they were somehow satanic had planned centennial observances, centennial celebrations. During the late spring and summer of 1969, the Sierra Club would sponsor a series of river trips that would serve as reenactments of Powell's pioneering journey down the river, and in Page, the doors of the chamber of commerce's new John Wesley Powell Museum would be opened to the public, a six-cent Powell-centennial stamp would be issued by the postmaster general, and—as Lady Bird Johnson had done at the dam itself three years before—Floyd Dominy and assorted western officials would christen the great spreading body of water that bore the explorer's name.

Down near the dam in the middle of June, standing on a sandstone spit that reached a little way into the water, a large crowd gathered to affirm that this reservoir indeed was Powell's proud legacy. Brown and blue bunting—symbolic of the silty old red river and the new blue, silt-settled lake—hung from the speaker's platform; the Navajo Tribal Band provided musical entertainment, as it had for more than a decade of such ceremonies; motorboats moved in formation in front of the assemblage. The upper 130 feet of the dam stood high above the water in the southern distance, and Navajo Mountain crowned redrock upland in the east.

"The Sierra Club to the contrary, I like dams," intoned Arizona Governor John Williams, a former radio broadcaster who had reported live from a shelter near this spot when the first blast blew out of the canyon wall back in 1956.

"The Sierra Club notwithstanding, this is a beautiful lake," Utah Governor Calvin Rampton affirmed, the large audience in complete agreement.

"A conservationist is one who is content to stand still forever," said Navajo Tribal Council Chairman Raymond Nakai, derisively defining the term. "Major Powell would have approved of this lake. May it ever be brimming full."

Finally, fifty-nine-year-old Floyd Dominy took his turn at the podium, thanking the speakers, as he always did, for their kind comments about his lake, telling all who heard him that they were welcome for his efforts, telling them to enjoy the lake for many years, then suddenly shocking them right out of their shoes when he announced that Dave Brower himself—the Sierra Club's Mephistopheles, the butterfly chaser with a net so big he had snared the Bureau's Grand Canyon dams, the most single-minded, obsessive, and vilified of all the radical conservationists—was standing right there among them, on hand as his special guest. What Dominy didn't say, and what none of them probably knew, was that by now Brower held no more sway with the sinister Sierra Club than did the smiling commissioner himself.

During David Brower's tenure as executive director of the Sierra Club, membership in the conservation group had risen from 7,000 in 1952 to more than 77,000 early in 1969. The club in that time had evolved from a little-known collection of mountain enthusiasts in northern California to the most powerful, resourceful, and renowned conservation organization in the United States, and probably the world. In less than two decades, Brower had led the club in successful opposition to federal plans to build dams in Dinosaur National Monument and the Grand Canyon; he had orchestrated the club's successful lobbying in support of the creation of Redwood and North Cascades national parks; his exhibit-format books had created a special and still-swelling niche in the publishing industry, and sales of the nineteen books in the series now totaled $10 million; the Brower-produced films *Glen Canyon* and *Grand Canyon* had had thousands of showings around the country; and *Life* magazine lately had defined him as "his country's number one working conservationist." "Even his enemies," added the *Nation*, "regard him with a respect tinged with awe."

Yet ironically, the ranks of Brower's enemies had grown large within the Sierra Club as well as without. As early as 1959, the club's board of directors had become concerned about Brower's militancy and about his maneuvering the organization into major roles in controversial national conservation issues, and although the board had stood squarely behind Brower at the outset of the Grand Canyon dams battle, and publicly had supported him in the face of the Internal Revenue Service's threat to revoke the club's tax-exempt status, many within the organization had grown increasingly un-comfortable with Brower's center-stage role in the national debate, as well as with what they believed was his cavalier use of the club's funds to further the fight. Brower had spent large sums on newspaper ads in opposition to the Grand Canyon dams, on a series of fruitless appeals of the IRS's ultimate decision to strip the club of its tax-exempt status, as well as on an expensive ad in the *New York Times* in the fall of 1968 that did no more than pat Brower and the club on the back for the fine books they'd been publishing—all expenditures lacking the board's consent. When the club's board voted 7 to 6 soon thereafter to remove Brower's spending power, it seemed clear to many within the organization that his command could not continue indefinitely.

Animosity between pro- and anti-Brower factions, outright anger between club members who had been long-standing comrades, came to a head in April 1969, during board elections punctuated by a labored internal debate over whether the organization should support plans to build a nuclear power plant at California's Diablo Canyon, Brower and his allies ultimately being soundly defeated. Two months later, right before Brower joined Floyd Dominy on the shore of Lake Powell, the Sierra Club's board met at San Francisco's Sir Francis Drake Hotel to decide whether Brower's tenure as the club's executive director should come to an end.

Wallace Stegner had lately written a letter to the *Palo Alto* (California) *Times* saying that his old friend and ally had been "bitten by the worm of power"; Ansel Adams, whom Brower had met on a trail in the Sierras back when he was still a teenager, openly averred that the club's fiscal ills and its changings-of-mind on the Diablo Canyon issue, both of which he blamed on Brower, had seriously damaged the club's reputation as well as its future effectiveness; Dick Leonard, Brower's old climbing partner and the man who had sponsored him for membership in the Sierra Club back in 1933, now stridently led the group favoring Brower's ouster, complaining that Brower seemed to think that any and all means were justified in the pursuit of his vision of conservation's ends.

When the board was polled after only a few minutes of open debate, only directors Martin Litton, the Colorado River dory man, and Eliot Porter, who had photographed five of the most successful exhibit-format books, plus three others supported Brower's staying on the job. Ten board members were opposed, the motion carried, and that was that. The white-haired but somehow still cherubic conservationist rose, made a short, emotional, and conciliatory farewell speech, then made his way to the edge of Glen Canyon—whose demise, he was certain, remained far more disastrous than his own.

Floyd Dominy and David Brower had made their rendezvous in the Arizona desert at the request of *New Yorker* magazine staff writer John McPhee. An ardent outdoorsman himself, McPhee long had been intrigued by the larger-than-life image Brower had assumed during the two decades of his conservation struggles, fascinated by the way in which a shy mountain climber from Berkeley had become the nation's archetypal defender of the natural world, a fellow capable of sparking fear and fundamental loathing among timber executives, mining-company presidents, water-management bureaucrats, and engineers by the bucketful. As he made initial plans to write a long profile of the then Sierra Club leader, it had struck McPhee that the best way to do so would be to journey with him out into wild country, and as well, if possible, to witness Brower in personal confrontation with people who opposed his views, who believed that humankind's proper role in relation to the natural world was to husband it, to put it to use, and not to leave it alone. If the idea was rather formulaic, McPhee knew, at least it ought to provide some fireworks, verbal fisticuffs, and, via these inevitable conflicts between the several parties, perhaps it would offer a bit of insight not only into who this character Brower *was* but also into what were the true dimensions of the current, burgeoning environmental debate.

Once Brower had agreed to the scheme, McPhee arranged trips into Washington's Cascade range, where the two had hiked with a minerals engineer and fierce advocate of the multiple use of public lands; to Cumberland Island, Georgia, where they had walked the longest undeveloped beach on the Atlantic coast with a resort developer convinced that people can and should live in fragile places; and finally to Glen and Grand canyons, where they would share powerboats on Lake Powell and pontoon rafts down the rugged rapids of the Colorado with the commissioner of Reclamation, a man already convinced that Brower wasn't "the sanctified conservationist that so many people think he is," Floyd Dominy always willing, no, *eager* to take on the opposition. It had been Charles Fraser, Brower's nemesis on the Georgia coast, who had defined him as a druid, one of the "religious figures who sacrifice people and worship trees." Yes. Brower was the *archdruid,* in fact, and these three encounters, McPhee hoped, would provide him engaging opportunities to defend his cathedrals, his pristine houses of prayer.

Prior to the Powell Centennial ceremony that Thursday morning in June, Dominy had escorted Brower and McPhee on a tour through the cool, cavernous innards of the dam, showing the two men the eight great generators down in the powerhouse, the dynamo housings painted bright yellow now, a low oceanic roar emanating from them as it always did, the faint vibration from the enormous machines rising up through the soles of their shoes. He had shown them the softly lighted, circular control room deep in the bowels of the powerhouse, where banks of dials, gauges, monitors, and flashing lights were vague evidence of the fact that, at that moment, 4,356 cubic feet of water per second were leaving the reservoir, en route to becoming a river again. They had seen the flat expanse separating the powerhouse from the dam itself, planted in grass now instead of Lady Bird Johnson's suggested bluebonnets— the "football field" according to Dominy, but the "golf course," if you had asked the employees who scurried about in plastic hardhats—and then Dominy had taken them down, way down, into the dam's foundation gallery 130 feet below the riverbed, 700 feet of concrete standing on top of them, to show the two men an example of how Glen Canyon Dam had cracked, the commissioner joking in the rainforest atmosphere of the passageway—water dripping from the ceiling and falling from the wall and running away in narrow gutters—that the situation demanded some Dutch boys, explaining that minerals in the water plus the Bureau's own injected grout eventually would slow much of the flow, but certainly not all of it. "You just cannot completely stop the Colorado River," Dominy explained. Then he offered his adversary a gift. "Dave, just to cement our friendship, I'm going to have a pair of bookends made from some of those old core samples for you," he said, referring to the test cores that had been taken as the concrete was being placed six, seven, eight years before. "Nothing could support a set of Sierra Club books better than a couple of pieces of Glen Canyon Dam. Would you accept that?"

"I'll accept the bookends," Brower told him. "Thank you very much, Floyd."

The two men did kind of *like* each other despite their differences, McPhee was somewhat surprised to discover. Riding the reservoir in Glen Canyon for three days, then the river in the Grand the following week, they drank beer and bourbon together, shamelessly flattered each other, spun yarns and told tall tales, argued about the other's mistaken plans and principles, and occasionally screamed and shouted. Dominy was either the fine fellow with the Irish wit and good humor—or the man who willingly would destroy nature for the sake of a few thousand kilowatts; Brower a hell of a nice guy you'd enjoy traveling with anywhere—or a pointy-headed elitist who would stoop to anything to get his way.

Motoring up the blue boulevard of the lake, traveling 20 miles an hour in a Bureau of Reclamation boat captained by Zug Bennett, a high-scaler who had gone to work on the dam project back in the autumn of 1956 and who now led these VIP tours in semiretirement, the dam man and the river defender marveled at the scenery, the horizontal blues of water and summer sky, the vertical buffs and browns, oranges and reds of the rock. "Who but Dominy would build a lake in the desert?" queried the commissioner himself, shouting to be heard above the growling engine. "Look at the country around here! No vegetation. No precipitation. It's just not the setting for a lake under any natural circumstances. Yet it is the most beautiful lake in the world."

Brower agreed. It was beautiful. "You can't duplicate this experience—this lake—anywhere else," Brower said. "But neither can you enjoy the original experience. That's the trouble. I camped under here once. It was a beautiful campsite. The river was one unending campsite. The ibis, the egrets, the wild blue herons are gone."

At the head of Clear Creek Canyon, a few flat-water miles up the arm of the Escalante where the Cathedral in the Desert was deep in water, Dominy chatted with the people in a nearby boat while Brower privately grieved over the loss of the most wondrous place he had ever seen.

"This lake is beautiful," said the one woman in the boat.

"Thank you," the commissioner told her in reply.

Near Rainbow Bridge, down-lake two dozen miles, the two men began to argue, carping about what it meant to save a region. "For every person who could ever have gotten in here when this place was in its natural state, goddamn it, there will be hundreds of thousands who will get in here, into all these side canyons, on the water highways. It's your few against the hundreds," Dominy contended, his voice rising with emotion. "Before I built this lake, not six hundred people had been in here in recorded history."

"By building this lake," Brower countered, "mankind has pre-empted [thousands of] acres of habitat for its own exclusive use. . . . A thousand people a year times ten thousand years will never see what was there."

"Read *Desert Solitaire*," said Dominy, explaining that some guy named Abbey didn't seem to think the dam would be around that long, the commissioner seemingly unperturbed by the notion that the dam that held this water back someday might be blown to smithereens.

People *would* see the rough, rock-strewn, narrow river bottom in the Grand Canyon for hundreds of years to come, it seemed, and now Brower and Dominy were two of those fortunate sojourners, riding the fat, nickel-colored rafts of Jerry and Larry Sanderson—former high-scalers themselves, dam builders now turned into river rats—downstream from Lee's Ferry, the smooth and mounded sandstone giving way to striated Redwall limestone, the sluggish current now tumbling into the same rapids that had terrorized Powell and his men, the river still a river despite the dam that controlled its flow.

"Notice that light up the line now, Floyd," Brower instructed late one afternoon, pointing to a cliff face glowing in the glancing light. "Look how nice it is on the barrel cactus."

"Gorgeous," Dominy agreed, his cigar sending smoke into the canyon air, the commissioner wearing long pants and a flannel shirt, a gold-braided cap emblazoned with the words *Lake Powell*, while Brower in turn wore only shorts and sneakers.

Sharing the bourbon Dominy had brought along in his briefcase, enjoying Jerry Sanderson's steaks at this fourth of several campsites, the two talked about the day they'd spent on the wave-rippled river while McPhee quietly listened, Dominy delighting in the rapids, Brower admitting they made him mighty nervous, before Dominy realized that the spot where they sat might have been covered by water held back by Hualapai Dam.

"Covered 168 feet deep," Brower noted derisively.

"It would be beautiful," Dominy insisted, still unwilling to acknowledge it as a dam that never would be built, "and, like Lake Powell, it would be better for *all* elements of society."

"Lake Powell is a drag strip for powerboats," Brower argued. "It's for people who won't do things except the easy way. The magic of Glen Canyon is dead. It has been vulgarized. Putting water in the Cathedral in the Desert was like urinating in the crypt of St. Peter's. I hope it never happens here."

"Don't give me the crap that you're the only one who understands these things," Dominy shouted back. "I'm a greater conservationist than you are, by far. I do things. I make things available to man. Unregulated, the Colorado River wouldn't be worth a good goddamn to anybody. You conservationists are phony outdoorsmen. I'm sick and tired of a democracy that's run by a noisy minority. I'm fed up clear to my goddamned gullet!"

The two men, both in late middle age now, both having suffered painful recent defeats—Dominy losing his dams, Brower the job that had defined him—railed on for a while in the dying and diffuse light of the canyon. But then they were buddies again, bolstered by bourbon, the writer recording their words, Brower telling Dominy they'd have to take a trip like this again sometime, Dominy telling Brower, by God, he'd have to come visit him at his Virginia farm, the shy and sensitive mountaineer and the blustery fellow who called himself the nation's water boy improbably side by side there on the banks of the coveted Colorado.

12
A ROCK-ENCIRCLED SEA

Balance had been the byword during his tenure at Interior, this much Stewart Udall was sure of by the time the Nixon Republicans took the helm at the beginning of 1969 and Udall returned to Arizona, and although the legislation wasn't signed into law until after he had left office, the Arizona Democrat took partial and proud credit for the creation of the new National Environmental Policy Act, as he saw it, surely the most consequential piece of conservation legislation ever enacted.

Henceforth, NEPA demanded, before any federally sponsored natural-resource project could proceed with construction, before any private project using or affecting public lands or water could proceed, a major study identifying its environmental consequences and proposing alternatives to construction had to be completed. The legislation guaranteed individuals, organizations, and governmental entities the right to comment formally on the environmental merits or demerits of each proposal as well. The law wasn't without its shortcomings—the "environmental impact statement" on a proposed federal dam, for instance, would be drafted by the Bureau of Reclamation, the very agency that wanted to build it—but it seemed certain that it would have a sweeping effect nonetheless, mitigating projects' impacts, delaying projects until they were deemed acceptable, occasionally blocking unworthy ones altogether.

Back in 1956, just two months had passed between the passage of the CRSP and the opening of the first incidental construction bids on the Glen Canyon project, the first spending of federal money; in only three more months, heavy construction was started. But from now on, in comparison, things would be very different. Once Congress had authorized a dam or a power plant, an airport or an atomic-testing ground, a lengthy review and comment process would begin, culminating with a document that would sanction the project, recommend changes, or spell its certain collapse. Had the CRSP passed Congress in 1970 or in the years thereafter, work in Glen Canyon—assuming it ever began—wouldn't have commenced for two years or three or maybe more.

Had he now been a congressman faced with the issue of whether to build a dam in Glen Canyon, Stewart Udall couldn't be sure how he might have voted. In the spring of 1956, his vote in favor of the CRSP had been enthusiastic. The story of the time on the floor of the House when he drank a glass of water with a piece of Glen Canyon sandstone immersed in it long since had entered the Capitol folklore, and his support of western water development always had been accompanied by similar bravado. Yet in the intervening years, his own trips into the canyon, into the rocky chaos of the country surrounding Rainbow Bridge, his efforts—one successful, one not— to turn the Canyonlands and Rainbow Bridge regions into national parks, and more recently, the angry debate over the fate of the Grand Canyon, all had combined to make him wonder whether he had been on the right side back in those seemingly simple and innocent fifties.

Were there better, simpler ways to guarantee the states their shares of the river without turning it into a plumbing system? Did irrigation projects that needed the infusion of hydropower revenues to appear economically sound deserve to be built in the first place? Water had to be delivered to the people and put to use, yes, but how in the world should you responsibly, carefully go about it? In the case of the Grand Canyon dams, Udall was certain now that they would have been difficult to justify for any purpose, least of all the simple powering of pumps to lift Phoenix-bound water over the Buckskin Mountains. And the alternative to them that he and his department had devised did seem satisfactory: the Navajos were anxious to sell their coal and to secure for their people a share of the jobs the mine and power plant would provide. But were the scars of strip-mining and the gray staining of the southwestern skies that much more benign than standing water, than the loss of rapids and riverine habitat? Too often, it seemed to him, politics was the business of selecting the lesser of several evils.

As far as Glen Canyon was concerned, Udall couldn't help but wonder if the conservationists hadn't let the big fish get away back in the days of the battle over Echo Park. Perhaps they should have let the government build its dam in Dinosaur in exchange for an agreement at long last to act on Harold Ickes's proposal for an Escalante National Monument; wouldn't a monument, better yet a national park, encompassing all of Glen Canyon have been worth the inundation of those protected canyons on the Green?

Whatever the answer to this question was, a dam did indeed rise out of Glen Canyon now, looking something like the cupped hand of God, backing up the river for almost 150 miles these days, and the former Interior secretary had to admit that Lake Powell *was* strangely, impossibly, unbelievably beautiful. No, it wasn't natural—nothing else on earth was remotely like it—but neither were gardens and parks and planted fields natural. People inarguably loved this man-made, rock-encircled sea. Nearly a million of them a year were visiting it already, and if few knew what lay leagues beneath its bright blue surface, they seemed to delight in it just the same. It was true, the river-cut Glen Canyon was gone, and Udall's vote a decade and a half before had helped to speed it on its way, but a kind of Glen Canyon remained, he was convinced, truncated and changed certainly, but somehow it still survived.

Stewart Udall, not quite fifty years old yet, had gone home to Arizona to practice law knowing that elective office in his home state might not be open to him again, his decision to oppose the Grand Canyon dams—despite the fact that the Central Arizona Project ultimately had been approved and was at long last leaving the drawing boards—still proof of his treason as far as more than a few of his fellows were concerned. Floyd Dominy, however, remained at his desk in Washington as the Nixon era commenced. Employed by the Bureau of Reclamation during the terms of five presidents now, Dominy had served as commissioner under four of them— Eisenhower, Kennedy, Johnson, and now another Republican whom he little liked or understood—remaining each successive president's choice to head up Reclamation because Floyd Dominy *was* Reclamation. Since the spring of 1959, he had been the indomitable, arrogant, energetic, bold, bellicose Kmish, and the Bureau's achievements and failures had been his own. He had acquired airplanes and assorted accoutrements of power; he had built a multistoried monument to himself and his agency in the foothills outside Denver; he had presided over the grand unfolding of the Colorado River Storage Project, and similarly had shepherded huge waterworks in California and in the Missouri River basin high in the grain-belt plains. He was the nation's water boy, he joked, a builder of dams since he was barely a man, a guy who headed the governmental agency that built things, one who'd traveled the world lending a hand at halting water, and, he readily acknowledged, one who'd been a controversial bastard for more years than he could begin to remember.

Dominy, too, was a fellow whose dam-divert-pump-and-pipe vision of Reclamation's role in western water development had failed as often as it had succeeded. As many as a dozen major projects conceived and planned under his administration had never materialized; his hope to expand Reclamation's efforts eastward into the domain of the Army Corps of Engineers had come to naught; his stupendous Pacific Southwest Water Plan had been reduced to a single long canal and a stake in a smoking power plant. With his senatorial cohort Carl Hayden retired now, and with Wayne Aspinall growing a little long in the tooth in the House, Dominy's days of cozy assistance from Capitol Hill were drawing to a close, and the conservationists seemed to be getting better all the time at convincing the public that dams somehow were destructive.

On top of all that, it seemed lately that Richard Nixon had asked the FBI to investigate the possibility of improprieties among hold-over, high-level federal officials, and the investigative agency had reported back to the president with a sheaf of eye-opening details concerning Dominy's intimidations of Bureau employees, his rants and ravings and well-tested devious tactics, as well as his seemingly insatiable appetite for extramarital romantic liaisons. There was nothing downright *indictable* in the FBI's collection of rumor, invective, and simple fact, but the Nixon administration was going to be a clean one, and the president and Interior Secretary Walter Hickel soon selected an Interior deputy-assistant named James Watt—a thirty-year-old from Wyoming who long had been in awe of the storied Commissioner Dominy—to inform him that a grateful government no longer required his services.

Floyd Dominy's battered leather briefcase, with the seal of the Bureau of Reclamation color-embossed on the side, had barely dried out from its splashings by the Grand Canyon's rapids before he had retired to his farm beside the Shenandoah, his dam-building days all done.

David Brower, the sometime friend of Stewart Udall and onetime chum of the now-retired commissioner of Reclamation—nemesis of both men throughout most of the 1960s—spent few days in the retirement the Sierra Club's board of directors had forced upon him. "We cannot be dilettante and lily-white in our work," he had commented in the wake of his resignation from his position as the club's seventeen-year executive director. "Nice Nellie will never make it. We cannot go on fiddling while the earth's wild places burn in the fires of our undisciplined technology." His personal solution to the alternative of simply fiddling was to create a new group, not a conservation club this time but rather an *environmental* organization, that word now sparked with currency, commitment to the cause, and a clear kind of militancy. Old-style conservation groups, it seemed, were content to whine politely about isolated losses of forests, seashores, canyons. The new environmentalists, on the other hand, understood that what was at stake was the very survival of the planet, and *ecology*, the subdiscipline of biology concerned with the relationships between living organisms and their surroundings, lately had become their catchphrase.

The ridiculous space program—what they viewed as the idiotic and spendthrift attempt to put men on the moon—ironically had offered the environmentalists rhetorical support for their contentions in recent days. As the crew members of *Apollo 8* had orbited the moon at Christmas 1968, they had photographed the distant earth, white and blue and beautiful, surrounded by the blackness of space.

Subsequent moon missions throughout 1969 had sent back similar images, pictures of a seemingly fragile planet, an island, a sphere that somehow appeared to be a single living organism. It was that notion, the idea that humankind's planetary home was alive yet exceedingly delicate, that Brower had wanted to convey when he and his wife, Anne, struck upon "Friends of the Earth" as the name of the new environmental organization they and a corps of allies were busily founding.

The Grand Canyon struggle had proven to the white-haired environmentalist that reason and honesty weren't enough. You had to be prepared to get tough, and stay tough, and continue the fight for as long as it took. Friends of the Earth, based in New York City in part to separate it clearly from the Sierra Club, wasn't even going to seek tax-exempt status. Right from the start, the organization was going to lobby for all it was worth, and it didn't make sense to pretend otherwise. It was going to work hard, stridently if it had to, for parks and wilderness expansion, for clean-air and clean-water legislation, for curbs on toxic chemicals, and even for an end to the ongoing war in Vietnam, Brower commenting in a Friends of the Earth ad that until now conservationists wrongly had operated "as though war is not as destructive as dams" or air pollution or DDT. "It is not true," he continued. "They are all of equal order, deriving as they do from a mentality which places all life and its vital sources in a position secondary to politics or power or profit."

But dams, of course, did remain fundamental and frightening symbols of evil. To his mind, Glen Canyon Dam remained the most demonic of all the dams, and Brower continued to hold himself largely responsible for the fact that it and its reservoir had blocked and largely obliterated a splendid system of desert canyons. Yet there was one slender hope, one buoyant bit of news out on the Arizona-Utah border: six years of power production and continuing downstream water commitments had meant that Lake Powell still was filling very slowly. The reservoir briefly had reached 3,600 feet above sea level in July 1970—only 100 vertical feet from the crest of the dam—but those final 100 feet would represent fully half of its storage capacity. Lake Powell was half empty still; its water still hadn't reached the boundary of Rainbow Bridge National Monument (although six more feet of rise would put them there), and more than 80,000 acres, roughly 130 square miles of sloping slickrock, of winding creeks and riverbed that were scheduled for inundation had yet to feel the relentless rise of the water.

The three barrier dam sites the Sierra Club had proposed early in the 1960s as locations for protecting Rainbow Bridge from the reservoir long since had been inundated, yet perhaps there still might be a way to "preclude impairment," as the CRSP legislation had stipulated long ago. If the government could be required to obey its own laws by limiting Lake Powell to a maximum height of 3,606 feet above sea level or less, the monument never would be encroached upon, the bridge itself wouldn't be endangered by water lapping at its base, and a lot of wonderful territory at the fringes of the current reservoir pool forever would be prevented from going under. The Bureau of Reclamation clearly had betrayed its legal obligation by its failure to construct a barrier dam, Brower contended in the Friends of the Earth newspaper, *Not Man Apart*, "however, because of gains that can be foreseen as a result of stabilizing Lake Powell at 3,606 feet or lower, we can be grateful for the Bureau's breach of promise."

But the Bureau never willingly would agree to limit Lake Powell, Brower realized; it always would ride on a raft of excuses and explanations why it simply had to have its maximum pool. Neither did it make any sense to try to get Congress to mandate the limited height; it was Congress, after all, that had refused to spend the money to build the barrier dam in the first place. No, the only thing to do was to sue. In this particular instance, what was required wasn't so much political and media-related hardball as simple, straightforward legal redress.

On November 4, 1970, five months after a widespread, celebratory "Earth Day" at least nominally, at least for a little while, had turned everyone in the country into an environmentalist, the Friends of the Earth, the Salt Lake City–based Wasatch Mountain Club, and Ken Sleight, the peripatetic river runner who long had claimed the Glen Canyon country as his rightful address, filed suit in U.S. District Court in Washington, D.C., asking the court to require the commissioner of Reclamation and the secretary of the Interior to obey the law specifically and immediately by limiting the maximum level of Lake Powell to 3,600 feet.

When the National Parks Association had tried a similar tactic back in 1962 to prevent Stewart Udall from proceeding with the closing of Glen Canyon Dam's diversion tunnels, the same court had ruled that that organization did not have the legal standing to sue the government. This time, however, with the members of the Wasatch Mountain Club claiming that the invasion of water into the national monument would impair their enjoyment and appreciation of the great stone arch as had been guaranteed them by President Taft when he created the monument back in 1910, and with Ken Sleight similarly asserting that his livelihood would be hindered if he no longer could guide tourists to a pristine Rainbow Bridge, the court agreed to hear the case, transferring it, however, to the jurisdiction of the federal district court in Salt Lake City—a venue, the environmentalists feared, that might not be nearly as conducive to granting them relief, but a venue nonetheless.

Sensing the potential for disaster, sensing one more conservationists' mess on their hands, Reclamation and Interior lawyers quickly succeeded in getting the court to agree that the reservoir could rise as planned until the suit was settled, then they began to prepare their case. Meanwhile, Reclamation officials in Washington and water men from the several states that effectively owned Lake Powell began a public-relations counterattack.

As far as the federal government and the states of the upper Colorado River basin were concerned, if they ever were required to limit Lake Powell to the 3,600-foot level, the consequences would be catastrophic. The first and most obvious detriment to them would be that Glen Canyon's reservoir would be able to store only about 14 million acre-feet of water instead of the 27 million planned for for more than a decade. And with the Colorado River Compact stipulation that more than half of *that* amount—7.5 million acre-feet—had to be sent through the dam each year to meet the upper basin's supply commitment to the lower basin, only a paltry supply of water would be left in long-term storage for the inevitable unrainy days, months, and years that would lie ahead. Scheduled upper basin irrigation and industrial water development would have to be curtailed drastically in order to be sure of supplying the proprietary rights and demands downstream.

Perhaps worst of all from a water-management point of view would be the fact that Glen Canyon Dam's twin spillways, their intakes 48 feet above the 3,600-foot level, never would feel the torrential rush of water, never could control a momentous flood. With the dam's eight penstocks and four outlet tubes capable of releasing no more than about 30,000 cubic feet of water per second, any time upstream runoff in excess of that amount was forecasted, the level of Lake Powell would have to be dropped quite low to accommodate the incoming water without exceeding the 3,600-foot maximum. In those rare years when a truly monumental flood was anticipated, there would be no alternative but to drop the lake as low as it would go—down to the level of the gates of the river outlets at elevation 3,374—then to cross every finger in the upper Colorado basin.

Throughout the many months during which the initial complaint was answered, companion briefs were filed, depositions were dutifully taken, motions were granted and motions in turn were denied, federal and state water officials and more than a few interested members of Congress remained quietly optimistic that the case would be decided in their favor. Not that they took David Brower and his hotshot lawyers for lightweights—they had learned well by now that that could be a fatal mistake—but there was this fundamental fact that seemed to support their cause: despite the fact that he belonged to the judicial branch, the case would be decided by a man who was part of the governmental system—one of the fellows, in effect—and even better, U.S. District Judge Willis W. Ritter was a Utahn, a native of the Mormon state of Deseret, a Saint himself and hence a water man, and surely he would understand both the calumny of David Brower and the calamity of a needlessly low Lake Powell. But somehow this jurist didn't.

The reservoir had crept upward to a height of 3,622 feet above sea level by February 27, 1973, the day Judge Ritter ordered the Department of the Interior and its Bureau of Reclamation forthwith to keep its reservoir from further invading Rainbow Bridge National Monument and to remove the water that so far had ebbed up the bed of Rainbow Creek inside the monument's boundary. ''It was pretty sneaky of Congress to pass a law and then ignore it completely,'' the judge told reporters after he had issued his ruling, a ruling the government immediately appealed, a ruling the region's water men nearly could not believe.

Within days after Ritter had refused the Interior department's request that he stay his injunction to limit Lake Powell's height, Utah's Senator Frank Moss and Representative Gunn McKay each had introduced bills that would do what Stewart Udall had hoped Congress would do as long ago as 1960—specifically rescind the language in the CRSP act that had assured protection of the monument, and specifically sanction the flooding of Rainbow Creek.

But the bills had barely begun their labyrinthine journeys through their respective houses of Congress in the summer of 1973 before the federal court of appeals in Denver overruled the Ritter decision, the higher court agreeing with the contention of the government's lawyers that Congress's several refusals to allocate funds for a barrier dam had taken precedence over the provisions of the CRSP, and noting as well the chaos that a permanently low Lake Powell would have created in the byzantine world of western water law.

Lawyers for Ken Sleight, for the Friends of the Earth and the Wasatch Mountain Club in turn petitioned the Supreme Court to hear the case, the environmentalists hopeful, if not quite confident, that the nation's highest court would see things Judge Ritter's way. But in the meantime, and for the foreseeable future, Lake Powell—half full now and gaining a thousand acre-feet with every inch—continued to creep up the canyon walls and spread into blue and broadening bays.

In town squares and in city parks, at 1,500 colleges and universities and at 10,000 high schools around the country, April 22, 1970, had been the day the earth began to be saved. But in Page, Arizona, with a cold spring wind blowing in from the west and with the sand swirling on Antelope Mesa a few miles southeast of town, it had been a day for another kind of commencement. With shiny-bladed shovels in their hands and ceremonial hard hats on their heads, the chiefs of five electrical utilities, the chairman of the Navajo Nation, plus local representatives of the project contractor and the Bureau of Reclamation had gathered that day to break ground for the construction of the Navajo Generating Station, a $350-million coal-fired power plant to be located on land leased from the Navajos.

Just before leaving office in January 1969, Stewart Udall, the man most responsible for forging this unusual and decidedly delicate coalition, had signed contracts committing the Department of the Interior, via the Bureau of Reclamation, to advancing $81 million in construction funds for the project in exchange for roughly 25 percent of the plant's long-term electrical output, as well as guaranteeing the plant up to 34,000 acre-feet of water a year to be drawn from nearby Lake Powell. Under the terms of the federal agreement, the plant would be owned and operated by the Salt River Project, a Phoenix-based public utility, and four other

WEST-member utilities, its coal supplied by the Peabody Coal Company's Black Mesa strip mines on land jointly controlled by the Navajo and Hopi tribes. When completed, the Navajo plant would be capable of producing 2.5 million kilowatts of electricity, as much power as Glen Canyon Dam and Hoover Dam *combined* could produce, and it seemed somehow fitting that it would be built by the Bechtel Corporation, the giant company that was the latter-day legacy of Dad Bechtel and his boys, co-builders of the Southwest's first big power project back in the 1930s, back in the days when hydropower—water pulsing through the penstocks of a stupendous dam—was surely the coming thing.

Residents of Page had worried about air pollution back when plans for the Navajo plant had been announced late in the 1960s, but Salt River Project officials had assured them that, although the project would get off the ground before the pending National Environmental Policy Act would mandate a lengthy environmental-impact process, the utilities nonetheless understood the swelling public mandate for clear skies, explaining that they would spend nearly $9 million to install state-of-the-art electrostatic precipitators and wet-lime flue-gas scrubbers at the plant, which together could capture more than 90 percent of its air-borne contaminants. Still, more than a hundred tons of fly ash and toxic gases a day would escape into the air. But at least the plant's towering stacks—taller than Glen Canyon Dam itself and rising above Lake Powell like strange new wind-eroded pinnacles—would lift the contaminants up to high-altitude winds that would carry them east and out of mind.

At age eighteen, it was time to push Page out of the nest. The government camp had grown up by now, its residents beginning to resent the government's control over every curb-and-gutter job, every sewer-fee increase, the government in turn getting tired of supporting the place. Since the first frowsy days in 1957, the Bureau of Reclamation had spent more than $4 million to turn Manson Mesa into a town, and for the past several years, despite a municipal-service charge assessed against all property owners, the community had run up annual debts averaging over $300,000. But now the faucet of federal funding was going to be turned off. Congressional legislation officially had withdrawn the town of Page from Reclamation's control, transferring to it 16.7 square miles of land, Lake Powell water rights that could supply domestic service to as many as 12,000 people, a guarantee of ongoing electrical service, and $500,000 in cash to help ease the pain of emancipation.

Residents voted in favor of home rule in December 1974, and by March 1975 Page was out on its own, Salt River Project officials congratulating the community on the occasion of its festive incorporation by proclaiming that Page had become "Arizona's most power-full town," and so it had. Combined power production down in Glen Canyon and out on Antelope Mesa soon would total well over three million kilowatts, enough to supply almost as many people, and the steel-lattice transmission towers that marched away to the south and east were vivid reminders to the townspeople that large chunks of Arizona, Nevada, and California now were plugged into Page and were sustaining it in return.

By the time the Navajo Generating Station was completed—dedicated to the prosperity of Page and of all the people who would feel the pulse of its power, on a sun-soaked day in June 1976—the town's second boom already had slipped into history. More than 700 people were employed at the mammoth power plant, 100 or so down at the dam, but Page had begun to tighten its belt again, the population receding to half what it had been at the chaotic height of Navajo's construction. But even with fewer than 5,000 people at home on the mesa now, Page remained the second largest town in sprawling Coconino County. Retail shops were proud of their longevity, boat sales and repair businesses were thriving, Holiday Inn lately had opened a lodge, and more than a million people a year were passing through Page en route to the lake that lapped at the orange cliffs.

In 1972, Congress had expanded the boundaries of the Glen Canyon National Recreation area to include 1.2 million acres, 1,875 square miles of federal preserve, reaching from Lee's Ferry all the way up river and lake and river again to Horsethief Bottom in Labyrinth Canyon of the Green, 50 miles above its confluence with the Colorado; reaching up the San Juan as far as its fabled Goosenecks, within a river bend or two of tiny Mexican Hat; encompassing the southern headlands of the Kaiparowits Plateau and the network of wild canyons that spilled into the Escalante. Once Lake Powell was full—when and if it climbed clear to the top of the dam—its surface area would comprise only 13 percent of the recreation area, the rest of the parkland made up of waterless, largely roadless, and decidedly wonderful desert.

And it had begun to look by now as though the reservoir actually would reach the 3,700-foot level before too many more years were out. The Supreme Court finally had upheld the appeals court decision, finding against Salt Lake City's Judge Ritter and the environmentalists he had sided with, ruling that Congress indeed could ignore the laws, regulations, and requirements of its own making whenever it chose to do so, its most recent actions and decisions always taking precedence over those that came before; the court took care to ignore the issues of whether the base of Rainbow Bridge should get wet or what level should be the reservoir's proper limit. There were only 40 feet left to rise now, but as the lake crawled on up a hundred side canyons and spread into ever bigger bays, its vertical progress would grow very slow, receding in fact in the summer months when every air conditioner south of Flagstaff was working overtime, the turbines at the base of the dam spinning wildly with the discharge of every possible drop of cold, clear, no-longer-captive water.

But Reclamation hydrologists were confident that they wouldn't have to wait till the end of the century to see the lake lapping at the crest of the dam. It would reach its watery zenith in ten years or less, they contended, perhaps in as few as five. As far as the career boys at the Bureau were concerned, Lake Powell was performing beautifully. It hadn't emptied into underground aquifers, as a few so-called experts had argued it would; all of it hadn't soaked into the spongy sandstone; the rock itself hadn't grown weak from the water and collapsed in a muddy, disastrous mess; and despite Dave Brower's dire predictions, the lake hadn't yet filled with silt.

A Reclamation study completed in 1962, the year before water first was impounded, had concluded that the long-term sediment accumulation rate in the reservoir would be 85,400 acre-feet a year, meaning that the entire reservoir pool would be full of silt—not water—in 316 years. But a second study in 1976, undertaken independently by a group of universities funded by the National Science Foundation, had found that only 27,000 acre-feet a year actually had accumulated since the reservoir was created early in 1963. At that rate, sediment wouldn't fill Lake Powell for a thousand years. Although the differences in the pre- and post-1963 sedimentation rates were surprising, Reclamation hydrologists assumed that they could be accounted for, first of all, by the fact that upstream reservoirs built as part of the CRSP now trapped some of the silt that previously had been carried downstream, as well as by land-use controls that had begun to restrict overgrazing, and by a variety of climatic factors. But whatever the mix of causes, it looked like good news indeed. Conservationists and Bureau critics had argued early on that Lake Powell, assuming it successfully held water, might not last a century before it silted up; at best, they guessed, it could survive two hundred years before it was nothing but a delta of dirt, before the river wound lazily across it, then poured over the crest of the dam in a waterfall that would shame Niagara. Yet at least for now, and based on the best available data, it appeared that Lake Powell would store water—although in annually smaller amounts—for quite a few years to come. It seemed that you'd have to stick around until 2900 or so to see that spectacular plunge.

Ed Abbey, for one, wasn't sure about Reclamation's optimistic projections of how soon the lake would fill with silt. Anyone who had ever glanced at the chocolate runoff of the Green and the Colorado, anyone who had ever seen the San Juan's storied sand waves, seen it rolling through its canyon looking more like liquid earth than simply a muddy river, knew that awesome amounts of sediment were being deposited everywhere the rivers and side streams slowed and then stopped as they met the reservoir's motionless water. And what did it matter if it took two hundred or two thousand years for Glen Canyon to fill with the soil of the upper Colorado River watershed? On either timetable, the result would be equally tragic. And there *was* an alternative, wasn't there?

In *Desert Solitaire*, Abbey had fantasized about the leveling of the dam with a stupendous explosion, the rapid its rubble would create fittingly named Floyd Dominy Falls. Yet as Abbey's reputation as a writer and an eloquent defender of desert wilderness had grown in the years following the publication of the book, he had begun to imagine an enlightened alternative to sending the dam sky-high. Writing this time in *Slickrock*, a 1971 Sierra Club exhibit-format book (the picture books edited by John G. Mitchell since David Brower's departure) combining Abbey's text with Philip Hyde's color photographs of the canyonlands, Abbey had contended that, as a power producer, Glen Canyon Dam already was sorely obsolete, such a technological dinosaur that surely the thing to do was to open the diversion tunnels and drain the reservoir.

Although the tunnels were, in fact, impossible to open—filled as they were with concrete—this too was a terrific fantasy: the dam would stand forever as a monument to humankind's folly, and the river would run around it. "Within the lifetime of our children," Abbey wrote, "Glen Canyon and the living river, heart of the canyonlands, will be restored to us." It was a wonderful image—Glen Canyon brought back to life, the reservoir only a fleeting aberration, a brief mistake, the serpentine river channel, the grottoes and glens, Music Temple, Mystery Canyon, the Cathedral in the Desert, and the Crossing of the Fathers all alive again, pristine except for the high, horizontal calcium-carbonate line marking the former reservoir's summit.

In 1975, Abbey's novel *The Monkey Wrench Gang* appeared, this time a kind of comic extravaganza that invented a band of environmental terrorists intent on chopping down the billboards, disabling the bulldozers, crippling the strip mines, uranium mills, and power plants that were turning the Southwest into some sort of industrial sacrifice zone. And oh how Abbey's marauders had their sights set on Glen Canyon Dam! In an author's note at the beginning of the novel, Abbey cautioned that although the book was fiction, it was "based strictly on historical fact. Everything in it is real and actually happened. And it all began just one year from today."

By the end of the 1970s, Abbey's novels, nonfiction books, and his fragrant and fulminating essays had turned him into a kind of high priest of what long ago he had called the cult of the wild. His willingness to be brash, venomously satiric, openly distrustful of government men promising that they would look after the land, and hateful of developers glibly promising easy street had struck a resonant chord with millions of Americans who were in the process of forming their own environmental consciousnesses. Although he sometimes seemed uncomfortable in the role, Abbey became a symbol not only of outrage but of hope, of the brave possibility that the slickrock deserts—and fragile country everywhere—somehow could survive. "Resist much, obey little," Abbey exhorted, his message clear enough to draw thousands into committed comradeship, bold enough to scare similar numbers to death.

On the first full day of spring in 1981, with the desert air already warm and the winter winds settling to barely a breeze, Ed Abbey watched from the bridge that spanned Glen Canyon while six men climbed over a gate blocking the access road that curled down to the crest of the dam, the group carrying a fat and heavy roll of black plastic. Reaching the viewpoint between the dam's twin elevator towers, the conspirators held tight to one end of the roll, then let it unfurl down the concave face, 300 feet of tapering plastic meant to resemble a crack, a fatal fissure in this, the most terrible of American dams. "Earth first!" shouted Abbey and the seventy people who watched with him from the bridge. "Free the Colorado!"

Ronald Reagan recently had become the president; James Watt, the minion who had fired Floyd Dominy a decade before, had become the secretary of the Interior; antienvironmental "sagebrush rebels" from far afield in the canyon country were claiming that they were going to seize the federal lands for their own exploitative purposes, and it was time to respond in kind. Organized by a group of zealous and experienced activists, former Wilderness Society and Friends of the Earth staff members, sixties radicals, and even—so they said—an anonymous congressman, this was the debut, the coming-out announcement of a coalition of people who no longer were going to play by the rules. "Earth First!" they were calling themselves—the exclamation point vitally important—and the newsletter they distributed contended that they were going to be passionate, confrontational, and very creative in their efforts to save western wilderness.

When he had founded Friends of the Earth back in 1969, David Brower had asserted that a conciliatory approach to environmentalism no longer would suffice. Now here were people—most of whom still held up Brower as the movement's patron saint—proclaiming that neither was Brower's brand of activism any longer enough. From now on the cause demanded action—whatever might capture the media's attention or worry the opposition—from a little guerrilla theater (such as the symbolic crack) to subterfuge and sabotage. "It's time for us to fight," declared Earth First! organizer Dave Foreman, himself a former representative of the Wilderness Society. And where better to declare war than beside the dam that for twenty years already had symbolized the West's tragic destruction?

Once the group had reassembled in the parking lot adjacent to the visitors' center, it was Ed Abbey's turn to speak. "Oppose the destruction of our homeland by those alien forces from Houston, Tokyo, Manhattan, D.C., and the Pentagon," intoned the writer, his gravelly voice and grizzled beard making him seem like some sort of environmental Isaiah. "And if opposition is not enough, we must resist. And if resistance is not enough, then subvert, delay until the empire begins to fall!"

Just then, police cars pulled into the parking lot, their lights flashing, the men who got out of them remaining constrained, asking questions about permits, about whom these people represented, about who was making the speech.

"That's Edward Abbey," the officers were told.

"Hey, Edward Abbey," exclaimed Officer E. C. White. "I've read his books."

White made his way through the crowd to the back of the pickup from which Abbey had been speaking, and from which petitions were being distributed urging the Department of the Interior to drain Lake Powell and "to allow the Colorado and San Juan rivers to cleanse their canyons and begin to recreate their wilderness." The policeman stretched out his hand. "Mr. Abbey, nice to meet you," he said. "I've been a fan of yours for a long time."

Lem Wiley, truly retired now, had returned again to Glen Canyon in the summer of 1980, the man who had built the dam, whose hat resided inside it, invited once more to sit on a dam-top dais as an honored guest, to take part in another ceremony. Yet this time, Wiley had returned for a special and personal reason. The construction veteran had wanted to see it with his own eyes, to see it once at least, something he wouldn't have missed for the world—the waters of Lake Powell wetting the 3,700-foot gauge on the upstream face of the dam, the lake spreading away to the rocky upland that separated the old channels of the river and Wahweap Creek, that high ground now an island, the lake spreading into the spillway-approach channels high on the canyon walls, pressing against the spillways' giant radial gates.

It had taken seventeen years for the reservoir to rise 560 feet, the final feet coming quickly and only recently as the runoff from the western Rockies tumbled down to the deserts. Reclamation workers at the dam had marked the time as 9:42 P.M., Mountain Daylight Time, June 22, 1980, when the cool, clear water had edged upward to 3,700 feet, Lake Powell finally impounding 27 million acre-feet of water, its surface area an ocean encompassing 252 square miles, the reservoir reaching 186 miles up the ancient bed of the Colorado River, 75 miles up the snaking San Juan, the lake touching 1,960 miles of shore now, longer than the Pacific coastline from Seattle to San Diego.

Two weeks after that long-awaited and lovely moment, Wylie had journeyed to Glen Canyon from his home in southern California to take part in Bureau-sponsored ceremonies celebrating the filling— he and Louis Puls, an octogenarian now, on hand as Reclamation relics, as physical evidence of the early days, proof that this monumental plug had been built by ordinary people. Wylie had listened politely to the speeches on the hot afternoon of July 11, had warmly acknowledged the praise that was heaped upon him, but at the end of the ceremony, he had laughed out loud, his eyes bright with delight, as the spillway gates were opened for the first time since the two tunnels had been dug twenty years before—water rushing down through the canyon walls, then roaring into Glen Canyon again.

Three summers hence, Lem Wylie was weak and growing infirm, but he nonetheless was back on the red canyon rims he first had surveyed in 1956, this time not to participate in festivities, but rather to observe a dam in crisis.

The snowpack high in the central Rockies had been only a little greater than its long-term average during the fall of 1982 and the winter of 1983. But heavy high-country snowstorms in May, followed by suddenly warmer weather, had pushed creeks and upper-basin tributaries out of their banks, swelling the downstream rivers to flood levels they hadn't reached in many years, the rivers sending 90,000 cubic feet of water per second into Lake Powell by early June. Not anticipating a runoff of such proportions, the Bureau of Reclamation hadn't substantially drawn down the level of Lake Powell over the course of the long winter in order to accommodate the inflow, and now there was no alternative but to open the penstock gates to full capacity and to open the four steel-lined river outlets that dropped through the dam as well, their jet-valves spraying spectacular foamy plumes into the downstream channel, making possible a combined, through-the-dam discharge of about 30,000 cubic feet per second. But the runoff didn't subside, and the lake continued upward, filling fast enough now that the situation was being monitored very closely.

Under the direction of Tom Gamble, Reclamation's Page-based Colorado River power-operations manager, Bureau officials conferred and finally decided to open the east spillway gate slowly and to begin bypassing large quantities of water around the dam, the first time either spillway ever had been pressed into flood-control service. For more than a week, water poured into the 41-foot-diameter tunnel, its volume steadily increased until—at 32,000 cubic feet per second—the water exiting the tunnel, pouring over the deflector bucket at the tunnel outlet and spewing into the riverbed, began to turn orange, began to spit out sandstone grit, pebbles, whole boulders even, the tunnel's concrete lining obviously torn away somewhere inside it, the water's pressure trying to gouge an even wider tunnel now, threatening to destroy it.

Gamble quickly ordered the spillway closed. If too much of the rock surrounding the tunnel were eroded away, particularly as it passed near the abutment of the dam itself, the result could be disastrous. It seemed unlikely that damage that extensive already had been done, but it was too great a risk to take to continue to operate the spillway without first seeing how badly it had been impaired. The following morning, a team of engineers in foul-weather suits, wearing hard hats mounted with miner's lamps, were lowered inside a cable-suspended cart down the 60-degree slope descending from the top of the tunnel, 600 feet down to the elbow where the tunnel became nearly horizontal. There they were shocked by the sight they encountered. In many places, the concrete lining simply had disappeared, the reinforcing bar that had been inside it twisted, snapped like a matchstick, gone. But far more than the concrete had been eaten away. Water now was ponded inside gaping holes in the sandstone itself, holes so big they prevented the further progress of the cart, the biggest gash, right at the elbow, as wide as the tunnel, 150 feet long, and so deep their probes didn't reach its nadir.

What the engineers witnessed in that wet cavern that morning was the result of a process called cavitation, something hydrologists had only dimly understood back when the dam was designed and built, the increasing effect of shock waves created when small vacuums or vapor cavities were formed on the downstream sides of minute bumps and irregularities in the tunnel's concrete lining, the rushing water effectively bouncing over the bumps and crashing into the lining below them, pounding at it, tearing it away, doing the same to the rock that was underneath it. The way to combat the cavitation process, Reclamation engineers now knew, was to introduce air into the flow of water, the air bubbles mixing with the water, cushioning its blows, dampening the disastrous effects of the shock waves. But for the moment, at least, trying to devise a means of repairing this tunnel, plus preventing future cavitation damage to it and its neighboring tunnel across the canyon, was premature. For the moment, the critical issue was going to be how in the world the dam's operators could wait out the continuing runoff, creating a reservoir of unprecedented size in the process, without having to resort to the spillways that were too suspect now to use.

Inflow into Lake Powell was estimated to be averaging 120,000 cubic feet per second by the end of June; with releases from the dam down to 30,000 cubic feet again, the reservoir would reach its maximum height of 3,700 feet above sea level almost immediately. The crest of the dam itself reached 3,715 feet—there was no danger of overtopping it, specifically because as the water level reached precisely 3,700 feet, it would begin to spill over the top of the radial gates protecting the two spillway entrances, then would fall into the tunnels despite the damage already done to one and the presumably precarious condition of the other.

With time a critical factor now, Gamble and his workers installed temporary flashboards on top of the gates, fashioned out of plywood procured from a lumberyard, and the plywood worked!—successfully holding the reservoir at bay until steel flashboards had been fabricated and were ready to be installed on the Fourth of July. Ten days later, on July 14, 1983, Lake Powell—nearly 300 square miles of water surface now—edged upward to an elevation of 3,708.4 feet, held that level for long, agonizing hours, then slowly, almost imperceptibly at first, began to drop.

The runoff had peaked, and within days it had subsided dramatically. The reservoir had successfully captured the flood; the dam hadn't been compromised, the power plant never impeded. One spillway tunnel now was in need of massive repair, and both would have to be retrofitted with some sort of air injection system to prevent similar crises in subsequent springs. Lem Wylie, staying out of the way but watching the work of the past days with passionate interest, was proud of Gamble and his crews, yet he couldn't help but second-guess some of the operation, wondering why in the hell the Bureau wasn't better than it was at weather forecasting, wondering why the hydrologists hadn't anticipated cavitation problems in the tunnels. But the bottom line for the old fellow who had seen those tunnels get their concrete coating in the first place was that the people who were managing this dam had responded to unforeseen and very serious events in a way that had resulted in their satisfactory conclusion. Wylie knew that the conservationists and these "environmentalists" that he'd been reading about had been hoping in recent weeks that all hell would break loose in Glen Canyon; they had wanted to prove that the river was still the boss, and, in a way, Wylie had to admit that indeed the river was. But unlike those people who did nothing but depend on their crazy-headed, let-nature-take-its-course kind of religion, Wylie was proudly in the camp with the people who resorted instead to ingenuity, to trial and error, to science. Wylie knew he belonged with the boys who built things.

Sitting in Tom Gamble's living room in Page, its windows offering a panoramic view of the canyon, the dam, and the blue and brimful sea, Wylie joked on an evening after the tension had eased that in his day the Bureau hadn't been so poor. "When I heard you'd put plywood flashboards on top of the gates, I just about jumped out of my pants," Wylie told Gamble, who had been in charge of the Glen Canyon operations since 1974, and his wife, Claudine, the daughter of Wylie's long-time colleague Norm Keefer.

Keefer had passed away recently; people said Louis Puls up in Denver was getting awfully frail; Floyd Dominy had been put out to pasture, and Wylie guessed Stew Udall still was down in Phoenix; Vaud Larson, Ruben Gaulke, Elmer Urban, Harry Gilleland, himself, and a dozen others all were alive and variously well, but all of them long since had retired from Reclamation. Their dam-building days were done, and now the work belonged to the bright young fellows he had been watching over the course of the past few days, and, well, that was how it should be.

Sitting in an overstuffed chair, shutting his eyes as he searched to remember names and dates, Wylie reminisced that evening about his life with Reclamation—about his days on the survey crew down at Hoover Dam, about the night Will Rogers came down into a diversion tunnel to buck up the spirits of the boys; about trying to build a hydro-dam in the midst of the Alaskan permafrost, that stuff worse than any rock you ever encountered; about coming to Glen Canyon. This place had changed so completely in the—what?—twenty-seven or twenty-eight years since he first had seen it. Where there had been nothing but sand and a hole and a lot of hope, good God what working people had been able to accomplish here! This place, this job had been the highlight of his career, Wylie said, but even if he was young and strapping again, he wouldn't want to do it over. There was too much in the way of politics now, too much of a byzantine system, far too many papers to shuffle and reports to issue in triplicate, too many people who looked at building something that would last and called it rape and ruin, too little confidence in what you could do if you simply pressed your nose to it.

"You see, I lived in an era in which these things could be done," Wylie explained, sounding wistful if unperturbed. "Nowadays, it's a little different."

EPILOGUE

In the eighty-two years since the Bureau had come to life as the piddling
little Reclamation Service, it had grown
into a government agency with a
billion-dollar budget and no trifling
amount of clout, and inarguably, it had
transformed the American West. By the
time the Glen Canyon spillways were
being repaired and retrofitted early in
1984, Reclamation could count itself
the builder and operator of 355 dams in
seventeen western states. Since the turn
of the century, the agency had constructed 17,000 miles of water-delivery
canals and pipelines, fifty hydroelectric
power plants, and it owned a piece of a
massive coal burner outside Page, Arizona. Sixteen million people now consumed Reclamation water in fields or at
kitchen faucets; ten million people depended on Bureau power. The Southwest had been watered, it had been
made habitable for huge numbers of
people, floods had been quelled and
disastrous droughts avoided, all for a
price of about $11 billion over the

course of those eight energetic decades. By now, the Granite Reef Aqueduct was 300 percent over budget but at last was nearing completion, almost ready to deliver water to the profligate city of Phoenix; two of Wayne Aspinall's five pet projects were under construction in western Colorado; the Bureau was tinkering here and fiddling there, but its halcyon days were done. Dams already stood at the optimal mountain and desert dam sites; irrigated agricultural lands were in surplus around the West; water supplies were adequate for the short term and would be far into the future if people could learn to conserve; and an environmentally cognizant era had opened, creating a political climate in which dams were no longer considered unmixed blessings. For eight years now, not a single new dam-and-reservoir project had been authorized, Congress gone skeptical, grown stingy on matters of western water. The Bureau of Reclamation would remain a potent force for years to come, of course, but within its ranks, even the thirstiest administrators and most ardent engineers agreed that the dam-building days that had consumed the careers of men like Lem Wylie never would come round again.

Reclamation's legacy, writ in rock and billions of tons of concrete as much as in moving water, was destined to be vigorously debated in the years of the agency's reassessment, the coming years of its outright decline. Had Reclamation truly succeeded in making much of the West habitable and hospitable for generations still to come? Had it actually accomplished its goals of forever harnessing menacing rivers, irrigating fallow fields, providing secure supplies of water in a region that otherwise was fatally arid? Or had it simply constructed monuments to itself—massive projects wasteful of money, materials, and wilderness as well as water— only to create fragile and necessarily short-lived oases, only to postpone the inevitable day when a populous, industrial West would run out of water once and for all?

Fifty years after Hoover Dam blocked Black Canyon and began to dole out the Colorado's water downstream, that dam remained in the minds of most people what it had always been—a symbol of the nation's resolve and resourcefulness in the midst of disastrously troubled times. But two decades after Glen Canyon Dam had climbed out of its canyon, there still was no consensus about *its* merits, nothing akin to agreement about whether it should ever have been built. The federal agency that had planned and constructed both dams was the bountiful fountain of all that was good in the country's arid quarter. It was also the single most glaring example of government gone awry, of rivers heedlessly and needlessly altered. Or, more accurately perhaps, it was something of both extremes.

Entering the nineties now, opinion likewise remains sharply divided about Glen Canyon's fate. Politicians and water managers are convinced that the dam and its power plant have facilitated municipal and industrial development stretching from Denver to San Diego, improving the lives of all the people who live in the sere Southwest. National Park Service officials point out that Lake Powell serves the recreational needs of millions of people a year and contend that the reservoir has made readily accessible canyons and desert landscapes that once were all but impossible to visit. But environmentalists and people who remember a simpler and emptier West, on the other hand, still view the damming of Glen Canyon as a disaster, as politics and technology gone mad, as the unconscionable destruction of what may well have been the most beautiful place on earth.

Twenty-five years after the last concrete was poured atop the dam, thousands still passionately defend the otherworldly beauty of Lake Powell and the project's myriad benefits; a quarter-century after the reservoir rose and Glen Canyon began to go under, others are still equally emphatic that one day the dam must be destroyed or dismantled and the flowing Colorado allowed to begin to re-create its canyon wilderness. The rhetorical struggle for Glen Canyon continues today because it poses such fundamental American questions: Should we let our technologies drive us, should we do whatever we're capable of doing in the certainty that science and engineering will enrich us? Or are we often wisest and most foresightful when we restrain ourselves, when we recognize natural limits, when we dare to leave the earth alone?

It seems certain in retrospect that Glen Canyon Dam was built, in part, because building is a fundamental response to human needs; we build, in a very literal way, because we have no alternative. What we construct can succor and sustain us, but it also can leave us diminished, the poorer for having destroyed what stood in its place. As we settled the American West, we hacked and hammered for two centuries—confident pioneers putting our marks on this place— before we began to notice the consequences; for fifty years we erected great dams before we began to revere the quiet canyons. We initiated Glen Canyon Dam at the end of an era in which we had proved beyond any doubt our ability to construct and to create. We completed the dam and pressed it into service as a new age opened, one in which still we would build, but in which caution and care and a new kind of collective humility would begin to be our bywords.

The high spine of the Rocky Mountains was buried in snow again in 1984, and when the weather warmed, creeks and rivers swelled to overflowing as they had done the year before, rushing through the piedmont and the flat farming valleys, roaring into the distant desert canyons. The level of Lake Powell had been drawn far down during the winter in anticipation of another heavy runoff, and in the months since Glen Canyon Dam's east spillway tunnel had been severely damaged, crews had worked round the clock to repair and reline the tunnel wall and to construct an air-injection system in both the east and west tunnels that would dampen the shock waves the next time a hundred-mile-an-hour jet of water raced down and around the dam.

The repair was finished by the time the river rose in Cataract Canyon, by the time the San Juan similarly swelled and a peak of 148,000 cubic feet of water per second poured into Lake Powell. But this time the reservoir never threatened to crest the tops of the spillway gates. The lake filled, but this time its operators weren't forced to let it overflow. In early August, however, with the heavy runoff ended and the lake level subsiding, it was time to test the $20-million spillway repair and retrofit. Divers worked underwater to chip out the epoxy that had sealed the spillway gates since the previous summer, then the 200-ton radial gate that protected the east tunnel was slowly hoisted open, and for five days, 50,000 cubic feet per second of foamy, air-cushioned water escaped to the river below. On the inspection that followed the test, the tunnel appeared intact, its concrete walls still silky, its cavitation-control system pronounced successful, the dam entirely back in business now, prepared again to capture and store the capricious Colorado, capable again of letting it go.

Floyd Dominy, raising bulls beside the Shenandoah now, well into his federal retirement and well shed of a billion bureaucratic headaches, still didn't agree for a second that the dam and the lake were some sort of mistake. He could—if you pressed him—sermonize for hours still about Glen Canyon's continuing benefits, but he really didn't care to; he was a seed stockman now, a rancher who took his bovines as seriously as he once had taken his dams, and his career at Reclamation somehow was ancient history.

Dominy initially hadn't been included among the special guests the Bureau had planned to bring back to Page for the 1980 ceremony marking the filling of the reservoir. In the years he had been away from Reclamation, although he remained something of a legend, his reputation had been soured instead of enhanced, and within the constrained, belt-tightened Bureau of the Carter administration, few were willing to hold up Floyd Dominy as a reminder of Reclamation's pugnacious past. Dominy finally was invited to the ceremony out in Arizona almost as a kind of guilty afterthought, but understanding the snub, he declined in the end to attend. At the close of the 1980s, the storied Kmish hadn't returned to Glen Canyon in twenty years, not since he had toured the dam and Lake Powell with Dave Brower in the summer of 1969, and although he still could wax poetic about the place, he didn't seem concerned about whether he ever would see it again.

Arizona voters returned Morris Udall to Congress for his fifteenth term in the fall of 1988, the lanky raconteur a longtime favorite among his congressional colleagues on Capitol Hill and the chairman of the House Interior Committee since that post had been vacated by Wayne Aspinall in 1972. Looking back at the end of the 1980s on his role in the political fight over the fate of the Colorado River, Mo Udall was frankly critical of his efforts two decades before to secure dams in the Grand Canyon. "I found myself in a bind," he remembered. "I was caught between my Mormon upbringing, my environmentalist leanings, and my constituents' near unanimous support for the dams. My decision finally turned on one irrefutable fact: water is life in the desert. . . . I came out in favor of the dams, and threw my whole energies into getting them approved. In retrospect (and this is something I wish I had understood then), two deeply rooted but antithetical American traditions—development of our natural resources and preservation of our wild landscapes—were about to collide. After two hundred years, the balance was about to shift. The wheel of history was turning—and I was in the way."

Wayne Aspinall, a man who proudly had championed his own district's interests above all else since he first had come to Washington in 1949, a man who'd never met a water project he didn't like during all those years, had been defeated in the Democratic party primary in 1972. His district's boundaries had been redrawn prior to the election, and his new constituents considered him old and out of touch with the burgeoning environmentalism he openly disdained. Nevertheless, the people who remembered his glory days still wistfully referred to him in his retirement as "Mr. Chairman," as "Mr. Reclamation," as a schoolmaster who'd become a legend. "The river means the West to me," Aspinall explained at the end of his life. "My whole ambition was to see western water made available for the development of the West." He died at his home overlooking a languid stretch of the Colorado in the autumn of 1983.

Carl Hayden, still a hero throughout his home state, died at age ninety-four the same year that voters sent Aspinall home to Colorado. Hayden had spent fifty-eight years in the U.S. House of Representatives and in the Senate, longer than anyone in history; he had served during six decades of overwhelming change in his native Southwest—the region transformed from an empty desert on the Mexican frontier into a brightly lit, well-watered oasis growing by startling leaps and bounds during the decades he spent at the center of Arizona's political scene. Although he had been pleased that the visitors' center at Glen Canyon Dam was named in his honor when it was completed in 1967, Hayden's true monument had been the Central Arizona Project. It had been authorized just months before he reluctantly retired from public service in January 1969, and he had done his best to live long enough to witness its bounty filling a beautiful concrete canal. But at his death in 1972, the Granite Reef Aqueduct still was 200 miles short of Phoenix.

Barry Goldwater, Hayden's colleague during the fifties and parts of the sixties, had been reelected to the Senate in 1968, his presidential aspirations long since squashed, but his service to Arizona continuing for three more terms. Looking back on his career at his retirement at the end of 1986, the seventy-seven-year-old said the single vote he regretted casting during his thirty years in the Senate was his vote in favor of the Colorado River Storage Project and the construction of Glen Canyon Dam. "While Glen Canyon has created the most beautiful lake in the world and has brought millions and millions of dollars into my state and the state of Utah," Goldwater said, "nevertheless, I think of that river as it was when I was a boy. And that is the way I would like to see it again."

Twenty-five years after the dam had begun to store water, Goldwater still owned land on the slope of Navajo Mountain and still talked about retreating there one day, where he could look out over the redrock until the end.

Lem Wylie died on August 21, 1984. His body was cremated, and his old friends Mack Ward, the Page pharmacist, and Lake Powell Air Service's Royce Knight took his ashes aloft and distributed them over Antelope Island—around a bend of the blue lake from the dam he had built back in the days when you could do things. Or so the two men tried to do. Each time Ward attempted to sprinkle Wylie's gritty remains out the window of the plane, they blew back inside. It was just like Lem, the two men agreed, to get the last word, to end a story on his terms.

Across a wide expanse of water from Antelope Island, Art Greene, his children, and their families had financed several expansions of their Wahweap Lodge and Marina by the time they sold their park service concession and lakeside property to the Del Webb Corporation, a developer of southwestern retirement communities, in 1976, the corporate concessionaire later acquiring the Bullfrog Bay marina, the marina that Gaylord and Joan Nevills Stavely had operated at Hite, and the Hall's Crossing marina run by Frank Wright, the river outfitter who had taken the reins of Norm Nevills's company early in the fifties.

Gay Stavely, based in Flagstaff now and running the Grand Canyon for a living, looked back on his early days—on the river and then the lake—and was certain he'd happily surrender every dime he made from Lake Powell if he could get Glen Canyon back. His former wife, Joan, in Page since 1977, when she became director of the Page–Lake Powell Chamber of Commerce, somehow was sure she had lived in two very different worlds during her lifetime, even though Page and her hometown hamlet of Mexican Hat were separated only by the high Kaibito Plateau and the many pinnacles of Monument Valley. "I don't equate the lake with the river," she would respond when people regularly asked her how she reconciled the two. "The river is someplace I was. It's tucked away in not only my memory, but in my heart. But the lake is lovely. This is still a rugged and beautiful place, still the only place I want to be."

Joan Stavely's old friend Greg Crampton, retired now from his professorship at the University of Utah and living in St. George, Utah, on the western edge of the canyonlands, continued to write about the rock-studded region that had shaped most of his professional and personal life; his 1986 book, *Ghosts of Glen Canyon*, had been designed for Lake Powell vacationers as a means of giving them a hint of the human and natural history that lay beneath that impossibly blue expanse of water. "Well, there it is," Crampton would say when asked about whether the lake ever should have come into existence, declining to join the still-raging argument. "If we regret that it is there, it doesn't seem to me that it needs to get in the way of our enjoyment of it. I enjoy the lake, but it's not the complete pleasure that came from floating the river."

Dave Foreman, the focal figure in the radical organization Earth First!—his group estimated to be made up of as many as 15,000 people at the end of the 1980s—still believed that action, rather than merely regret, was the only correct response to what he perceived as dozens of environmental calamities that had spread across the West like running sores. Boldly committed to what he termed "eco-defense," Foreman needed no time to ponder when asked what he could imagine doing in a final heroic act of sabotage. "I would hijack a semi truck," he responded full of cheerful bravado, "fill it full of nitrogen fertilizer and soak it down with diesel fuel and drive it onto Glen Canyon Dam."

Dave Brower, the mountaineer with the steel Sierra Club cup forever attached to his belt, the most radical conservationist of the 1950s, 1960s, and 1970s, now ironically looked like a kind of constrained and measured moderate in comparison with men like Foreman. Brower had no personal interest in demolishing the dam or in surreptitiously monkey-wrenching the destructive development of other parts of the West, yet he remained committed to the causes of wilderness and the protection of natural beauty, and increasingly, he was working hard in support of international disarmament, viewing the possibility of nuclear annihilation as the fundamental threat to the island earth.

Still feisty, often stubborn, still utterly independent as he entered his eighth decade, Brower by now had made a public peace with the Sierra Club, first having been elected an honorary vice-president and lifetime member of the organization, then returning to a nonadministrative seat on the board. The white-haired evangelist was back in the bosom of the club that had nurtured his zeal and his dedication to conservation beginning in the 1930s, the organization that he had made renowned throughout the nation, Brower perceived as a kind of prophet in his advancing age, a staunch defender of the faith, Brower himself still sadly convinced that his role in the decision to dam Glen Canyon was "the greatest sin I have ever committed."

Stewart Udall had spoken at a testimonial dinner for Dave Brower back in 1982, one celebrating his sometime nemesis's seventieth birthday, calling him then the "preeminent fang" of conservation's cutting edge, and in his 1988 update of his book *The Quiet Crisis*, Udall looked back on the battles that had pitted the two men against each other in the 1960s, confident that "if one believes, as I do, that the American people altered their thinking about their environment in the 1960s, in due course there will be laurels aplenty for David Brower."

Udall, now an environmental consultant, a writer, and occasional legal hired hand, was certain that his eight years as secretary of the Interior had spanned a fundamentally decisive era, a watershed of change, and he remained convinced that competing interests in the Colorado River basin had been brought into a tenuous balance during the years he kept the public watch. And his dream of a contiguous chain of preserves along the Colorado reaching upstream from the high and haughty wall of Hoover Dam to the river's canyon-bottom confluence with the Green finally had been achieved—the recreation area surrounding Lake Mead stretching upstream to the Grand Wash Cliffs to meet the expanded western boundary of Grand Canyon National Park, the park reaching all the way to Lee's Ferry now, beyond it the recreation area surrounding Lake Powell and encompassing the canyons at its flanks, then still farther upstream, the crowning sanctuary of Canyonlands National Park—more than 600 miles of publicly protected lands and waterways, reservoirs and rivers in almost equal measure.

In the one hundred and twenty years since Powell and his men first floated the Colorado's canyons, the region had changed to a degree the major wouldn't have been able to imagine. While he guessed that dams someday would block its course, he couldn't remotely have envisioned the sheer numbers of people who one day would live in the dry Southwest, sustained by the Colorado's water and power. And in the canyons themselves—so deep and empty and desolate that, it seemed to Powell, only adventurers like him ever would see them—vacationers now flocked in stunning numbers.

In 1986 for the first time, the annual number of people riding the river through the Grand Canyon reached more than 20,000, an average of more than a hundred people a day embarking from the park service's boat ramp near the spot where John D. Lee's ferry had crossed the river a century before, where Powell once had cached his boats. In 1966, in comparison, the year-long number of river runners had totaled 1,067; back in 1946, only Norm Nevills and three companions had floated the Grand's hushed stillwater stretches and ridden its rugged and punctuating rapids. Although regulated by the National Park Service, river running in the Grand Canyon had become so popular at the end of the 1980s that the very presence of people, combined with the wildly fluctuating water levels that were the product of Glen Canyon Dam's power plant, plus the utter absence of floods, seemed to stretch the definition of wilderness to its most liberal limits, the Colorado through the Grand Canyon now only an intimation of the marauding river it once had been, of the pristine place where people once seldom ventured.

A few miles upstream during the spring, summer, and golden fall of 1988, more than three *million* people set sail in houseboats and fishing dinghies, in speedboats and rusted skiffs onto the broad waters of Lake Powell, more than visited Yellowstone, more than peered into the nearby Grand Canyon, the numbers vindicating Floyd Dominy's seemingly outlandish prediction from nearly thirty years before. Despite the politicians who had come to rue the reservoir, despite the environmentalists who still despised it, Glen Canyon National Recreation Area had become the most popular federal preserve in all the interior West, its sixty-nine peak-season rangers plainly too few to monitor and adequately patrol the lake's 250 square miles of water surface, its nearly 2,000 miles of shoreline, or its estimated 16,000 feasible campsites. The park service by now had removed the floating marina from its former anchorage at the mouth of Bridge Canyon in an effort to mitigate the sometimes terrible congestion in the straight-walled little gorge, but still more than 100,000 people annually journeyed by boat to the base of Rainbow Bridge, young Navajo rangers stationed there trying in vain to keep swimmers from diving off its abutments, trying to limit the trash, trying as best they could to explain to the hordes of visitors that the bridge was more than simply a curiosity and that it deserved a measure of reverence, that it was Nonnezoshi, the rock from whence rainbows climbed into the summer sky.

Navajos had been at home in the Glen Canyon country more than a century now; seven hundred years had passed since the Anasazi had departed and wandered away to the south, never to return to farm in the canyon bottom or to build dams from sandstone blocks. But just twenty-five years after Glen Canyon Dam had been completed and its reservoir had begun to rise, people in the newest society in the American West—one still struggling to take shape—believed it was time to commemorate the events of a quarter-century before, some of them praising the triumph of the dam's construction, others still lamenting the loss of a wild river and the canyon that had contained it.

In Salt Lake City on January 15, 1988, a standing-room-only crowd of young people in parkas and hiking boots, some carrying babies on their backs, paying homage to a place most of them also had been too young to see, gathered in a high-school auditorium to mourn the death of the pristine canyon. And they came to hear Ed Abbey, the sixty-year-old whose writings had turned him into an unlikely kind of hero—one who himself would live for only another year. That night, Abbey encouraged his audience to have heart— Glen Canyon would be wild again one day. Sponsored by the Utah chapter of the Sierra Club, this was the third annual "archdruid lecture"—John McPhee's appellation for David Brower by now having become a generic label for anyone who staunchly professed the wilderness faith—the evening including folk music, a showing of Brower's old and grainy still-montage film *Glen Canyon*, and culminated by the bard from Moab, from Oracle and mythical Wolf Hole, reading the chapter from *Slickrock* in which he decried the "damnation" of the place he was sure had been the spiritual heart of that wonderful region of rock. The thousand or more people present were on their feet and cheering when he assured them that, even if centuries passed before it happened, someday the dam would be reduced to a concrete-studded rapid, and the river would be back at work resculpting its awesome canyon.

It was early May before the people who still fervently believed in the *rightness* of the dam and the long and twisting ribbon of water that stood behind it gathered in Page to celebrate the achievement of that earlier era—old-timers who had lived in the construction camp but who hadn't returned in twenty-five years surprised to discover that Page now looked a lot like American communities everywhere, its commercial streets lined with franchised restaurants and shopping plazas, its park and its golf course green and shaded now, new homes spreading across the mesa top, the three impossibly tall exhaust towers of the Navajo Generating Station—its power output dwarfing the electrical generation down at the dam—now the predominant landmarks.

The often-incessant spring winds were quiet on Saturday morning, May 7, the day dawning clear and warm, helicopters fluttering overhead early to keep an eye out for saboteurs, eighty Arizona and Utah state troopers on duty out of Reclamation's concern that kooks or crackpots might try to disrupt the midday ceremonies, that they might attempt to assault the dam. Security at the visitors' center was tight, but still tourists could view the exhibits, the photographic displays of how the gargantuan dam had been constructed, the Normal Rockwell painting of a Navajo family staring in wonder at the dam from the red canyon rim, the bright-lighted, clicking monitor, its numbers changing every second, that recorded the income the dam's power plant had generated since September 1964—$899,417,239 for only a moment and then a few dollars more. From a broad and arcing bank of windows, visitors could look down on the dam itself, seeing the great monolith that had transformed this place—its downstream face sweeping up out of the canyon in an arresting, somehow stunning curve; the enormity of the dam, its unadorned design and its very plain purpose separating a sea of bright blue water from a sheer-walled canyon where a river continued to run.

Down on the crest of the dam at noon, five hundred people, perhaps half of them Navajos, had assembled by the time the Page High School band struck up "The Star-Spangled Banner" and crews from Phoenix and Salt Lake City television stations turned on their Mini-cams, Mormon Bishop Leo Larson offering an invocation after the national anthem. Security men had searched all morning for suspicious characters, for signs of disruptive intent, but by the time the ceremony started, the only potential troublemaker they had encountered was peripatetic Kenneth Sleight, the veteran Glen Canyon river runner whose shock of hair was graying now, wearing a snap-button cowboy shirt and sneakers, sitting among the Navajos and the townspeople of Page. With dozens of eyes trained on him, Ken Sleight sat with his head in his hands while a succession of dignitaries went to the podium and lauded the giant structure beneath their feet.

"I'm filled with pride in those far-thinking Reclamation folk who came before me," current Bureau Commissioner C. Dale Duvall told the gathering. "For a quarter of a century, lives of people throughout this region have been improved by Glen Canyon Dam." Then one by one, representatives and senators from the two adjoining states, the governor of Arizona, and the chairman of the Navajo tribe rose to praise what had transpired in Glen Canyon beginning those many years before, to affirm the technological triumph of the dam and the beauty of the lake that spread into the slickrock north, the National Park Service's Regional Director Lorraine Mintzmyer finally assuring the crowd that all of them would continue to work together, "to see that this jewel of the West never loses its luster."

At the end of the ceremony, Sleight slowly made his way to the dais, security men trailing him, ready to grab him if he made an irregular move. But all that the legendary desert rat wanted to do was to shake hands with the special guests, to smile politely, and to give them a copy of a statement he had written, its point of view differing drastically from the sentiments just expressed, a missive describing his personal ordeal in watching Glen Canyon go under, urging the politicians to help protect what remained of the desert wilderness.

His statement distributed, his simple demonstration done, Sleight walked off the crest of the dam, pausing for only a moment as he went, scanning the curling lake that so many people somehow seemed to love, then peering downstream into the part of the canyon that still resembled the way it had appeared when he first encountered it—a river running in Glen Canyon as it had half a lifetime before, as it had when the dam-building Anasazis survived here and the world was young and surely easy to understand.

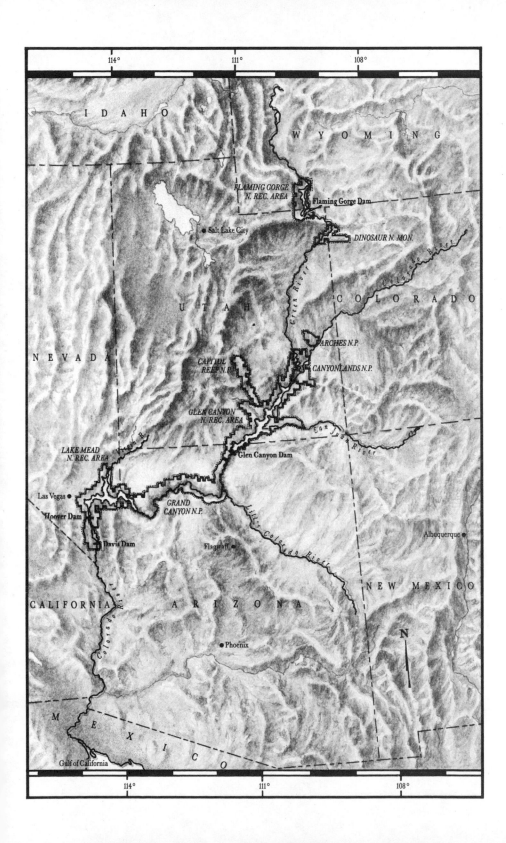

IDAHO

WYOMING

FLAMING GORGE
N. REC. AREA

Flaming Gorge Dam

• Salt Lake City

DINOSAUR N. MON.

Green River

Colorado River

UTAH

COLORADO

NEVADA

ARCHES N.P.

CAPITOL
REEF N.P.

CANYONLANDS N.P.

GLEN CANYON
N. REC. AREA

San Juan River

LAKE MEAD
N. REC. AREA

Glen Canyon Dam

Las Vegas •

Hoover Dam

GRAND
CANYON N.P.

Little Colorado River

Davis Dam

Flagstaff •

Albuquerque •

CALIFORNIA

ARIZONA

NEW MEXICO

N

Phoenix •

Colorado River

M E X I C O

Gulf of California

114°

111°

108°

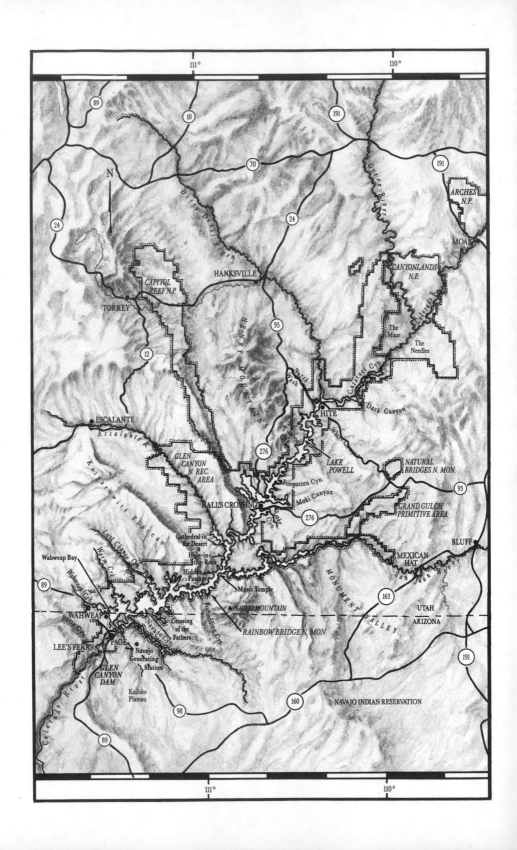

Acknowledgments

Many people contributed to this book, and I'm indebted to them for their invaluable contributions. Pamela Jones made available to me her exhaustive research into the short and lively history of Page, Arizona, and was wonderfully supportive. David Strang and Jerry Jones of the Salt River Project allowed me to view dozens of hours of videotaped interviews, archival film, and television footage, as well as their own excellent documentary film, *Page, Arizona: A New Beginning*. Tom Gamble, former Bureau of Reclamation power-operations manager at Glen Canyon Dam, pointed me in several important directions in the beginning, and he and his wife, Claudine Morrow Gamble, always were cordial and encouraging. Gamble's successor, Blaine Hamman, and his staff also offered me assistance on several occasions. Julia Betz at the John Wesley Powell Museum kindly allowed me to peruse that institution's archives. The irrepressible Harry Gilleland took me on terrific, eye-popping tours, compiled videotapes, introduced me to people I needed to meet, and answered a thousand questions with wit and warmth and enthusiasm. My friend Stephen Trimble offered me early encouragement, numerous suggestions, and shared his intimate knowledge of and affection for the canyonlands. C. Gregory Crampton, Jean Duffy Nydegger, W. L. "Bud" Rusho, Joan Nevills Stavely, Ken Sleight, Stewart Udall, and Elmer Urban graciously offered me their time and their recollections about the years of the 1950s and 1960s.

I encountered much of the source material for this book in libraries, and for their very specific contributions, I'm grateful to the staffs of the American Heritage Center at the University of Wyoming, Bancroft Library at the University of California, Hayden Library at Arizona State University, Marriott Library at the University of Utah, the Special Collections section of the University of Arizona Library, and the library at the Bureau of Reclamation's western regional headquarters in Denver. Near my home, Reed Library at Fort Lewis College and the Cortez Public Library regularly were helpful.

Once again, I offer my heartfelt thanks to my agent, Barney Karpfinger, for believing that this could become a book, and to my editor, Marian Wood, for so deftly shaping it into one. I'm grateful to Karen, of course, for riding out the rough water as well as the calm, and for living out west with me.

BIBLIOGRAPHY

Without the rich material provided by the histories, analyses, and memoirs that preceded it, this book couldn't have been written. Of particular help to me were C. Gregory Crampton's *Ghosts of Glen Canyon*, David Lavender's *River Runners of the Grand Canyon*, *Cadillac Desert* by Marc Reisner, the Bureau of Reclamation's *Technical Record of Design and Construction: Glen Canyon Dam and Powerplant*, and Susan Schrepfer's exhaustive interviews with David Brower for the Oral History Program of the University of California's Bancroft Library, compiled in typescript under the title *David R. Brower: Environmental Activist, Publicist, and Prophet*.

I similarly made use of hundreds of letters, pamphlets, reports, and newspaper and magazine articles. Although they are not listed below, I will be glad to answer queries directed to me through the publisher about specific sources. Throughout the text, I have referred to articles and letters (as well as books) that played important roles in the events that occurred in Glen Canyon, in addition to reporting and commenting on them. Although not necessarily noted in the text, issues of the *Congressional Record*, *High Country News*, the *Page Signal*, the *Sierra Club Bulletin*, and *Western Construction* also offered particularly valuable information.

Abbey, Edward. *Desert Solitaire: A Season in the Wilderness*. New York: McGraw-Hill, 1968.

———. *The Monkey Wrench Gang*. Philadelphia: Lippincott, 1975.

Abbey, Edward, and Hyde, Philip. *Slickrock*. San Francisco: Sierra Club Books, 1971.

Brower, David. *David R. Brower: Environmental Activist, Publicist, and Prophet*. Berkeley: Bancroft Library Oral History Program, University of California, 1980.

Bureau of Reclamation. *The Colorado River: A Comprehensive Report on the Development of Water Resources*. Washington, D.C.: U.S. Department of the Interior, 1947.

———. *Lake Powell: Jewel of the Colorado.* Washington, D.C.: U.S. Department of the Interior, 1965.

———. *Lake Powell 1986 Sedimentation Study.* Washington, D.C.: U.S. Department of the Interior, 1988.

———. *Preliminary Geological Report: Glen Canyon Dam-Site.* Washington, D.C.: U.S. Department of the Interior, 1949.

———. *Technical Record of Design and Construction: Glen Canyon Bridge.* Washington, D.C.: U.S. Department of the Interior, 1959.

———. *Technical Record of Design and Construction: Glen Canyon Dam and Powerplant.* Washington, D.C.: U.S. Department of the Interior, 1970.

Clark, Georgie, and Newcomb, Duane. *Georgie Clark: Thirty Years of River Running.* San Francisco: Chronicle Books, 1977.

Cohen, Michael P. *The History of the Sierra Club: 1892–1970.* San Francisco: Sierra Club Books, 1988.

Collier, Michael. *An Introduction to Grand Canyon Geology.* Grand Canyon: Grand Canyon Natural History Association, 1980.

Crampton, C. Gregory. *Ghosts of Glen Canyon.* St. George: Publisher's Place, 1986.

———. *Land of Living Rock.* New York: Knopf, 1972.

———. *Standing Up Country.* New York: Knopf, 1964.

Edmonds, Carol. *Wayne Aspinall: Mr. Chairman.* Lakewood: Crown Point, 1980.

Everhart, Ronald E. *Glen Canyon–Lake Powell: The Story Behind the Scenery.* Las Vegas: KC Publications, 1983.

Fox, Stephen R. *John Muir and His Legacy: The American Conservation Movement.* Boston: Little, Brown, 1981.

Fradkin, Philip. *A River No More: The Colorado River and the West.* New York: Knopf, 1981.

Goldwater, Barry. *Delightful Journey Down the Green and Colorado Rivers.* Phoenix: Arizona Historical Foundation, 1970.

Grey, Zane. *Tales of Lonely Trails.* New York: Harper and Brothers, 1922.

High Country News. *Western Water Made Simple.* Covelo: Island Press, 1987.

Hyde, Philip. *A Glen Canyon Portfolio.* Flagstaff: Northland Press, 1979.

Jennings, Jesse D. *Glen Canyon: A Summary.* Salt Lake City: University of Utah, 1966.

Jones, Stan. *Glen Canyon Dam Souvenir Guide Book.* Page: Sun Country Publications, 1984.

Lavender, David. *River Runners of the Grand Canyon.* Grand Canyon: Grand Canyon Natural History Association, 1985.

Leydet, François. *Grand Canyon: Time and the River Flowing.* San Francisco: Sierra Club Books, 1964.

Long, Paul V., Jr. *Archaeological Excavations in Lower Glen Canyon, Utah, 1959–1960.* Flagstaff: Museum of Northern Arizona, 1966.

Luckert, Karl W. *Navajo Mountain and Rainbow Bridge Religion.* Flagstaff: Museum of Northern Arizona, 1977.

Mann, Dean, et al. *Legal-Political History of Water Resource Development in the Upper Colorado River Basin.* Los Angeles: Lake Powell Research Project, University of California at Los Angeles, 1974.

McPhee, John. *Encounters with the Archdruid.* New York: Farrar, Straus and Giroux, 1971.

Nash, Roderick. *Wilderness and the American Mind.* New Haven: Yale University Press, 1982.

National Park Service. *A Survey of the Recreational Resources of the Colorado River Basin.* Washington, D.C.: U.S. Department of the Interior, 1950.

Palmer, Tim. *Endangered Rivers and the Conservation Movement.* Berkeley: University of California Press, 1986.

Porter, Eliot. *The Place No One Knew: Glen Canyon on the Colorado River.* San Francisco: Sierra Club Books, 1963.

Powell, John Wesley. *The Exploration of the Colorado and Its Canyons.* Mineola: Dover, 1961.

Reisner, Marc. *Cadillac Desert: The American West and Its Disappearing Water.* New York: Viking, 1986.

Richardson, Elmo. *Dams, Parks, and Politics: Resource Development and Preservation in the Truman-Eisenhower Era.* Lexington: University of Kentucky Press, 1973.

Rusho, W. L. *Everett Ruess: A Vagabond for Beauty.* Salt Lake City: Gibbs M. Smith, 1983.

Rusho, W. L., and Crampton, C. Gregory. *Desert River Crossing: Historic Lee's Ferry on the Colorado River.* Salt Lake City: Gibbs M. Smith, 1981.

Stavely, Gaylord. *Broken Waters Sing.* Boston: Little, Brown, 1971.

Stegner, Wallace. *The American West as Living Space.* Ann Arbor: University of Michigan Press, 1987.

———. *Beyond the Hundredth Meridian: John Wesley Powell and the Second Opening of the American West.* Boston: Houghton Mifflin, 1954.

———. *The Sound of Mountain Water.* New York: Dutton, 1980.

———, ed. *This Is Dinosaur: Echo Park Country and Its Magic Rivers.* New York: Knopf, 1955.

Stevens, Joseph E. *Hoover Dam: An American Adventure.* Norman: University of Oklahoma Press, 1988.

Terrell, John Upton. *War for the Colorado.* Volume 2. Glendale: Arthur H. Clark, 1965.

Udall, Morris. *Too Funny to Be President.* New York: Henry Holt, 1988.

Udall, Stewart. *The Quiet Crisis*. New York: Holt, Rinehart & Winston, 1963.

———. *The Quiet Crisis and the Next Generation*. Salt Lake City: Gibbs M. Smith, 1988.

Vélez de Escalante, Fray Silvestre. *The Dominguez-Escalante Journal*. Provo: Brigham Young University Press, 1976.

Warne, William E. *The Bureau of Reclamation*. New York: Praeger, 1973.

Watkins, T. H., ed. *The Grand Colorado*. Palo Alto: American West Publishing Company, 1969.

Wiley, Peter, and Gottlieb, Robert. *Empires in the Sun: The Rise of the New American West*. New York: Putnam, 1982.

Worster, Donald. *Rivers of Empire: Water, Aridity, and the Growth of the American West*. New York: Pantheon, 1985.

Young, Robert W. *A Political History of the Navajo Tribe*. Tsaile, Ariz.: Navajo Community College Press, 1978.

Zaslowsky, Dyan, and the Wilderness Society. *These American Lands*. New York: Henry Holt, 1986.

INDEX